COPPER TRANSPORT AND ITS DISORDERS
Molecular and Cellular Aspects

ADVANCES IN EXPERIMENTAL MEDICINE AND BIOLOGY

Editorial Board:
NATHAN BACK, *State University of New York at Buffalo*
IRUN R. COHEN, *The Weizmann Institute of Science*
DAVID KRITCHEVSKY, *Wistar Institute*
ABEL LAJTHA, *N. S. Kline Institute for Psychiatric Research*
RODOLFO PAOLETTI, *University of Milan*

Recent Volumes in this Series

Volume 445
MATHEMATICAL MODELING IN EXPERIMENTAL NUTRITION
Edited by Andrew J. Clifford and Hans-Georg Müller

Volume 446
MOLECULAR AND CELLULAR MECHANISMS OF NEURONAL PLASTICITY:
Basic and Clinical Implications
Edited by Yigal H. Ehrlich

Volume 447
LIPOXYGENASES AND THEIR METABOLITES: Biological Functions
Edited by Santosh Nigam and Cecil R. Pace-Asciak

Volume 448
COPPER TRANSPORT AND ITS DISORDERS: Molecular and Cellular Aspects
Edited by Arturo Leone and Julian F. B. Mercer

Volume 449
VASOPRESSIN AND OXYTOCIN: Molecular, Cellular, and Clinical Advances
Edited by Hans H. Zingg, Charles W. Bourque, and Daniel G. Bichet

Volume 450
ADVANCES IN MODELING AND CONTROL OF VENTILATION
Edited by Richard L. Hughson, David A. Cunningham, and James Duffin

Volume 451
GENE THERAPY OF CANCER
Edited by Peter Walden, Uwe Trefzer, Wolfram Sterry, and Farzin Farzaneh

Volume 452
MECHANISMS OF LYMPHOCYTE ACTIVATION AND IMMUNE REGULATION VII:
Molecular Determinants of Microbial Immunity
Edited by Sudhir Gupta, Alan Sher, and Rafi Ahmed

Volume 453
MECHANISMS OF WORK PRODUCTION AND WORK ABSORPTION IN MUSCLE
Edited by Haruo Sugi and Gerald H. Pollack

Volume 454
OXYGEN TRANSPORT TO TISSUE XX
Edited by Antal G. Hudetz and Duane F. Bruley

A Continuation Order Plan is available for this series. A continuation order will bring delivery of each new volume immediately upon publication. Volumes are billed only upon actual shipment. For further information please contact the publisher.

COPPER TRANSPORT AND ITS DISORDERS
Molecular and Cellular Aspects

Edited by

Arturo Leone
University of Salerno
Salerno, Italy

and

Julian F. B. Mercer
The Murdoch Institute for Research into Birth Defects
Royal Children's Hospital
Parkville, Australia

KLUWER ACADEMIC / PLENUM PUBLISHERS
NEW YORK, BOSTON, DORDRECHT, LONDON, MOSCOW

```
Library of Congress Cataloging-in-Publication Data

Copper transport and its disorders : molecular and cellular aspects /
  edited by Arturo Leone and Julian F.B. Mercer.
      p.   cm. -- (Advances in experimental medicine and biology ; v.
448)
   "Proceedings of a satellite meeting of the European Human Genetic
Society on copper transport and its disorders: molecular and
cellular aspects, held May 21-25, 1997, in Sestri Levante, Italy"-
-T.p. verso.
   Includes bibliographical references and index.
   ISBN 0-306-46045-9
   1. Copper--Metabolism--Disorders--Congresses.  2. Copper-
-Physiological transport--Congresses.  3. Copper--Metabolism-
-Congresses.   I. Leone, Arturo.  II. Mercer, Julian F. B.
III. Series.
RC632.C64C67  1998
616.3'99--dc21                                              98-46056
                                                                CIP
```

Proceedings of a Satellite Meeting of the European Human Genetic Society on Copper Transport and Its Disorders: Molecular and Cellular Aspects, held May 21 – 25, 1997, in Sestri Levante, Italy

ISBN 0-306-46045-9

© 1999 Kluwer Academic / Plenum Publishers, New York
233 Spring Street, New York, N.Y. 10013

10 9 8 7 6 5 4 3 2 1

A C.I.P. record for this book is available from the Library of Congress.

All rights reserved

No part of this book may be reproduced, stored in a retrieval system, or transmitted in any form or by any means, electronic, mechanical, photocopying, microfilming, recording, or otherwise, without written permission from the Publisher

Printed in the United States of America

PREFACE

This book is a compilation of presentations at the first meeting devoted to the molecular and cellular biology of copper transport. When we first considered the possible program for the meeting, we felt that a forum to integrate the recent advances in molecular understanding of copper transport with the older knowledge of copper metabolism was needed. In addition we wished to have a strong emphasis on the diseases of copper including the genetic diseases, Menkes and Wilson, and other possible health aspects of this metal seen from a molecular perspective. Overall we were very happy with the success of the meeting, and most participants were very enthusiastic. Unfortunately we were not able to obtain manuscripts from every contributor, but the selection in this book covers most of the topics discussed.

The history of biological research into copper dates from the latter half of the last century when the presence of copper as a component of living systems was first noted, but it was not until the 1920s that the essential role of copper was first recognized. J.S. McHargue found that plants and animals needed copper for optimal growth and health and proposed that copper was needed for life (McHargue, 1925). Other groups soon confirmed these observations in plants. In animals the requirement of copper for hematopoiesis was discovered in 1928 (Hart et al.,1928) ; this observation was the first link between the nutrition of copper and iron, which is still being investigated, and recent results from yeast which have added to the understanding of this interaction which appears to be mediated by iron oxidases such as ceruloplasmin. A number of the contributions to the conference dealt with this interesting metabolic interaction.

The role of copper in the demyelinating disease, enzootic neonatal ataxia was discovered by Bennetts and Chapman in 1937. Although this was a very early observation, the reason for the demyelenation is still not fully understood. Nevertheless many studies have shown that many of the effects of copper deficiency could be traced to the deficient activity of copper-dependent enzymes. One of the fundamental enzymes required for energy metabolism is, of course, cytochrome oxidase, and the reduction of this enzyme may underlie the demyelination found in copper deficiency.

In humans, demonstration of copper deficiency has been more difficult and is manifest mainly in childhood when the demand for copper needed for growth is highest. Preterm infants are most at risk, and infants recovering from malnutrition can develop iron-unresponsive anemia, neutropenia and connective tissue abnormalities. The copper intake for optimal health is still not clear, and some of the issues are discussed by Olivares et al. Overviews of copper physiology are provided by the Maria Linder and Harry McArdle. It is hoped that the increasing sophistication of our knowledge of the molecular basis of copper transport will in turn be converted into new ways of assessing copper status in the population.

The most dramatic example of copper deficiency in humans is the genetic disorder Menkes disease. This is an X-linked condition first described by John Menkes in 1962 (Menkes et al., 1962) as a complex syndrome characterized by retarded growth, neurodegeneration and unusual hair. The biochemical basis of this disease remained a mystery until David Danks realized that the symptoms of children with Menkes disease resembled many of the features of copper-deficient animals (Danks et al.,1972). For instance, copper deficient sheep produce wool with a defective structure rather like the hair of Menkes patients. Danks also observed that the arterial abnormalities in Menkes disease were reminiscent of the defective arteries in copper-deficient pigs. His knowledge of the symptoms of copper deficiency came from the extensive research that had been carried out on trace element deficiencies in Australia.

Menkes disease has proven important for the rapid advances in understanding of the molecular basis of copper homeostasis. Using positional cloning strategies, the gene affected in this disease was isolated by three groups (Chelly et al.,1993; Mercer et al.,1993; Vulpe et al.,1993). The identification of this gene and the identification of its product as a copper transporting ATPases have stimulated great interest in the area of copper metabolism, and many of the papers in the conference are related to the consequences of this discovery.

Further studies on the properties of the Menkes protein have revealed the unexpected finding that it normally resides in the *trans*Golgi network (TGN) of the cell, but in the presence of copper, it relocalized to the plasma membrane (Petris et al.,1996). This property appears to be consistent with the requirement for the Menkes protein to deliver copper to lysyl oxidase in the TGN, but to export copper when cytoplasmic levels become excessive. Thus the studies of copper transport are moving into the area of cell biology and necessitate an understanding of the role of protein trafficking in copper homeostasis.

Recently the Cu-induced trafficking of the Wilson protein has been reported (Hung et al., 1997), and this suggests a common mechanism regulating the intracellular location and hence function, of the mammalian copper ATPases. The mechanism regulating the copper trafficking is unknown, but clearly of great interest. To clarify the amino acid motifs on the Menkes and Wilson proteins which direct the intracellular location of the protein and control the copper trafficking will require *in vitro* mutagenesis studies. These studies have been hampered by the difficulty of expressing the Menkes protein; however, these problems have now been solved by a number of groups, and we have a chapter describing the expression of a functional Menkes protein from a cDNA construct (La Fontaine et al). Since the meeting, a report on the mutation of the N-terminal metal binding sites has shown that they are important in the copper transport function of the molecule (Payne and Gitlin, 1998), but no information is yet available on the motifs needed for the copper trafficking response. These studies are underway in a number of laboratories (including JFM's) and results should be available in the near future.

Alternative splicing of both the Menkes and Wilson transcripts has been reported, but the functional significance of these alternate transcripts is unknown. Some forms are likely to produce non functional products, but others may be active and have a distinct role in copper transport; e.g. Menkes gene transcript lacking exon 10 is produced in most tissues and maintains an open reading frame, the predicted protein lacking transmembrane domains 3 and 4. Data was presented at the meeting that this form localizes to the endoplasmic reticulum, rather than the TGN. Whether the protein has a function in this location remains unclear. The chapter by Ed Harris et al. discusses some of the multiple forms of the Menkes protein. Interestingly, an alternate form of the Wilson disease protein, lacking exons 6, 7, 8 and 12 has been found to be located in the cytoplasm (Yang et al., 1997), so it may be that we have yet to uncover all the possible forms of these ATPases.

The potential toxicity of copper has also been recognized for many years. Acute toxicity of copper is well recorded from accidental or deliberate poisoning. Chronic toxicity is mainly found in the genetic copper toxicoses such as Wilson disease, in which copper accumulates to very high levels in the liver and causes copper damage to this organ and to the central nervous system. Although Wilson disease has been known as a copper disorder for much longer than Menkes disease, the isolation of the gene was facilitated by the discovery of the Menkes gene. Both are copper ATPases, and current research is beginning to unravel some of the functional differences between the two molecules. The session on Wilson disease presented recent data on the types of mutations, current strategies of treatment, animal models, and the interesting possibility that another copper transport gene in the liver may underlie Indian Childhood Cirrhosis and Tyrolean Childhood Cirrhosis.

The possible role of copper in generating free radicals, which in turn could cause disease, is suggested by the mutations of superoxide dismutase in amyotrophic lateral sclerosis. Further interest has been raised by the finding that the amyloid precursor protein binds copper. These issues were discussed in the session on copper-related diseases and toxicity, and there are three chapters in the book which cover these topics. The recent finding that the prion protein is a copper-binding molecule (Brown et al.,1997) increases the evidence, although still circumstantial, that free radical generation by copper could be a common factor in a number of important neurological disorders. We await with interest further developments in this area.

Talks on various animal models of copper deficiency and toxicity were presented at the meeting. The mottled mice are good models for Menkes disease and its variants and provide an interesting challenge for phenotype genotype correlations (see chapter by Mercer et al.). Detailed analysis of the various sites of expression of the Menkes and Wilson proteins is important to understand the physiological effects of the absence of these proteins. Saymour Packman presents the elegant *in situ* hybridization analyses from his group which showed the widespread distribution of expression of both genes during development. The results suggest some roles for the Wilson protein during develoment which have yet to be clarified. Carl Keen's work on the consequences of nutritional copper deficiency during development were of interest and importance for furthering understanding of the effects of Menkes disease *in utero*. There are animal models of Wilson disease and a comparison of the features of these was presented by John Howell. There are interesting differences between the phenotypes of the models; two rodent models have been shown to have mutations in the Wilson disease gene homologue, namely the LEC rat and the toxic milk mouse. Nevertheless, even these two rodent models are somewhat different in their presentation. There may be species differences in copper metabolism which cause these distinct phenotypes, or as is proposed for Wilson disease, different mutations may also result in distinct clinical conditions. The molecular basis for the other animal models of Wilson disease (e.g. the Bedlington Terrier and the sheep) has not been established, but it is conceivable they may have alterations in other genes involved in the copper pathway.

Both Menkes and Wilson proteins have six copper binding sites in their N-terminal regions. The function of these sites is still unclear, although recent data suggests they are needed for copper transport. The characterization of the structural properties of these regions presents an interesting challenge to groups interested in metal binding motifs. We included a session describing some of the copper binding properties of these regions, since an understanding of the biological function and the reason for the presence of six such sites will require studies of this type.

Soon after the discovery of the Menkes gene, bacterial homologues were described, illustrating the extreme conservation of this copper-transporting ATPase. Recent information on the *Enterococcus hirae* copper transporters is presented in Marc Solioz's chapter. Subsequently, yeast homologues have been reported, and their function was discovered by means of their iron-dependent phenotype. In an exciting analogy with mammalian systems, the yeast Menkes homologue (or perhaps more appropriately in this context, Wilson homologue) was found to deliver copper to a ceruloplasmin homologue (Fet3) in a late Golgi compartment (for a recent paper see (Yuan et al., 1997)). This illustrated another link between copper and iron metabolism. The power of yeast genetics to identify genes important in copper homeostasis is graphically illustrated in the paper by Val Culotta. Her elegant work has identified molecules involved in the intracellular movement of copper. These molecules were suspected to exist to move copper around to various locations in the cell but were not identified by classical biochemical techniques. Interestingly, the discoveries have come, not from the studies directed at copper transport, but indirectly by isolation of genes involved in resistance to oxidative stress. Human homologues have been found (Klomp et al., 1997), and it is clear that these studies have completed one of the missing links in the copper transport pathway and have led to the proposal to call the intracellular carriers "copper chaperonins" (Culotta et al., 1997).

Regulation of the genes involved in copper homeostasis at a transcriptional level was discussed by a number of speakers. Dennis Winge discussed the detailed knowledge that has been obtained on the copper regulation of genes in *E. coli* and yeast. Arturo Leone's group has been working on the nuclear factors interacting with the metallothionein gene promoter elements. Apart from this paper, there was not much mention of the enigmatic metallothioneins and their role in copper homeostasis. These molecules need to be fitted into the growing picture of intracellular copper movement.

A workshop on Menkes disease provided a forum of the discussion of issues of treatment. This proved somewhat controversial since strongly held divergent opinions exist as to the efficacy of copper therapy. Although copper histidine treatment has been quite effective in treating the disease in four cases, a number of other Menkes patients have not responded to therapy. We had hoped the reason for the different clinical outcomes would receive rational consideration; however, it is clear that insufficient data is available for this issue to be resolved. We decided that it would not be particularly useful to include any written contributions from participants in the workshop, and as editors, we will make some overall comments based on our perspective.

The key issue in treatment seem to be whether all Menkes patients will respond to copper histidine therapy, or only those patients who have residual Menkes protein activity. For those interested in this issue, there is some correspondence in the *Annals of Neurology* (Sarkar, 1997). Characterization of the patients at the protein level has only been carried out in one case of successfully treated Menkes disease that we know of, and this patient does have residual protein, however, the activity of this protein has not been determined (JFM unpublished data). It does seem important that treatment be commenced as early as possible and if possible, *in utero*, because reduced copper availability in the last trimester is likely to have deleterious effects on brain development. Another area of research would be to investigate other copper complexes to determine if they have advantages in delivering copper to the brain. John Sorenson presented some of his ideas on lipophilic copper complexes which may have applications in therapy. As functional tests for the effect of mutations in the Menkes disease protein are developed and more patients are studied, the controversial aspects of treatment should be resolved.

We gratefully acknowledge that the conference in Sestri Levante would not have been possible without Leonardo Melò, a Menkes patient in Italy. Sadly, Leonardo died in December, 1991, but his parents have shown great courage and determination in establishing the Menkes disease association in Italy, " Amici di Leonardo" (Friends of Leonardo). It was Mirella Melò who first suggested the idea of a conference to discuss the latest issues of research into Menkes disease, and the organization has provided substantial funding to make the meeting possible. Her efforts to promote research on Menkes disease and on copper metabolism are a reminder for us as scientists that the data we obtain could provide new strategies to cure copper-dependent diseases in the future. Following the original idea of Mirella, we plan to continue to meet as the International Copper Research Group, and our next meeting has been scheduled in Australia.

We are also grateful for the support to the meeting of the following Istitutions: Associazione "Amici di Leonardo", Bourroughs Wellcome Fund, Consiglio Nazionale delle Ricerche, Istituto Italiano del Rame, Istituto Nazionale della Nutrizione, International Copper Association, Università degli Studi di Salerno; and Companies: Bibby, Bio-Rad Laboratories, Eppendorf, Beckman Analytical, Corning Costars Italia, Dia-Chem, Microglass, Millipore.

Arturo Leone
Julian F. B. Mercer

REFERENCES

Bennetts, H. W., and Chapman, F. E. (1937). Copper deficiency in sheep in Western Australia: a preliminary account of the etiology of enzooic ataxia of lambs and an anemia of ewes. Aust. Vet. J. *13*, 138–149.

Brown, D. R., Qin, K., Herms, J. W., Madlung, A., Manson, J., Strome, R., Fraser, P. E., Kruck, T., von Bohlen, A., Schulz-Schaeffer, W., Giese, A., Westaway, D., and Kretzschmar, H. (1997). The cellular prion protein binds copper in vivo. Nature *390*, 684–687.

Chelly, J., Tumer, Z., Tonnerson, T., Petterson, A., Ishikawa-Brush, Y., Tommerup, N., Horn, N., and Monaco, A. P. (1993). Isolation of a candidate gene for Menkes disease that encodes a potential heavy metal binding protein. Nature Genet. *3*, 14–19.

Culotta, V. C., Klomp, L. W. J., Strain, J., Casareno, R. L. B., Krems, B., and Gitlin, J. D. (1997). The copper chaperone for superoxide dismutase. J. Biol. Chem. *272*, 23469–23472.

Danks, D. M., Campbell, P. E., Stevens, B. J., Mayne, V., and Cartwright, E. (1972). Menkes's kinky hair syndrome:an inherited defect in copper absorption with wide-spread effects. Pediatrics *50*, 188–201.

Hart, B. A., Steenbock, H., Waddell, J., and Elvehjem, C. A. (1928). Iron in nutrition VII. Copper as a supplement to iron for hemoglobin building in the rat. J. Biol. Chem. *77*, 797–812.

Hung, I. H., Suzuki, M., Yamaguchi, Y., Yuan, D. S., Klausner, R. D., and Gitlin, J. D. (1997). Biochemical characterization of the Wilson disease protein and functional expression in the yeast *Saccharomyces cerevisiae*. J. Biol. Chem. *272*, 21461–21466.

Klomp, L. W., Lin, S.-J., Yuan, D. S., Klausner, R. D., Culotta, V. C., and Gitlin, J. D. (1997). Identification and functional expression of HAH1, a novel human gene involved in copper homeostasis. J. Biol. Chem. *in press*.

McHargue, J. S. (1925). The occurrence of copper, manganese, zinc, nickel and cobalt in soils, plants and animals, and their possible function as vital factors. J. Agric. Res. *30*, 193–196.

Menkes, J. H., Alter, M., Stegleder, G. K., Weakley, D. R., and Sung, J. H. (1962). A sex-linked recessive disorder with retardation of growth, peculiar hair, and focal cerebral and cerebellar degeneration. Pediatrics *29*, 764–779.

Mercer, J. F. B., Livingston, J., Hall, B. K., Paynter, J. A., Begy, C., Chandrasekharappa, S., Lockhart, P., Grimes, A., Bhave, M., Siemenack, D., and Glover, T. W. (1993). Isolation of a partial candidate gene for Menkes disease by positional cloning. Nature Genet. *3*, 20–25.

Payne, A. S., and Gitlin, J. D. (1998). Functional expression of the Menkes disease protein reveals common biochemical mechanisms among the copper-transporting ATPases. J. Biol. Chem. *in press*.

Petris, M. J., Mercer, J. F. B., Culvenor, J. G., Lockhart, P., Gleeson, P. A., and Camakaris, J. (1996). Ligand-regulated transport of the Menkes copper P-type ATPase efflux pump from the Golgi apparatus to the plasma membrane: a novel mechanism of regulated trafficking. EMBO J. *15*, 6084–6095.

Sarkar, B. (1997). Early copper histidine therapy in classic Menkes disease. Annal. Neurol. *41*, 134–136.

Vulpe, C., Levinson, B., Whitney, S., Packman, S., and Gitschier, J. (1993). Isolation of a candidate gene for Menkes disease and evidence that it encodes a copper-transporting ATPase. Nature Genet. *3*, 7–13.

Yang, X., Kawardada, Y., Terada, K., Petrukhin, K., Gilliam, T. C. a., and Sugiyama, T. (1997). Two forms of Wilson disease protein produced by alternative splicing are localized in distinct cellular compartments. Biochem. J. *326*, 897–902.

Yuan, D. S., Dancis, A., and Klausner, R. D. (1997). Restriction of Copper Export in Saccharomyces cerevisiae to a Late Golgi or Post-Golgi Compartment in the Secretory Pathway. J. of Biolog Chem *Vol. 272, No. 41*, 25747–25793.

CONTENTS

1. Copper Transport in Mammals 1
    Maria C. Linder, Norma A. Lomeli, Stephanie Donley, Farrokh Mehrbod,
        Philip Cerveza, Stephen Cotton, and Lisa Wooten

2. Models to Evaluate Health Risks Derived from Copper Exposure/Intake in
        Humans 17
    Manuel Olivares, Ricardo Uauy, Gloria Icaza, and Mauricio González

3. Cu Metabolism in the Liver 29
    Harry J. McArdle, Michelle J. Bingham, Karl Summer, and T. J. Ong

4. Multiple Forms of the Menkes Cu-ATPase 39
    Edward D. Harris, Manchi C. M. Reddy, Yongchang Qian,
        Evelyn Tiffany-Castiglioni, Sudeep Majumdar, and John Nelson

5. The Cell Biology of the Menkes Disease Protein 53
    Michael J. Petris, Julian F. B. Mercer, and James Camakaris

6. Functional Analysis of the Menkes Protein (MNK) Expressed from a cDNA
        Construct 67
    Sharon La Fontaine, Stephen D. Firth, Paul J. Lockhart, Hilary Brooks,
        James Camakaris, and Julian F. B. Mercer

7. Mutation Spectrum of *ATP7A*, the Gene Defective in Menkes Disease 83
    Zeynep Tümer, Lisbeth Birk Møller, and Nina Horn

8. Animal Models of Menkes Disease 97
    Julian F. B. Mercer, Loreta Ambrosini, Sharon Horton, Sophie Gazeas, and
        Andrew Grimes

9. Developmental Expression of the Mouse Mottled and Toxic Milk Genes 109
    Yien-Ming Kuo, Jane Gitschier, and Seymour Packman

10. The Treatment of Wilson's Disease 115
    George J. Brewer

11. Indian Childhood Cirrhosis and Tyrolean Childhood Cirrhosis: Disorders of a Copper Transport Gene? 127
    M. S. Tanner

12. Animal Models of Wilson's Disease 139
    J. McC. Howell

13. Copper-Binding Properties of the N-Terminus of the Menkes Protein 153
    Paul Cobine, Mark D. Harrison, and Charles T. Dameron

14. Expression, Purification, and Metal Binding Characteristics of the Putative Copper Binding Domain from the Wilson Disease Copper Transporting ATPase (ATP7B) 165
    Michael DiDonato, Suree Narindrasorasak, and Bibudhendra Sarkar

15. Structure/Function Relationships in Ceruloplasmin 175
    Giovanni Musci, Fabio Polticelli, and Lilia Calabrese

16. Autoxidation of Amyloid Precursor Protein and Formation of Reactive Oxygen Species 183
    G. Multhaup, L. Hesse, T. Borchardt, Thomas Ruppert, R. Cappai, C. L. Masters, and K. Beyreuther

17. Copper-Zinc Superoxide Dismutase and ALS 193
    Joan Selverstone Valentine, P. John Hart, and Edith Butler Gralla

18. A Study of the Dual Role of Copper in Superoxide Dismutase as Antioxidant and Pro-Oxidant in Cellular Models of Amyotrophic Lateral Sclerosis ... 205
    M. T. Carrì, A. Battistoni, A. Ferri, R. Gabbianelli, and G. Rotilio

19. The Effect of Copper on Tight Junctional Permeability in a Human Intestinal Cell Line (Caco-2) 215
    Simonetta Ferruzza, Yula Sambuy, Giuseppe Rotilio, Maria Rosa Ciriolo, and Maria Laura Scarino

20. Metal Regulation of Metallothionein Gene Transcription in Mammals 223
    P. Remondelli, O. Moltedo, M. C. Pascale, and Arturo Leone

21. Copper-Regulatory Domain Involved in Gene Expression 237
    Dennis R. Winge

22. Intracellular Pathways of Copper Trafficking in Yeast and Humans 247
    Valeria Cizewski Culotta, Su-Ju Lin, Paul Schmidt, Leo W. J. Klomp, Ruby Leah B. Casareno, and Jonathan Gitlin

23. Copper Homeostasis in *Enterococcus hirae* 255
    Haibo Wunderli-Ye and Marc Solioz

Index 265

COPPER TRANSPORT AND ITS DISORDERS
Molecular and Cellular Aspects

COPPER TRANSPORT IN MAMMALS

Maria C. Linder, Norma A. Lomeli, Stephanie Donley, Farrokh Mehrbod, Philip Cerveza, Stephen Cotton, and Lisa Wooten

Department of Chemistry and Biochemistry and Institute for Molecular
  Biology and Nutrition
California State University, Fullerton, California 92834-6866

1. OVERVIEW OF COPPER TRANSPORT AND FUNCTION

Copper is an important essential trace element for all living organisms needed for the activity of a variety of enzymes involved in critical areas of metabolism (Linder, 1991a; Linder and Hazegh-Azam, 1996). These include cytochrome c oxidase, the terminal enzyme of respiration; lysyl oxidase, essential for cross linking of elastin; dopamine-monooxygenase, required for synthesis of catecholamines; and several enzymes involved in anti-oxidant defense, notably ceruloplasmin, Cu/Zn superoxide dismutase, and copper-thionein. In most creatures including mammals, copper enters cells and organisms primarily or exclusively via absorption from the digestive tract. Adult humans consume in the range of 0.6–1.6 mg Cu per day, and a large proportion of this (60–70%) is absorbed. However, in considering the daily absorption, transport and metabolism of dietary copper, it is important to note that this occurs in the context of an even larger flux of copper from the body itself into the gastrointestinal tract, and back. This is surmised from the data on copper concentrations and daily volumes of GI fluids shown in Table 1. From these, it seems that almost 5 mg of Cu are secreted daily into the adult digestive tract as part of these various fluids. With the addition of dietary copper, a total of about 6 mg Cu daily thus enters the GI tract. Clearly, most of this is reabsorbed: only about 1 mg is lost in the feces; and dietary copper is only a small proportion of the total fluxing in and out of the digestive tract. Absorption and reabsorption of copper occur primarily in the small intestine, and the copper absorbed is in ionic form.

A portion of the copper secreted into the GI tract is not available for resorption. This is particularly so for copper secreted in bile (Linder, 1991a; Linder et al., 1997a), indicating that biliary secretion is the main excretory route from the body, being lost in the feces. [Very little (about 50 ug) is lost through the urine.]

After intestinal absorption, copper enters the blood of the mammalian organism in ionic form. Once in the blood, distribution appears to occur in two stages. In the first,

Table 1. Concentration and daily secretion of copper in human gastrointestinal fluids. (Summarized from Linder et al., 1997a.) The volume of duodenal fluid has been assumed

Fluid	Volume (mL/day)	Copper Concentration (ug/mL)	Daily secretion (ug/day)
Salivary	1500	0.22	330
Gastric	2500	0.39	975
Bile	625	4.0	2500
Pancreatic	1500	0.6	900
Duodenal	1500?	0.17	250
			Total 4900 per day

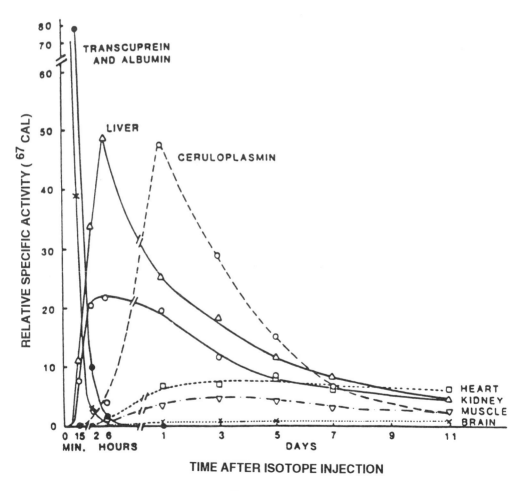

Figure 1. Time dependent distribution of tracer ^{67}Cu to blood proteins and tissues after administration to rats. Relative specific activity of ^{67}Cu associated with plasma proteins (albumin, transcuprein and ceruloplasmin) and with various rat tissues at various times after i.p. administration of ng quantities of radioactive copper.

most of the copper is taken up by the liver and kidney; in the second, a portion re-emerges in the blood plasma buried in ceruloplasmin. Following the appearance of dietary copper in ceruloplasmin there is uptake by other tissues. This process is illustrated by the data in Fig. 1 for rats treated with tracer ^{67}Cu(II) and has been extensively described and reviewed (Linder et al., 1997a; Linder, 1991a; Weiss et al., 1985). In the figure, one sees the initial high specific activity of copper-binding components in the blood plasma (albumin and transcuprein), which rapidly declines at the same time as accumulation of radioactivity is occurring in liver. This is followed by the appearance of radiolabeled ceruloplasmin in the plasma, and a decrease in its radioactivity as other tissues accumulate the tracer. In general, the data from rat studies agree with those obtained in humans using heavy isotopes of copper (Scott and Turnlund, 1994; Linder et al., 1997a). What emerges is that almost all the copper in extracellular body fluids is not transported in the form of free ions or chelates, but rather that it binds tightly to and is carried by specific proteins. These seem to target it to specific tissues. It seems likely that this targetting involves specific cell surface receptors and transporters, which have been identified for ceruloplasmin, but about which as yet relatively little is known. As already indicated, the three proteins that appear to be involved in blood copper transport are albumin, transcuprein and ceruloplasmin.

2. PROTEINS OF COPPER TRANSPORT IN BLOOD

Albumin (Table 2) was first identified as the plasma protein most likely to bind ionic copper in studies by Bearn and Kunkel (1954), using unphysiologically large amounts of ^{64}Cu. Not long thereafter, the presence of a high affinity binding site at the amino terminus of the protein was identified (Breslow, 1964), which has been the subject of repeated studies over the years (Lau and Sarkar, 1971; Masuoka et al., 1993; Linder, 1991a). It is noteworthy that although there is enough albumin in human plasma to bind as much as 30 mg Cu(II) per mL at the high affinity site alone, in fact only about 180 ng Cu per mL actually bind! This attests to the fact that (a) copper is truly a trace element (there is very little of it in the circulation, or in cells, for that matter), and (b) that there are other plasma components with an even higher affinity for copper than albumin, such as transcuprein and perhaps also some specific peptides (see Linder, 1991a and Linder et al., 1997a for details). One can speculate that the presence of this "excess" albumin might, at least theoretically, provide some protection against the formation and circulation of copper chelates that would catalyse formation of free radicals, should there be a sudden influx of copper ions. Albumin does not appear to be necessary for the normal distribution of incoming dietary copper (to liver and kidney), as demonstrated in genetically analbuminemic rats (Vargas et al., 1994).

Transcuprein was the name given to a much larger plasma protein labeled with tracer ^{67}Cu in parallel with albumin after administration of the isotope to rats, or upon mixing it directly with rat plasma (Weiss et al., 1985; Linder et al., 1991a, 1997a). Fig. 2A

Table 2. Albumin and copper transport

- Most abundant plasma protein (66-70% of plasma protein)
- Multiple functional transporter (from fatty acids to tryptophan and metal ions)
- Most albumins have a very high affinity Cu(II) binding site at the N-terminus involving a histidine
- About 17% of the copper in human plasma is bound to albumin
- Because of its abundance, serum albumin could bind 30,000 ug Cu/mL
- Does not appear to be required for normal distribution of copper entering blood from the digestive tract (Nagase analbuminemic rats)

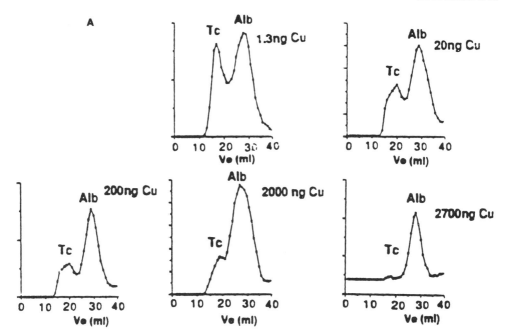

Figure 2. Copper associated with transcuprein (Tc) and albumin (Alb) in blood plasma, after separation by size exclusion chromatography. Copper elution prfiles were obtained by applying 1.0 mL samples to 50 mL columns and collecting 1 mL fractions. On these columns, transcuprein elutes in the void volume and albumin in the included volume (Weiss et al., 1985). Ceruloplasmin (Cp) elutes inbetween. A and C are reprinted from Linder, 1991a, with permission. A. Changes in labeling distribution associated with the addition of increasing quantities (ng) of actual copper (as ^{67}Cu-nitrilotriacetate; Cu-NTA) to samples of rat plasma, as indicated. B. Distribution of radioactivity after fractionation of samples of serum from rabbits (open circles) and humans (closed circles) pretreated with 5 ng portions of ^{67}Cu-NTA . Elution positions of Tc, Cp and Alb are shown. C. Elution of actual copper (determined by atomic absorption) when whole human plasma is fractionated. The main peak in the middle is copper in ceruloplasmin. Copper in the void volume and with albumin is seen as shoulders on either side of the ceruloplasmin peak.

shows the size exclusion chromatographic separation of ^{67}Cu-labeled transcuprein (Tc; first peak) and albumin (Alb; second peak) after tracer treatment of rat plasma, and how the proportion of radioactivity associated with Tc decreases and that with albumin rises as increasing amounts of Cu(II) are added. This indicates that (a) transcuprein can bind Cu even in the presence of large amounts of albumin; and (b) copper binding sites on transcuprein are easily saturated, and (as expected) the excess binds to albumin. With time after administration of ^{67}Cu to animals (Fig. 1), the radioactivity with transcuprein declines rapidly and in parallel with that on albumin, and at the same time as it is being absorbed by the liver and kidney. Since albumin is not required for normal copper distribution, it seems likely that the specific uptake by liver and kidney of copper entering from the intestine is mediated by transcuprein, although it readily exchanges Cu(II) with albumin (Weiss et al., 1985; Linder, 1991a), and both are normally part of the "exchangeable plasma copper pool".

Some of the characteristics of transcuprein are summarized in Table 3. It has an apparent molecular weight of 270 k and is composed of two subunits of about 200 and 60 kDa (Fig. 3), the larger containing some N-linked carbohydrate (Askary, 1996). Recent amino acid sequence data we have obtained strongly suggest that the 200 k subunit of

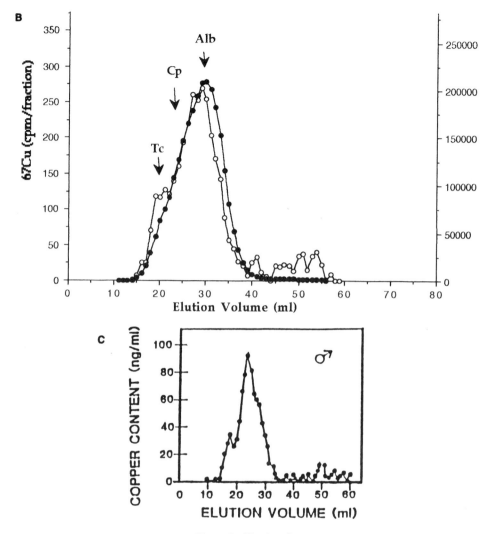

Figure 2. (*Continued*)

transcuprein is "alpha-1-inhibitor-3" (Table 4), a member of the family of macroglobulins (Schweizer et al., 1987; Eggertsen, et al., 1991; Regler et al., 1991). Plasma macroglobulins are known for their ability to inhibit proteases by trapping them. The 60 kDa subunit of transcuprein is albumin-like and has the N-terminal amino acids needed for high affinity copper binding. One possibility is that the macroglobulin is trapping the albumin-like protein (Osterberg et al., 1994), which in turn has the copper binding site. The macroglobulin itself may also provide copper binding: It is known that alpha-2-macroglobulin carries zinc (Adham et al., 1977; Osterberg and Malmensten, 1984), and there is preliminary evidence it can bind copper as well (Linder, 1991a). There are several plasma macroglobulins, and their distribution among mammalian species is not uniform. Thus, for example, rodents have alpha-1-macroglobulin (a 145 kDa subunit tetramer) along with alpha-1-inhibitor-3 (Rubenstein et al., 1993; Lonberg-Holm et al., 1987), but little or no alpha-2-macroglobulin. Indeed, alpha-1-I3 is a major rat plasma protein. Humans, on the

Table 3. Transcuprein and copper transport

- Discovered in rat studies with tracer doses of ^{67}Cu(II)
- Mr 270 k by size exclusion chromatography
- Two subunits (about 200 k and 60 k)
- 200 k subunit is lightly glycosylated (about 10 kDa)
- An alpha-$_1$-glycoprotein
- Has a higher affinity for Cu(II) than rat albumin
- Readily exchanges Cu(II) with albumin
- Accounts for about 12% of the copper in rat serum

other hand, do not have these two macroglobulins and have large amounts of alpha-2-macroglobulin. Thus, it seems likely that humans do not have transcuprein, per se, but employ another macroglobulin for the same (copper transport) purpose. We are currently investigating these possibilities and verifying in rats that alpha-1-I3 is truly transcuprein. Meanwhile, we have verified that the transcuprein (void volume) fraction of human plasma does not bind as much tracer copper as does that of rodents (Fig. 2B). Nevertheless, a significant portion of the copper in human plasma is associated with this fraction (Fig. 2C) (Wirth and Linder, 1985; Linder, 1991a; Barrow and Tanner, 1988; Evans and Fritze, 1969). From the data currently available, it would appear that about 10% of human plasma copper may be with transcuprein-like macroglobulins (Linder, 1991a).

3. CERULOPLASMIN IN COPPER TRANSPORT

3.1. Ceruloplasmin Structure and Expression

Ceruloplasmin has long been recognized as the major copper containing protein in the blood plasma, comprising about 65% of the copper in that fluid in the human (Wirth and Linder, 1985; Barrow and Tanner, 1988; Linder, 1991a). Human blood ceruloplasmin

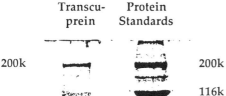

Figure 3. SDS-Polyacrylamide gel electrophoresis of purified transcuprein. Subunits of about 200 and 60 k are indicated. N-terminal and internal amino acid sequencing indicates that the 200 k subunit is alpha-1-inhibitor-3, while the 60 k subunit is highly homologous to albumin. The "subunits" of about 100 kDa appear to be degradation products of the 200 k subunit. From Tsai, 1990.

Table 4. Alpha-1-Inhibitor 3

- Belongs to the macroglobulin family of protease inhibitors (traps)(which includes alpha-2-macroglobulin)
- Monomeric protein of about 190 kDa (1446 amino acids)(Subunits of other macroglobulins are in the same size range)
- Has open and closed conformations (x-ray scattering) depending upon whether it is binding another protein
- [Alpha-2-macroglobulin can bind Cu(II) and is considered a major transporter of Zn(II); it also binds small basic polypeptides, including some growth factors]
- Not detected in human plasma and may be rodent specific among mammals
- Four rat genes for alpha-1-I3; at least two forms in rat plasma; one is negative acute phase reactant, the other is not; liver specific expression

is a single polypeptide of 1046 amino acids to which several carbohydrate chains are attached, with a total molecular weight in the range of 132,000. Recent x-ray crystallographic studies (Zaitseva et al., 1996) indicate that it is composed of 6 plastocyanin-like domains, numbers 2, 4 and 6 each containing a single Cu atom, with three additional Cu atoms between domains 1 and 6, presumably comprising its trinuclear copper center, for a total of 6 Cu atoms per molecule. Copper in ceruloplasmin is buried, non-dialysable, and not a part of the exchangeable plasma copper pool. Not all ceruloplasmin is fully loaded with copper, especially in copper deficiency (see Linder, 1991a; Middleton and Linder, 1993). Blood ceruloplasmin is thought to derive primarily from the liver (Linder et al., 1997a), but many other tissues also express ceruloplasmin and probably secrete it, particularly those involved in producing proteins for other body fluids, like Sertoli cells, mammary gland, and the choroid plexus of brain (Linder and Hazegh-Azam, 1996).

3.2. Functions of Ceruloplasmin in the Blood

Extensive research data indicate that ceruloplasmin is a multifunctional protein, particularly involved in radical scavenging and in copper as well as iron transport (Linder and Hazegh-Azam, 1996; Linder et al., 1997a). As concerns iron transport (Table 5-I), the focus of much current interest, it has long been known that chronically reduced levels of blood ceruloplasmin are associated with increased accumulations of iron in liver and some other organs, and that intravenous infusion of ceruloplasmin (but not ionic copper) results in an immediate release of iron from the liver to the blood (Linder,1991a and 1993; Osaki et al., 1971; Roeser, et al., 1970; Ragan et al., 1969). The same phenomenon has been observed with copper deficient intestine (Wollenberg et al., 1990). The prevailing concept, first postulated by Frieden and Osaki (Osaki et al., 1966; Frieden, 1986), is that iron is available for release from cells in the ferrous form, and that plasma ceruloplasmin helps to oxidize it to Fe(III) so it is able to attach to its transport protein in the plasma, transferrin. Recent findings for humans with the rare genetic abberration of aceruloplasminemia (Yoshida et al., 1995) are consistent with this concept, in that there is tissue iron accumulation. Also, a copper protein (Fet3) is required by yeast cells for transmembrane transport of iron (Askwith et al., 1994; Dancis et al., 1994), although in this case importation rather than release is involved. Nevertheless, ceruloplasmin cannot be essential for the flux of iron out of cells of the liver or other organs, as very large amounts of iron enter and exit these tissues daily but excess iron only accumulates very slowly even when ceruloplasmin is totally absent from the plasma.

There is also much evidence that ceruloplasmin plays a role in antioxidant defense (Table 5-II), particularly during inflammation when production of the protein is increased. Numerous *in vitro* and cell culture studies indicate that ceruloplasmin provides specific

Table 5 Serum ceruloplasmin functions. (for details, please consult Linder and Hazegh-Azam, 1996; Linder et al, 1997a; and Linder, 1991a.)

I. Ceruloplasmin has a role in iron transport across cell membranes
- in copper deficiency, when ceruloplasmin concentrations fall very low, there is a gradual accumulation of iron in liver and some other organs
- release of iron to the blood from copper deficient liver or intestine is immediately enhanced by adding ceruloplasmin to the perfusing blood
- only small quantities of ceruloplasmin are needed for this function (1-2% of normal levels)
- humans with aceruloplasminemia also accumulate Fe in some organs
- ceruloplasmin has ferroxidase activity [oxidizes Fe(II) to Fe(III)]
- Fe(III) [not Fe(II)] binds to the iron transport protein, transferrin
- Fet3, a copper protein with homology to laccase and ceruloplasmin, is required for transmembrane import of Fe into yeast cells

II. Ceruloplasmin has a role in inflammation when iron transport is reduced
- liver biosynthesis and plasma levels are increased in inflammation and infection; inflammatory cytokines stimulate this response
- the role in inflammation may be to protect normal cells against oxygen radicals produced in sites of inflammation by leukocytes:
  · ceruloplasmin neutralizes oxygen radicals and inhibits
  · damage induced by a variety of such radicals
  · ceruloplasmin receptors are present on most cells, including red cells
  · and would normally be expected to be saturated with ceruloplasmin (K_D is in the range of 10^{-8}M; blood ceruloplasmin is about 2 mM)

III. Ceruloplasmin has a role in copper transport
- ceruloplasmin-copper given i.v. is readily absorbed by most tissues
- ceruloplasmin-copper is readily absorbed by cells in culture
- specific tissues like heart and placenta are particularly avid for ceruloplasmin- versus non-ceruloplasmin-copper in the plasma
- ceruloplasmin-copper turns over fairly rapidly (half-life in rats is about 20 hours) and is enhanced by copper deficiency
- ceruloplasmin-copper is absorbed more rapidly by most tissues during copper deficiency

protection against oxygen radicals and is a scavenger of such radicals, although not a superoxide dismutase (see Linder, 1991a; Goldstein and Charo, 1982). It is noteworthy that this function of ceruloplasmin is clearly distinct from its function in promoting efflux and transport of iron. In inflammation, there is an increased expression of ceruloplasmin by the liver (Gitlin, 1988), resulting in increased levels of ceruloplasmin in the plasma (Fig 4). However, this does not result in an increased efflux of iron from the liver but rather the opposite (Konijn and Hershko, 1977, 1981): and levels of iron in plasma are markedly reduced. Indeed, it is a well known hallmark of inflammation that extracellular transport of iron diminishes (Fig. 4), presumably to make less of it available to infectious agents for whom iron is growth limiting (see Weinberg, 1978; Linder, 1993).

The evidence for blood ceruloplasmin as a copper transport protein is summarized in Table 5-III. As indicated, numerous studies, particularly from our laboratory, using cell culture as well as whole animals, have shown that ceruloplasmin-copper is readily absorbed by most cells or tissues, and that many cells or tissues show a marked preference for uptake of copper from ceruloplasmin as opposed to that in the exchangeable plasma pool (involving albumin and transcuprein) (Linder et al., 1997a). One example is given in Fig. 5, in which tissue uptake of tracer copper from ^{67}Cu-labeled ceruloplasmin or the exchangeable-copper fraction of the plasma was compared after intravenous infusions into

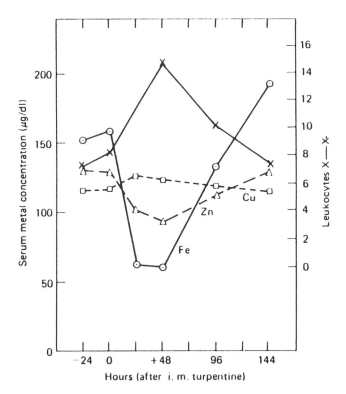

Figure 4. Changes in serum iron, copper and zinc in response to systemic inflammation in humans. Reprinted with permission from Linder, 1991b.

pregnant rats (Lee et al., 1993). Clearly, ceruloplasmin-copper was preferentially absorbed by all the tissues examined, although there was considerable variation in terms of the relative preference for the two copper sources: placenta (and therefore fetal tissues), as well as heart and brain had preferences in the range of 5–7-fold, whereas it was only twice as likely to be absorbed by the uterus, liver and kidney. In these same studies, the importance of ceruloplasmin as a source of copper for tissues like the placenta was corroborated by showing that inhibition by cycloheximide of ceruloplasmin synthesis (and incorporation of ^{67}Cu into ceruloplasmin) also markedly inhibited accumulation of ^{67}Cu in most tissues, after administration of ionic ^{67}Cu (Lee et al., 1993; Linder et al., 1997a).

More recently, to further demonstrate the importance of ceruloplasmin as a tissue copper source, rats maintained on low copper diets with and without copper added to their drinking water were compared in terms of their tissue uptake of ^{67}Cu from ceruloplasmin administered i.v. (Table 6). With the exception of the kidney, there was greater uptake of copper from ceruloplasmin by tissues of rats that had been on the copper deficient diets. This indicates that cells even mildly deficient in copper will enhance their uptake of the element from ceruloplasmin.

3.3. Ceruloplasmin and Copper in Milk

We and some others have established that ceruloplasmin is present not only in plasma and some other body fluids (Table 7) but also in the milk of many species, including that of humans, pigs and rats (Wooten et al., 1996; Linder et al., 1997a,b). This is con-

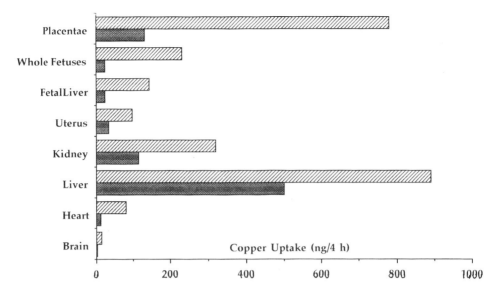

Figure 5. Tissue uptake of copper from ceruloplasmin or the exchangeable copper pool after I.V. administration to pregnant rats. Data are from Lee et al. (1993) in which pregnant rats near term were injected by tail vein with ng quantities of ^{67}Cu-labeled ceruloplasmin (slashed bars) or with ^{67}Cu-NTA mixed with rat plasma (dark bars). Mean values for actual uptake of copper (in ng over 4 hours) are shown. These were calculated based on the plasma specific activities of each copper source. All differences were statistically significant (p< 0.05 or less).

sistent with the finding of mRNA for ceruloplasmin expressed by mammary gland and mammary tumors, and with the finding (by in situ hybridization) that expression is in the ductal endotheliel cells involved in milk protein production (Jaeger et al., 1991; Linder and Hazegh-Azam, 1996). Quantities of ceruloplasmin in the milk vary with lactational stage in humans and pigs, being 2–4 fold greater in the first days after birth (post colostrum) than after one month or more of lactation. We have been examining the ceruloplas-

Table 6. Effect of mild copper deficiency on uptake of copper from ceruloplasmin

	Total uptake (cpm):		
	Cu-sufficient	Cu-deficient	% Change
Kidney	11500	7510*	-35
Liver	23280	32470*	+39
Heart	1640	1915	+17
Spleen	711	1253*	+63
Skeletal	34398	51156*	+49

Cu-ceruloplasmin was purified from the plasma of a donor rat and administered by tail vein. Blood samples (120 ul) were taken for counting of radioactivity at various times thereafter to determine half life (Fig. 6), and blood and tissue samples were obtained at sacrifice, 24 hours after ceruloplasmin administration. Tissue uptake was determined by gamma counting and corrected for residual blood contamination: Values (percent weight) for blood contents of organs used were kidney 6.4, liver 8.4, heart 3.0, spleen 6.4, skeletal muscle 1.3. Ceruloplasmin (measured as p-phenylene diamine oxidase activity) was 62 % lower in the copper deficient rats. [Body weights of these female Sprague Dawley rats were 205 ± 13 (Mean ± SD).] Values are means for groups of 6 rats. SDs averaged about 10%.
*Statistically significant difference as compared with Cu-sufficient ($p < 0.01$).[67]

Table 7. Ceruloplasmin and copper concentrations in other human body fluids

	Ceruloplasmin		
	Protein (mg/L)	Cu (ug/L)	Total copper (ug/L)
Plasma/Serum	320	750	1200
Synovial fluid	284	600	3500
Lymph	1.5	5	1200
Cerebrospinal fluid	0.9	3	5000
Amniotic fluid	(17)	(6)	175 (192)
Milk	4.7	16	1670

(Modified from Linder et al., 1991a.) Except for plasma and milk, for which values were determined directly, the ceruloplasmin-copper values shown were calculated based on the assumption of a Cu/protein ratio similar to plasma ceruloplasmin. The data in parentheses for amniotic fluid are for rats, near term.

min mRNA in biopsies of mammary gland from the pig (Linder et al., 1997b; Cerveza et al., 1998) and find (a) by Northern analysis, that the size of the message appears to be identical to that expressed in pig liver; (b) by RT-PCR with human-based primers, that all of the exons coding for the protein are present and of the expected size; and (c) by purification and Western blotting, that pig milk and serum ceruloplasmins are of the same size (about 141 kDa), and that only one major form is present. Moreover, there is a clear parallel between the degree of expression of ceruloplasmin mRNA by the mammary gland and daily production of the protein. Fig. 7 illustrates this relationship and shows that, while

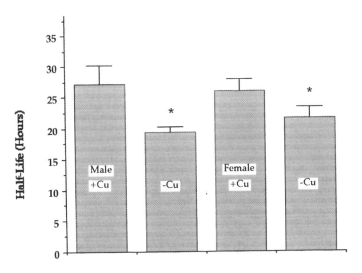

Figure 6. Effect of copper deficiency on the half-life of ceruloplasmin-copper. Groups of 6 adult male (left two bars) and female (two right bars) Sprague Dawley rats were infused via the tail vein with ug quantities of ^{67}Cu-labeled ceruloplasmin purified from the plasma of ^{67}Cu-NTA treated donor rats. All rats had been fed a copper deficient diet, but half of each set were supplemented with copper in their drinking water. The first order fall of blood radioactivity was monitored over about 24 hours, from which half-life was calculated. Data are Means ± SD. Stars (*) indicate a significant difference due to copper deficiency ($p< 0.01$).

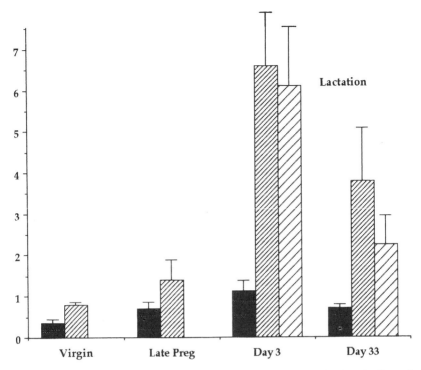

Figure 7. Differences in ceruloplasmin mRNA content of mammary gland and production of milk ceruloplasmin in pigs before and during lactation. The data show mean values ± SD for total RNA concentration (mg/g tissue) (black bars), total Cp mRNA (relative units) (grey slashed bars), and daily milk ceruloplasmin production (mg/day) (light slashed bars) in virgin animals, in late pregnancy, and on days 3 and 33 of lactation. From Cerveza et al., 1998.

there was some expression of ceruloplasmin mRNA also in conditions when there was no lactation (the virgin and late pregnancy states), expression increased dramatically with onset of lactation. Total RNA concentrations (per g mammary gland) increased already prior to lactation and increased further when lactation began. Production of ceruloplasmin and expression of its mRNA then decreased to the same degree over the next weeks of lactation. The smaller changes in *plasma* ceruloplasmin concentrations that occurred in these animals in no way paralleled the observed changes in *milk* ceruloplasmin.

Results of our current studies of lactating rats indicate that copper administered as ^{67}Cu-NTA in ng amounts i.p. or i.v. is rapidly absorbed from the circulation by mammary tissue (Fig. 8). what is particularly noteworthy is that initially, uptake by mammary gland is greater than that of the liver! Copper also very rapidly appears in the milk (Fig. 9) and in milk ceruloplasmin (determined by immunoprecipitation); and it seems that a constant proportion of tracer copper entering the milk (presumably secreted by mammary cells) is with ceruloplasmin. These preliminary results (Fig. 9) also suggest that in rats, a relatively large proportion of the copper in rat milk is associated with ceruloplasmin. Combined with our previous findings (Wooten et al., 1996) that ceruloplasmin-copper administered orally to suckling rats is taken up more rapidly than is ionic copper mixed with milk (Table 8), these findings suggest that ceruloplasmin plays a significant role in the nutritional transport of copper from mother to offspring during the suckling period. (As shown in the ta-

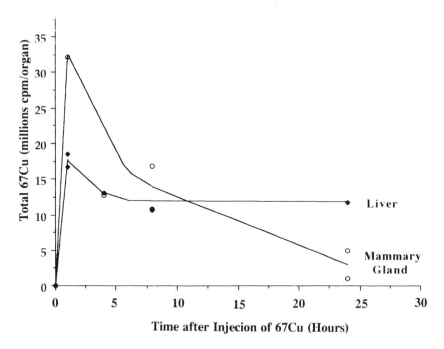

Figure 8. Course of incorporation of ^{67}Cu(ii) tracer into mammary gland and liver of lactating rats. Lactating rats were injected i.p. with ng quantities of radioactive copper, as the Cu-NTA chelate, and sacrificed at various times thereafter for determination of total radioactivity in each tissue. Points represent total cpm in livers (solid circles) and mammary gland (open circles) of individual animals. (Some points overlapped.)

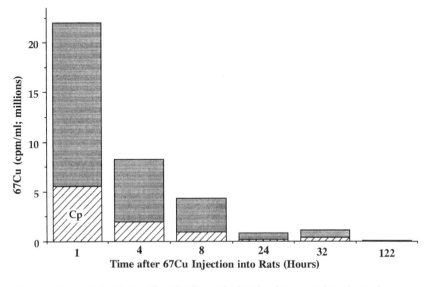

Figure 9. Radioactive copper in whole milk and milk ceruloplasmin of rats at various times after administration of ^{67}Cu-NTA. Values are averages for 3–4 rats at each time point after treatment as in Fig. 8. The overall percentage of copper in ceruloplasmin (Cp) was 27 ± 8 (Mean ± SD, N = 24). The latter was determined by counting washed immunoprecipitates.

Table 8. Preferential uptake of copper from ceruloplasmin after feeding to suckling but not weanling rats

	Radioactivity in			
	Liver		Other Sites	
Source	Cp	Cu-NTA	Cp	Cu-NTA
Newborn rats				
4 hours	1.6	0.5*	3.2	1.4 (carcass)
24 hours	12	5.4*	8.0	5.4*
Weanling rats				
4 hours	6.0	5.6	94	93 (stomach + intestines)
24 hours	23	23	70	75

Uptake (percent of dose) of ^{67}Cu from ceruloplasmin (Cp) or Cu(II)-nitrilotriacetate (Cu-NTA) at various times after oral administration to newborn and weanling rats. Data are mean values for ^{67}Cu in liver and carcass (minus the GI tract) or remaining in the stomach and GI tract after administration of equivalent amounts of Cu tracer as Cp or Cu-NTA. (Summarized from Lee et al., 1993.)
*significantly less uptake than from ceruloplasmin (p < 0.01)

ble, uptake of ceruloplasmin-copper is no longer preferred after weaning.) This would be in addition to the role of plasma ceruloplasmin in nutritional copper transport during gestation, already identified (Lee et al., 1993).

REFERENCES

Adham, N.F., Song, M.K., and Rinderknecht, H. (1977). Binding of zinc to alpha-2-macroglobulin and its role in enzyme binding activity. Biochim. Biophys. Acta 495, 212–219

Askary, S.H. (1995). Purification, copper binding an sequencing of transcuprein. Master's thesis, California State University, Fullerton.

Askwith, C.E., Eide, D., Van Ho, A., et al. (1994). The FET3 gene of S. cerevisiae encodes a multicopper oxidase required for ferrous ion uptake. Cell 76, 403–410.

Barrow, L., and Tanner, M.S. (1988). Copper distribution among serum proteins in paediatric liver disorders and malignancies. Eur. J. Clin. Invest. 18, 555–560

Bearn, A.G., and Kunkel, H.G. (1954). Localization of ^{64}Cu in serum fractions following oral administration: an alteration in Wilson's disease. Proc. Soc. Exp. Biol. Med. 85, 44–48

Breslow, E. (1964). Comparison of cupric ion-binding sites in myoglobin derivatives and serum albumin. J. Biol. Chem . 239, 3252–3259

Cerveza, P.J., Cotton, S.J., Mehrbod, F., Hazegh-Azam, M., Lomeli, N., Linder, M.C., Wickler, S., and Fonda, E. (1998). Ceruloplasmin expression in pig mammary gland, milk and serum during lactation. Submitted

Dancis, A., Yuan, D.S., Moehle, C., et al. (1994). Molecular characterization of a copper transport protein in S. cerevisiae: an unexpected role for copper in iron transport. Cell 76, 393–402

Eggertsen, G., Hudson, G., Shiels, B., Reed, D., and Fey, G.H. (1991). Sequence of rat alpha-1-macroglobulin, a broad-range proteinase inihibitor from the alpha macroglobulin-complement family. Mol. Biol. Med. 8, 287–302

Evans, D.J.R., and Fritze, K. (1969). The identification of metal-protein complexes by gel chromatography and neutron activation analysis. Anal. Chim. Acta 44, 1–7

Frieden, E. (1986). Perspectives in copper biochemistry. Clin. Physiol. Biochem. 4, 11–19

Gitlin, J.D. (1988). Transcriptional regulation of ceruloplasmin gene expression during inflammation. J. Biol. Chem. 263, 6281–6287

Goldstein, I.M. and Charo, I.F. (1982). Ceruloplasmin: An acute phase reactant and anti-oxidant. Lymphokines 8, 373–411

Jaeger, J.L., Shimizu, N., and Gitlin, J.D. (1991). Tissue-specific ceruloplasmin gene expression in the mammary gland. Biochem. J. 280, 671–677

Konijn, A.M., and Hershko, C. (1977). Ferritin synthesis in inflammation. I. Pathogenesis of impaired iron release. Brit. J. Haematol. 37, 7–16.

Konijn, A.M., and Hershko, C. (1981). Ferritin synthesis in inflammation. II. Mechanism of increased ferritin synthesis. Brit. J. Haematol. 49, 361–370.

Lau, S., and Sarkar, B. (1971). Ternary coordination complex between human serum albumin, copper (II) and L-histidine. J. Biol. Chem. 246, 5938–5943

Lee, S.H., Lancey, R., Montaser, A., Madani. N,, and Linder, M.C. (1993). Ceruloplasmin and copper transport during the latter part of gestation in the rat. Proc. Soc. Exp. Biol. Med. 203, 428–439

Linder, M.C. (1991a). The Biochemistry of Copper (New York, New York: Plenum Press)

Linder, M.C. (1991b). Nutrition and metabolism of the trace elements. In Nutritional Biochemistry and Metabolism, M.C. Linder, ed. (New York, New York: Elsevier Medical Publishers), pp. 215–276

Linder, M.C. (1993). Interactions between copper and iron in mammalian metabolism. In Metal-Metal Interactions, B. Elsenhans, W. Forth, and K. Schumann, eds. (Gutersloh, Germany: Bertelsmann Foundation Publishers), pp. 11–41

Linder, M.C., and Hazegh-Azam, M. (1996). Copper biochemistry and molecular biology. Am. J. Clin. Nutr . 63, 797S-811S

Linder, M.C., Wooten, L., Cerveza, P., Cotton, S., Shulze, R. and Lomeli, N. (1997a). Copper transport. Am. J. Clin. Nutr. Suppl., in press.

Linder, M.C., Cerveza, P., Cotton, S., Lomeli, N., Donley, S., Cotton, S., Shulze, R.A., Mehrbod, F., Dominguez, D., and Sridhar, A. (1997b). Ceruloplasmin in milk and its formation by mammary gland. FASEB J. 11, Abstract 1230

Lonberg-Holm, K., Reed, D.L., Roberts, R.C., Hebert, R.R., Hillman, M.C. and Kutney, R.M. (1987). Three high molecular weight protease inhibitors of rat plasma. J. Biol. Chem. 262, 438–445

Masuoka, J., Hegenauer, J., Van Dyke, B.R., and Saltman, P. (1993). Intrinsic stoichiometric equilibrium constants for the binding of zinc(II) and copper(II) to the high affinity site of serum albumin. J. Biol. Chem. 268, 21533–21537

Middleton, R.B., and Linder M.C. (1993). Synthesis and turnover of ceruloplasmin in rats treated with 17-beta-estradiol. Arch. Biochem. Biophys. 302, 362–368

Osaki, S., Johnson, D.A. and Frieden, E. (1966). The possible significance of the ferrous oxidase activity of ceruloplasmin in normal human serum. J. Biol. Chem. 241, 2746–2751

Osaki, S., Johnson, D.A. and Frieden, E. (1971). The mobilization of iron from the perfused mammalian liver by a serum copper enzyme, ferroxidase I. J. Biol. Chem. 246, 3018–3023

Osterberg, R., Boive, T., Wang, W., Mortensen, K., Saito, A., Sinohara, H., and Ikai, A. (1994). Small angle scattering study of alpha-1-inhibitor III from rat blood plasma. Biochim. Biophys. Acta 1207, 152–158

Osterberg, R., and Malmensten, B., (1984). Methylamine-induced conformational change of alpha-2-macroglobulin and its zinc(II) binding capacity. An x-ray scattering study. Eur. J. Biochem. 143, 541–544

Ragan, H.A., Nacht, S., Lee, G.R., Bishop, C.R., and Cartwright, G.E. (1969) Effect of ceruloplasmin on plasma iron in copper deficient swine. Am. J. Physiol. 217, 1320–1323

Regler, R., Sickinger, S., and Schweizer, M. (1991). Differential regulation of the two mRNA species of the rodent negative acute phase protein alpha-1-inhibitor 3. FEBS Lett. 282, 368–372

Roeser, H.P., Lee, G.R., Nacht, S. and Cartwright, G.E. (1970). The role of ceruloplasmin in iron metabolism. J. Clin. Invest. 49, 2408–2417

Rubenstein, D.S., Thogersen, I.B., Pizzo, S.V., and Enghild, J.J. (1993). Identification of monomeric alpha-macroglobulin proteinase inhibitors in birds, reptiles, amphibians and mammals, and purification and characterization of a monomeric alpha-macroglobulin proteinase inhibitor from the American bullfrog Rana catesbiana. Biochem. J. 2990, 85–95

Schweizer , M., Takabayashi, K., Geiger, T., Laux, T., Biermann, G., Buhler, J.M., Gauthier, F., Roberts, L.M., and Heinrich, P.C. (1987). Identification and sequencing of cDNA clones for the rodent negative acute phase protein alpha-1-inhibitor 3. Eur. J. Biochem. 164, 375–381

Scott, K.C., and Turnlund, J.R. (1994). Compartmental model of copper metabolism in adult men. J. Nutr. Biochem. 5, 342–350

Tsai, M.T. (1990). Transcuprein purification and characterization. Master's Thesis, California State University, Fullerton.

Weinberg, E.D. (1978). Iron and infection. Microbiol. Rev. 42, 45–66

Weiss, K.C., and Linder, M.C. (1985). Copper transport in rats involving a new plasma protein. Am. J. Physiol. 249, E77-E88

Wirth, P.L., and Linder, M.C. (1985). Distribution of copper among components of human serum. J. Nat. Cancer Inst. 75, 277–284

Vargas, E.J., Shoho, A.R., and Linder, M.C. (1994). Copper transport in the Nagase analbuminemic rat. Am. J. Physiol. 267, G259-G269

Wollenberg, P., Mahlberg, R. and Rummel, W. (1990). The valency state of absorbed iron appearing in the portal blood and ceruloplasmin substitution. Biol. Metals 3, 1–7.

Wooten, L., Shulze, R.A., Lacey, R.W., Lietzow, M., and Linder, M.C. (1996). Ceruloplasmin is found in milk and amniotic fluid and may have a nutritional role. J. Nutr. Biochem. 7, 632–639

Yoshida, K., Furihata, K., Takeda, S. (1995). A mutation in the ceruloplasmin gene is associated with systemic hemosiderosis in humans. Nature Genet. 9, 267–272

Zaitseva, I., Zaitsev, V., Card, G., Moshkov, K., Box, B., Ralph, A., and Lindley, P. (1996). The nature of the copper centres in human ceruloplasmin. J. Biol. Inorg. Chem. 1, 15–23

2

MODELS TO EVALUATE HEALTH RISKS DERIVED FROM COPPER EXPOSURE/INTAKE IN HUMANS

Manuel Olivares, Ricardo Uauy, Gloria Icaza, and Mauricio González

Institute of Nutrition and Food Technology
University of Chile

INTRODUCTION

Copper (Cu) is essential for the survival of plants and animals. Animal and human studies have shown that Cu is involved in the function of several enzymes (Linder and Hazegh-Azam, 1996). Different studies have demonstrated that Cu is required for infant growth, host defense mechanisms, bone strength, red and white blood cell maturation, iron transport cholesterol and glucose metabolism, myocardial contractility, and brain development (Danks, 1988).

Copper deficiency can result in the expression of an inherited defect such as Menkes disease or in an acquired condition. Acquired deficiency is a clinical syndrome that occurs mainly in infants, although it has also been described in children and in adults. This deficiency can be the consequence of decreased Cu stores at birth, inadequate Cu supply, inadequate Cu absorption, increased requirements, and increased losses (Olivares and Uauy, 1996a). Clinically evident Cu deficiency is a relatively infrequent condition in humans. The most frequent clinical manifestations of acquired Cu deficiency are anemia, neutropenia and bone abnormalities that include osteoporosis and fractures (Shaw, 1992; Olivares and Uauy, 1996a).

Many nutrients consumed in excess can be toxic. Acute Cu toxicity is infrequent in man, and usually is a consequence of ingesting contaminated foodstuffs or beverages (including drinking water), and from accidental or voluntary ingestion of high quantities of Cu salts. Acute symptoms include salivation, epigastric pain, nausea, vomiting and diarrhea. Intravascular hemolytic anemia, acute liver failure, acute renal failure with tubular damage, shock, coma, and death have been observed in severe Cu poisoning (USEPA, 1985; USEPA, 1987).

There are some reports in humans, suggesting that the consumption of beverages or drinking water contaminated with Cu results in nausea, vomiting, and diarrhea (Wyllie, 1957; CDC, 1975; Spiltany et al., 1984; Stenhammar, 1979; Berg et al., 1981; Knobeloch et al., 1994). However, the threshold for acute gastrointestinal adverse effects of Cu in drinking water has not precisely established.

The long term toxicity of Cu has been less studied. Chronic toxicity in humans is observed principally in patients with Wilson disease and from the occurrence of infantile cirrhosis in areas of India (Indian childhood cirrhosis), and isolated clusters of cases in other countries (idiopathic copper toxicosis) that have been also related to excessively Cu intake (Pandit and Bhave, 1996; Horslen et al., 1994). However, the most likely explanation for this condition appears to be a combination of a genetically determined defect in Cu metabolism, and a high Cu intake (Olivares and Uauy 1996b; Scheinberg and Sternlieb, 1996; Müller et al., 1996).

The need to develop specific approaches for the assessment of risk associated to the exposure of essential elements has been recently recognized. Traditional methods to assess health based risk used for non essential elements start by defining an intake/exposure level at which no observed adverse effects (NOAEL) of biological significance are found or the lowest level at which adverse effects are observed (LOAEL). In addition uncertainty (UF) or modifying factors (MF) are used to adjust NOAEL or LOAEL and define reference doses (RfD) which are chronic intake/exposure levels considered safe or of no significant health risk for humans (USEPA, 1993; Barnes and Dourson, 1988; Dourson, 1994). Uncertainty factors are used in deriving RfDs from experimental data of toxicity obtained from animal studies, or when data from humans is insufficient to fully account for variability of populations or special sensitivity sub groups of the general population, or when NOAEL has been obtained in studies of insufficient duration to assure chronic safety, or when the database on which the NOAEL is supported is incomplete, or when the experimental data provides a LOAEL instead of a true NOAEL (Dourson, 1994; Dourson and Stara, 1983). The usual value for each UFs is a 10 fold reduction in the acceptable exposure level for each of the considerations listed above and may be used in isolation or in combination depending on the specific element being assessed. Modifying factors are additional uncertainty factors which have a value of 1 or more but less than 10 and are based on professional judgment of the overall quality of the data available (Dourson, 1994). Given the limited human data available the limitations of animal models and the uncertainties of the interpretation the traditional toxicological approach to defining limits for exposure from essential elements summarized here may in fact lead to establishing limits which promote or even induce deficiency if followed by the population.

NEW APPROACH FOR THE ASSESSMENT OF HEALTH RISK FROM EXPOSURE/INTAKE TO ESSENTIAL TRACE ELEMENTS

A new approach considers the health risk assessment for essential elements taking into account risks associated to low intakes as well as high intakes. The relationship between intake/exposure level and risk for essential elements has a U of J shape, not perfectly symmetric but basically should include risk of deficiency associated to low intakes and risk of toxicity associated to high intakes (Goyer, 1994).

The steps that should be followed in applying the new risk assessment model are:

Step 1. Selection of Data from Human Studies and Relevant Animal Studies

The review of experimental studies in which several intake/exposure levels have been evaluated in order to estimate requirements and toxic effects is necessary. Since the aim is to develop population guidelines the estimate should include a measure of variability in order to define the distribution of requirements and toxicity for an ideal normal population defined as acceptable range of exposure/intake (AREI) for the specific element.

The nature of this review should include the full range of biological effects starting with the most extreme such as death, completing the list of effects with the most subtle such as taste.

Death and disease related to acute or chronic exposure/intake to essential elements can in fact occur and should be critically assessed. Intakes at which death from deficit or excess have been reported should be studied, considering that in some situations the biological effect of deficit or excess from the essential element may be due to other intervening factors, such as disease conditions or genetic predisposition. In these cases, it is particularly difficult to establish a causal relationship, unless controlled studies are available. The controlled clinical trial is the closest to proof of causality, eliminating the element (in the case of toxic effects) or incorporating the agent (in the case of deficit) in a group of subjects randomized to control or experimental conditions is necessary to test the hypothesis of a causal relationship. There are few elements where this type of data are available from chronic exposure or acute poisoning ; in the case of elements used for suicides, such as copper or iron. In the case of animals LD_{50} and LD_5 and LD_{95} have been defined for some elements, but are not available for most elements. The limitation of human data on lethality from deficit or toxicity is that they are not population based, since they are usually linked to single case reports or at best cluster of cases due to unusual circumstances. This is due to the fact that human populations in modern times are protected from death from deficient or toxic exposure/intake levels.

For most essential elements disease conditions associated to deficit have been well described (Fe, Cu, Zn, Mn, Se); much less is known of pathologic conditions linked to excess, few of these have been well characterized. In both cases, deficit and excess, disease has sometimes been linked to genetic conditions which favor the occurrence of deficiency or toxicity by modifying requirements, decreasing or increasing absorption or altering excretion. In many pathologic conditions specific data on chronic levels of exposure/intake leading to disease from excess or deficit may be lacking, but information on the actions required to cure the condition may be indicative on level of exposure/intake which can be considered safe. For example we know the level of zinc intake required to prevent growth failure in infants (NRC, 1989) or the level of iron excess which may induce hemolysis in susceptible individuals (Williams et al., 1975). On the other hand, in many cases the pathological effect of the metal may occur only in the presence of another concurrent factor, viral infection in the case of myocardial degenerative effects induced by selenium deficiency (Keshan disease) (Cousins, 1996) or copper toxicity in patients undergoing hemodialysis due to chronic renal failure (Bloomfield et al., 1971). The data from exposure/intake levels to prevent disease states, if available, represents the best option to define AREI since there is no question that preventing disease is relevant to the protection of human health.

The next level in evaluating data to estimate AREI is the assessment of studies that evaluate biologically significant effects of various exposure/intake levels. The difficulty here is establishing what will be considered biologically significant and what will be an effect with no significance. A suitable definition for biological significance in this context is the capacity of the biological indicator modified by E/I to predict the occurrence of deficiency or toxicity disease associated to the corresponding element. For example elevated ferritin may serve to predict liver damage induced by elevated iron intake in susceptible individuals (Beaumont et al., 1979). A functional assay, such as red cell resistance to peroxide stress may also predict the risk of hemolysis from excess iron (Farrell et al., 1977). On the other hand the elevation of superoxide dismutase in this same condition may represent an adaptive response to oxidant stress without pathological implication (Nielsen and Milne, 1993). Unfortunately most biochemical or functional biomarkers have not been validated in terms of their ability to anticipate the occurrence of disease. In this context, the most valid are those that relate to the limits of AREI, for example levels of E/I that indicate excessive or deficient retention of specific element. For example negative Zn balance over time will lead to disease state, on the other markedly positive copper balance may serve to indicate level of E/I which over time may lead to toxicity. For the purpose of quantifying AREI, a key biomarker may be the change in bioavailability induced by high or low E/I since this may be the most sensitive indicator of excess or deficit. For example high Cu E/I is associated to lower bioavailability while low Cu E/I leads to higher copper bioavailability, as measured by labeled Cu studies using stable isotopes which permit the separation of exogenous E/I from endogenous excreted Cu (Turnlund et al., 1989).

Following these concepts one could develop quantitative estimates of AREI on the basis of lethal effect, prevention of disease or assuring normal range of biomarkers with biological significance. The most stringent criteria will undoubtedly be this last one. In terms of feasibility, prevention of disease and normalcy of biomarkers are most likely to be used.

Step 2. Biokinetics and Interactions

A key consideration in defining AREI is the biological handling of the element which may affect the biological impact in terms of deficit or excess. This includes absorption, transport, metabolism, excretion and storage of the element and the possible interaction with other elements or factors which modify them. Consideration of these factors in adjusting up or down the level of AREI in a given context is important since they may modify the biological effect significantly. Factors which affect biokinetics differ for each element, but basically include interaction of elements amongst themselves, interactions with other dietary factors and the effect of environmental agents which may affect biokinetics. For example excess zinc will interfere with Cu absorption (van Campen and Scaifi, 1967), thus the higher limit of Cu AREI will be greater if there is concomitantly exposure to high zinc. Ascorbic acid will enhance the absorption of iron (Stekel et al., 1985), thus the lower limit for iron I/E may be smaller if diets are high in ascorbic acid, on the contrary if fiber intake is high iron absorption will be less (Simpson et al., 1981). Trace metals are not metabolized, thus excess E/I if not adequately excreted will accumulate and once storage capacity is saturated will lead to toxicity. The capacity to adapt to high E/I for most trace essential metals is dependent on the regulation of absorption. The capacity to store trace elements is present for some metals, for example, iron and Cu but not for all, i.e. zinc. Thus, deficiency of zinc induced by low intake may be apparent over a shorter time frame than for iron or Cu. Excretion of most essential trace elements is by the gastrointestinal route but it is not effective to regulate homeostasis since it is easily saturated leading to accumulation and possible toxicity. There is no efficient way to excrete iron except by blood loss (Bothwell et al., 1979), while Cu is mainly excreted by the biliary route (Linder and Hazegh-Azam, 1996). Thus excess E/I, once it saturates excretion and storage capacity may lead to toxicity. Storage of trace elements is mainly in the liver, thus this organ is potential subjected to toxic manifestations, this is the case for iron and Cu.

Step 3. Comparable Procedures to Define Critical Effects from Deficit as well as Excess

Another key point in the definition of an AREI is the assessment of the critical effects of deficiency and toxicity relevant to human health. The most sensitive indices of excess or deficiency may be biomarkers without clear functional or health significance. The other extreme, death associated to organ damage, induced by excess or deficit are clearly of health significance but are not relevant as sensitive indicators of health risks.

We will use Cu to exemplify the assessment of health risks associated to Cu E/I. Biochemical changes such as red cell SOD activity as an index of Cu deficit (Uauy et al., 1985) or changes in plasma Cu/Ceruloplasmin molar ratio as an index of Cu excess are sensitive but non significant indicators of health risks (Frommer, 1976). Biological indices of subclinical effects on specific function indicating potential adverse effects of Cu deficit or excess, for example decreased white blood cell phagocytic capacity in the case of deficit or increased serum aminotransferase or transpeptidase hepatic enzymes in response to excess Cu are often used in human studies (Heresi et al., 1985; JECFA, 1982). Clinical effects such as bone fractures in the case of Cu deficiency or liver dysfunction due to fibrosis in the case of excess are clearly significant in terms of health risk (Shaw, 1992; Sternlieb, 1980), but are difficult to assess in controlled studies since ethical principles in human investigation preclude a precise quantitative definition of these endpoints. Death from bacterial infection associated to neutropenia in the case of deficit or liver failure associated to excess are clearly not applicable endpoints to define ranges of AREI, although they are of unquestionable significance.

The concept that arises from the principle of comparable effects to define excess and deficit is that one should select effects of similar health significance to define upper and lower ranges of AREI, in general the range should be defined by effect response to exposure/intake levels that prevent the appearance of subclinical adverse effects. The review of these indices and the corresponding studies should be done by toxicologists and nutritionists familiar with health risk assessment. The combined effort should yield clearly defined critical endpoints for upper and lower ranges of AREI given the available information.

Step 4. Quantitative Evaluation of Critical Effects to Define AREI

The data to support the lower limit of AREI comes from mineral balance studies and/or from clinical studies indicating that these amounts are needed to prevent deficiency signs. On the excess side the data relevant to humans are quite limited, the upper end of the exposure range is usually derived from the limited studies of individuals taking mineral supplements, accidental or suicidal attempts, or exposure in the workplace or in the environment. It is virtually impossible to find population based evidence of toxic effects in adult humans consuming regular food and potable water, since all public health measures are directed to prevent this. In some cases there are potential rare genetically susceptible groups that may manifest toxicity at exposure/intake levels observed in normal diets (Scheinberg and Sternlieb, 1996; Fleet and Mayer, 1996).

Another approach to estimate AREI in quantitative terms is based on examining the range of exposure/intake of healthy populations, this assumes that if the population is in good health the exposure must be safe. This is particularly helpful if the experimental human information available is limited. A starting point in the definition of AREI can be the customary exposure/intake observed in "healthy" populations. If the upper or lower cut-offs of AREI are obtained extrapolating from animal studies the customary exposure/in-

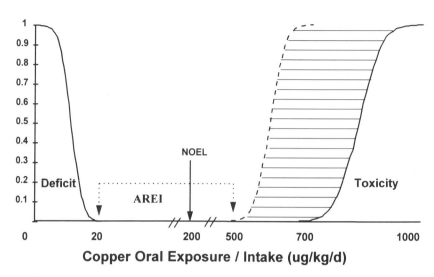

Figure 1. Safe range of oral exposure/intake for adults following the deficiency/excess risk model. The uncertainties in our knowledge of toxic manifestation in humans is depicted by the stripped area. At extremely high chronic exposure liver disease may occur, no specific data provide firm values for NOAEL, but based on WHO/FAO/IAEA recent recommendations the no observed effect level NOEL for copper is close to 200 ug per kg. The acceptable range of exposure/intake (AREI) is shown by dashed line.

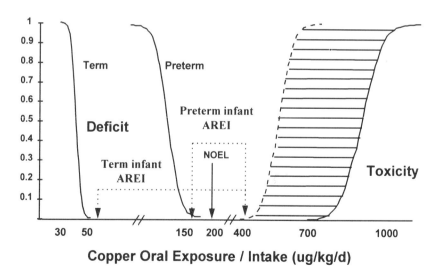

Figure 2. Safe range of oral exposure/intake for infants following the deficiency/excess risk model. The uncertainties in our knowledge of toxic manifestation in infants is depicted by the stripped area. At extremely high chronic exposure liver disease may occur, no specific data provide firm values for NOAEL. As can be noted from the comparison of this figure with figure 1, infants are more sensitive to copper deficit and excess relative to adults per unit of body weight. In the case of premature infants the need is 7.5 times greater than that of adults per unit of body weight. In the latter case the acceptable range or oral intake (AREI) is particularly narrow.

Models to Evaluate Health Risks Derived from Copper Exposure/Intake in Humans

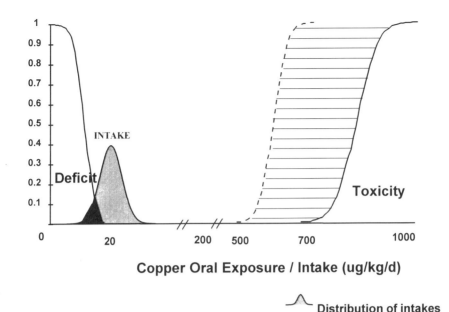

Figure 3. Safe range of oral exposure/intake (AREI) for adults compared to usual dietary intake (gray hatched area following the deficiency/excess risk model). The uncertainties in our knowledge of toxic manifestation in humans is depicted by the stripped area. Data on intake comes from recent WHO/FAO/IAEA data. As depicted in this figure a significant proportion of the population (20%), shown by the black hatched area may be receiving too little copper, while excess copper is not a problem unless food or water is contaminated with copper.

take of "healthy" populations should serve to validate the extrapolation. It would clearly be unwise to define AREI that are outside range of customary intakes of healthy populations (figures 1 and 2). As can be noted from these figures, infants are more sensitive to copper deficit (by a factor of 2.5) and to copper of excess (by a factor of 0.8) relative to adults. That is, they need 2.5 times as much copper as adults per unit of body weight to prevent deficit and tolerate 0.8 times less copper than adults per unit of body weight. The human data do not provide firm data to define a NOAEL but based on a WHO/FAO/IAEA expert group recommendations the no observed effect level for copper is 200 ug per kg body weight (WHO/FAO/IAEA, 1996). As depicted in figure 2 the case of premature infants is special, their need is extremely high 7.5 times greater than that of adults per unit of body weight, while we know little about their limit of tolerance if we consider the same value that adults the acceptable range would be particularly narrow.

The application of this approach is presented in figures 3 and 4. The safe range of oral exposure/intake AREI for adults compared to usual dietary intake is shown in figure 3. Data on intake comes from recent IAEA data collected in several developed countries (WHO/FAO/IAEA, 1996; IAEA, 1992). As demonstrated by these data there is a significant proportion of the population, nearly 20% that may be receiving too little Cu while excess Cu is not a problem unless food or water is contaminated with Cu. The AREI for infants compared to usual dietary intake reveals a similar problem (figure 4), a significant proportion of the infant population is at risk for deficit, while most preterm infants will develop clinical deficiency unless given formula supplemented with extra Cu to concentrations from 1.2 to 2.0

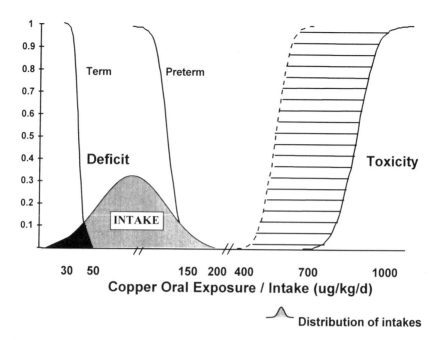

Figure 4. Safe range of oral exposure / intake (AREI) for infants compared to usual dietary intake (gray hatched area following the deficiency/excess risk model). The uncertainties in our knowledge of toxic manifestation in humans is depicted by the stripped area. Data on intake comes from recent WHO/FAO/IAEA data. As depicted in this figure a significant proportion of the term infant population, shown by the black hatched area may be receiving too little copper, while excess copper is not a problem unless food or water is contaminated with copper. In the case of preterm infants most will develop copper deficiency unless given formula supplemented with extra copper.

mg/l (Shaw, 1992). These illustrations may serve to orient the regulatory process since presently there are literally thousands of infants receiving Cu containing formulas to assure that their needs are fulfilled and millions of healthy individuals exposed to the oral intakes shown in the figures. As can be noted from the comparison of adults and infants, the latter are more sensitive to Cu deficit and to excess relative to adults per unit of body weight. The actual intake data demonstrates that on a population basis the risk of deficiency is much greater than excess. The recent Environmental Health Criteria for Copper meeting held in Brisbane in 1996 recognized this basic information in establishing its conclusions (Becking, 1996). In fact, except for contaminated food and drink or genetic defects in copper elimination toxicity is extremely uncommon in humans.

Step 5. Qualitative and Quantitative Assessment of Critical Effects to Define AREI for Normal Populations

Based on the data reviewed and adjusted under steps 1–3 a tentative exposure/intake level to prevent toxicity as well as deficiency for the general population should be derived. The upper and lower cutoff points for the range of acceptable oral intakes should be defined for population groups. The lower cutoff point should be sufficient to meet the requirements of most individuals in the population. This is usually called the recommended dietary allowance or intake (RDA or RDI). Based on criteria used to define RDAs or RDIs "most" usually implies 97.5 % of the population; if the mean and standard deviation for requirements are known this point is defined by the mean + 2 SD, if the SD is not known a CV of 15 % is customarily used to account for population variability in requirements. Similarly the upper end point should protect most individuals from the risk of toxicity, a statistical definition for "most" in this context is lacking, but should be derived based on the mean and distribution of toxic effect dose, using the traditional toxicology approach the ED_{50} should be defined and based on known variability or extrapolation from dose effect response an $ED_{2.5}$ should be defined. Special consideration of upper and lower cutoff points should be made in defining AREI for physiologic conditions affecting normal populations, this is infancy, pregnancy, lactation, aging, exercise, and changes in climatic conditions when pertinent. In the case of lower cut off points of AREI these are usually considered in setting RDAs but for the upper cut off point specific values for these categories should be given if the values are sufficiently different.

The acceptable range of exposure/intake AREI, that is the E/I levels sufficient to meet the requirement as well as prevent from risk of toxicity for most individuals in the population, may be extremely narrow if traditional uncertainty factors are used in defining the upper value of the acceptable range. The need to achieve a balanced and comparable approach to assess risk from deficit as well as excess is needed when evaluating essential elements (Bowman and Rishert, 1994).

A key issue is health risk assessment is the need to evaluate exposure/intake from all possible sources. For example for Cu the oral route is the main way of exposure to the element, food and water are the predominant sources of Cu (USEPA, 1985). Inhalation exposure from polluted air may occur in the workplace, specially in mining and in agricultural work where Cu salts are used as pesticides (USEPA, 1987). For all practical purposes the oral route is the only one of health significance as it relates to Cu exposure. The effect of Cu in air may be of importance because of the direct effects on the lung. Food may account for over 90 % of copper intake in adults if water has low Cu content(< 0.1 mg/l). If water Cu content is higher 1–2 mg/l it may account for close to 50 % of total intake. In the case of infants consuming Cu supplemented artificial formula the contribution of water may be less than 10 % whereas if the formula is not fortified with Cu water may contribute over 50 % of total Cu intake, specially when water Cu content is 1–2 mg/l.

In the case of Cu the defined AREI needs to meet not only the criteria for chronic safety but also should be tolerated acutely. For food, this is not an issue, since Cu in food will likely never induce acute adverse effects. On the contrary soluble Cu salts will induce acute vomiting if taken in sufficient amounts, ionic Cu is a powerful emetic agent as previously indicated (USEPA, 1985; USEPA, 1987). For drinking water most values are below 1 mg/l and very few over 2 mg/l. At this level no acute manifestations of intolerance except bitter taste for sensitive individuals are found. Above 5 mg/l an increasing proportion of subjects evidence nausea and abdominal pain.

In summary if one were to apply an upper limit of 0.5 mg/kg per day accepting that the contribution from diet is usually 0.05 or at most 0.1 mg per kg per day the safety margin for Cu intake derived from water is sufficient to support the present provisional guideline value of 2 mg/l . This analysis does not consider the issue of limited bioavailability of Cu and that of adaptation to varying levels of exposure/intake. If these are included the chronic safety margin would be even greater.

Step 6. Adjustment of AREI for Special Groups of Public Health Significance

Another guiding principle in defining AREI is that the acceptable oral exposure/intake range should be safe for the general population and not be expected to meet the requirements or prevent excess for special groups.

For example the zinc needs of patients with chronic diarrhea or of patients on chronic hemodialysis may fall outside the AREI. In both these cases intakes that are of health risk for the "healthy" population may be required to meet the special needs of these patients. Safe zinc exposure/intake levels for hemodialysis patients may lead to deficiency in healthy subjects while intakes required by patients with chronic diarrhea may be excessive for normal subjects.

In the case of essential elements it is clearly impossible to assure that all (100%) of subjects will be protected from deficit or excess. The range of AREI defined at the international level is not intended to address disease conditions or genetic alterations in trace element metabolism which determine special sensitivity to excess or deficit. These conditions should be addressed by national or regional regulating agencies based on the public health relevance of these special conditions. The AREI is clearly not intended to meet the special needs of population subgroups with genetic alterations of trace metal metabolism, for example achrodermatitis enterophatica a disease of zinc metabolism in which absorption is extremely low (Moynahan, 1974). In fact because of an intestinal transport defect AE patients need 2–3 times the usual zinc intake, in fact that level of E/I may interfere with Cu absorption increasing the risk of Cu deficit in normal subjects. Similarly Wilson disease patients will demonstrate marginal benefits of eliminating Cu from food and water which may trigger Cu deficiency in most healthy subjects.

REFERENCES

Barnes, D.G. and Dourson, M.L. (1988). Reference dose (RfD): description and use in health risk assessments. Regul. Toxicol. Pharmacol. 8, 471–486.

Beaumont, C., Simon, M., Fauchet, R., Hespel, J.P., Brissot, P., Genetet, B. and Bourel, M. (1979). Serum ferritin as a possible marker of the hemochromatosis allele. N. Engl. J. Med.. 301, 169–174

Becking, G. (1996). Copper: essentiality and toxicity. IPCS News 10, 4–5.

Berg, R., Lundh, S., Jansonn, G. and Rappe, A. (1981). Kopparforgiftning av dricksvattnet som orsak till diarre hos barn. Halsovardskontakt. 10, 6–8

Bloomfield, J., Dixon, S.R. and McCredie, D.A. (1971). Potential hepatotoxicity of copper in recurrent hemodialysis. Arch. Int. Med. 128, 555–560.

Bothwell, T.H., Charlton, R.W., Cook, J.D. and Finch, C.A. (1979). Iron Metabolism In Man. (Oxford: Blackwell), pp. 245–255.

Bowman, B.A. and Rishert, J.F. (1994). Comparison of the methodological approaches used in the derivation of recommended dietary allowances and oral reference doses for nutritionally essential elements. In: Risk Assessment Of Essential Elements, W. Mertz, C.O. Abernathy, S.S. Olin, eds. (Washington, DC: ILSI Press), pp. 63–73.

Center for Disease Control (CDC). (1975). Acute copper poisoning - Pennsylvania. Mortal. Morbid. Weekly Rep. 29, 99.

Cousins, R.J. (1996) Zinc. In: Present Knowledge In Nutrition. E.E. Ziegler, L.J. Jr. Filer, eds. (Washington, DC: ILSI Press), pp. 293–306.

Danks, D.M. (1988). Copper deficiency in humans. Annu. Rev. Nutr. 8, 235–257.

Dourson, M.L. (1994). Methods for establishing oral reference doses. In: Risk Assessment Of Essential Elements, W. Mertz, C.O. Abernathy, S.S. Olin, eds. (Washington, DC: ILSI Press), pp. 51–61.

Dourson, M.L. and Stara, J.F. (1983). Regulatory history and experimental support of uncertainty (safety) factors. Regul. Toxicol. Pharmacol. 3, 224–238.

Farrell, P.M., Bieri, J.G., Fratatoni, J.F., Wood, R.E. and Di Sant'Agnese, P.A. (1977). The occurrence and effects of human vitamin E deficiency. J. Clin. Invest. 60, 233–241.

Fleet, J.C. and Mayer, J. (1996). Discovery of the hemochromatosis gene will require rethinking the regulation of iron metabolism. Nutr. Rev. 54, 285–292.

Frommer, D.J. (1976). Direct measurement of serum non-caeruloplasmin copper in liver disease. Clin.Chim. Acta. 68, 303–307.

Goyer, R. (1994). Biology and nutrition of essential elements. In: Risk Assessment Of Essential Elements, W. Mertz, C.O. Abernathy, S.S. Olin, eds. (Washington, DC: ILSI Press), pp. 13–19.

Heresi, G., Castillo-Durán, C., Muñoz, C., Arévalo, M. and Schlesinger, L. (1985). Phagocytosis and immunoglobulin levels in hypocupremic infants. Nutr. Res. 5, 1327–1334.

Horslen, S.P., Tanner, M.S., Lyon, T.D.B., Fell, G.S. and Lowry, M.F. (1994). Copper associated childhood cirrhosis. Gut. 35, 1497–1500.

International Atomic Energy Agency (IAEA). (1992). Human Dietary Intakes Of Trace Elements : A Global Literature Survey Mainly For The Period 1970–1991. I. Data List And Sources Of Information. (Vienna: IAEA).

Joint FAO/WHO Expert Committee on Food Additives (JECFA). (1982). Toxicological Evaluation Of Certain Food Additives. World Health Organization Tech Rep Ser 683. (Rome: WHO), pp. 265–296.

Knobeloch, L., Ziarnik, M., Howard, J., Theis, B., Farmer, D., Anderson, H. and Proctor, M. (1994). Gastrointestinal upsets associated with ingestion of copper-contaminated water. Environ. Health Perspect. 102, 958–961

Linder, M.C. and Hazegh-Azam, M. (1996). Copper biochemistry and molecular biology. Am. J. Clin. Nutr. 63, 797S-811S.

Moynahan, E.J. (1974). Acrodermatitis enterophatica: a lethal inherited human zinc-deficiency disorder. Lancet ii, 399–400.

Müller, T., Feichtinger, H., Berger, H. and Müller, W. (1996). Endemic Tyrolean cirrhosis: an ecogenetic disorder. Lancet. 347, 877–80.

National Research Council (NRC). (1989). Recommended Dietary Allowances. 10th ed. (Washington, DC: National Academy of Sciences), pp. 205–213.

Nielsen, FH. and Milne, D.B. (1993). Oxidant stress effects on the clinical and nutritional deficiencies of trace elements. Int. J. Toxicol. Occupat. Environ. Health. 2, 59 (abstract).

Olivares, M. and Uauy, R. (1996a). Copper as an essential element. Am. J. Clin. Nutr. 63, 791S-796S.

Olivares, M. and Uauy, R. (1996b). Limits of metabolic tolerance to copper and biological basis for present recommendations and regulations. Am. J. Clin. Nutr. 63, 846S-52S.

Pandit, A. and Bhave, S. (1996). Present interpretation of the role of copper in Indian childhood cirrhosis. Am. J. Clin. Nutr. 63, 830S-5S.

Scheinberg, I.H. and Sternlieb I. (1996). Wilson disease and idiopathic copper toxicosis. Am. J. Clin. Nutr. 63, 842S-845S.

Scheinberg, I.H. and Sternlieb, I. (1994). Is non-Indian childhood cirrhosis caused by excess dietary copper?. Lancet. 344, 1002–1004.

Shaw, J.C.L. (1992). Copper deficiency in term and preterm infants. In: Nutritional Anemias. S.J. Fomon, S. Zlotkin, eds. Nestlé Nutrition Workshop Series Vol 30. (Vevey: Nestec Ltd, New York: Raven Press) , pp. 105--119.

Simpson, K.M., Morris, E.M. and Cook, J.D. (1981). The inhibitory effect of bran in iron absorption in man. Am. J. Clin. Nutr. 34, 1469–1478.

Spitalny, K.C., Brondum, J., Vogt, R.L., Sargent, H.E. and Kappel, S. (1984) Drinking water-induced intoxication in a Vermont family. Pediatr. 74, 1103–1106.

Stekel, A., Olivares, M., Pizarro, F., Amar, M., Chadud, P., Cayazzo, M., Llaguno, S., Vega, V. and Hertrampf, E. (1985). The role of ascorbic acid in the bioavailability of iron from infant foods. Int. J. Vitamin. Nutr. Res. 55 (Suppl 27), 167–175.

Stenhammar, L. (1979). Kopparintoxikation -en differentialdiagnos vid diarre hos barn. Lakartidningen. 76, 2618–2620.

Sternlieb, I. (1980). Copper and the liver. Gastroenterology. 78, 1615–1628.

Turnlund, J.R., Keyes, W.R., Anderson, H.L. and Acord, L.L. (1989). Copper absorption and retention in young men at three levels of dietary copper by use of the stable isotope ^{65}Cu. Am. J. Clin. Nutr. 49, 870–878.

Uauy, R., Castillo-Durán, C., Fisberg, M., Fernandez, N. and Valenzuela, A. (1985). Red cell superoxide dismutase activity as an index of human copper nutrition. J. Nutr. 115, 1650–1655.

US Environmental Protection Agency (USEPA), Environment Criteria and Assessment Office. (1985). Drinking Water Document For Copper (final draft). (EPA 600/X-84/190–1).(Cincinnati, OH: Environmental Protection Agency).

US Environmental Protection Agency (USEPA), Environment Criteria and Assessment Office. (1987). Summary Review Of The Health Effects Associated With Copper. (EPA 600/8–87/001).(Cincinnati, OH: Environmental Protection Agency).

US Environmental Protection Agency (USEPA), Environmental Criteria and Assessment Office. (1993). The Integrated Risk Information System (IRIS) (online). (Cincinnati, OH, Environmental Protection Agency).

van Campen, D.R. and Scaifi, P.U. (1967). Zinc interference with copper absorption in rats. J. Nutr. 91, 473–476.

WHO/FAO/IAEA. Copper. In: Trace Elements In Human Nutrition And Health. (Geneva:World Health Organization), pp. 123–143.

Williams, M.L., Shott ,R.J., O'Neal, P.L. and Oski, F.A. (1975). Role of dietary iron and fat on vitamin E deficiency anemia of infancy. N. Engl. J. Med. 292, 887–890.

Wyllie, J. (1957). Copper poisoning at a cocktail party. Am. J. Public. Health. 47, 617.

3

Cu METABOLISM IN THE LIVER

Harry J. McArdle,[1,*] Michelle J. Bingham,[2] Karl Summer,[3] and T. J. Ong[2]

[1]Rowett Research Institute
Greenburn Road, Bucksburn
Aberdeen AB21 9SY, Scotland
[2]Dept of Child Health, Ninewells Hospital and Medical School
Dundee, DD1 9SY, Scotland
[3]GSF-National Research Centre for Environment and Health
Institute of Toxicology
PO Box 1129: D-85758
Oberschleissheim, Germany

The liver is central to copper metabolism in mammals. Following transfer across the gut, about 40 % of the metal is taken up from the portal vein in each pass (see (Linder, 1991)). The liver also excretes Cu through the bile. The chemical characteristics of Cu in bile are poorly understood, but it is probably excreted as a Cu ion and complexes with bile salt in the canaliculi.

Cu levels in the liver in most mammals are high at birth and decrease to much lower levels in the adult (reviewed in (Linder, 1991)). Some of this Cu stored in the liver is excreted in the bile at birth or shortly thereafter, when the bile duct becomes patent, but it may also be important during the perinatal period, providing Cu for ceruloplasmin synthesis when the neonate is at risk of Cu deficiency due to poor supply from the mother's milk.

The liver synthesises and secretes the plasma cupro-protein ceruloplasmin. Ceruloplasmin is a glycoprotein with azide-sensitive ferroxidase activity and a molecular weight of 131,000. Each molecule of protein has six bound Cu atoms and another loosely associated (Zaitseva et al., 1996). In the plasma between 65 and 90 % of copper is present as ceruloplasmin with the remainder being associated with albumin, transcuprein and low molecular weight ligands. Ceruloplasmin is also an acute phase protein and serum levels increase during pregnancy and in response to oral contraceptives (Linder, 1991).

---

[*] Correspondence to: Dr Harry J. McArdle Rowett Research Institute, Greenburn Road, Bucksburn, Aberdeen AB21 9SY, Scotland. Tel: +44-1224-716628, Fax: +44-1224-716622, e.mail: hjm@rri.sari.ac.uk

Copper Transport and Its Disorders, edited by Leone and Mercer.
Kluwer Academic / Plenum Publishers, New York, 1999.

How Cu is delivered is beginning to become clear. In this review, we will consider how the different pumps operate, whether they are energy dependent, how specific they are and the form of Cu that they recognise. We have broken down the overall process into several different stages; (1) uptake across the hepatocyte membrane, which itself involves two different steps; (2) fractionation inside the cell; (3) transport across the Golgi by ATP7B and a brief examination of the effect of the LEC mutation. We have not considered excretion of Cu by the hepatocyte.

Cu UPTAKE ACROSS THE HEPATOCYTE MEMBRANE

After it is transferred across the gut membrane, Cu appears in portal serum attached to albumin, histidine, transcuprein or in a ternary complex of CuAlbHis (Lau et al., 1974; Weiss and Linder, 1985). Uptake across the hepatocyte membrane is carrier mediated (Ettinger et al., 1986; McArdle et al., 1988), but is not dependent on cell energy, is not coupled to the Na gradient and is specific for Cu ions (McArdle et al., 1988). Zn can compete, but only at a much higher concentration (McArdle et al., 1988; Weiner and Cousins, 1980). The data also indicates that the transporter is a dimer, coupled by two disulphide bridges, one inside and one outside the membrane (Fig 1).

Early data showed that the transporter recognises Cu as a CuHis$_2$ complex, although Cu and His are taken up separately, by distinct pathways (McArdle et al., 1988). This resulted in the hypothesis that Cu is transferred from albumin, which was thought to act as the initial carrier, to histidine and then to the transporter (Schmitt et al., 1983).

There were some data, however, which suggested that the process may be slightly different. Under the conditions prevailing in plasma, Cu is probably carried specifically bound to albumin as CuAlb or a CuAlbHis ternary complex (Lau et al., 1974). Whether this CuAlbHis complex has any physiological consequence was not clear.

Albumin has both specific and non-specific binding sites for Cu. It is possible, by incubating at different pHs, to manipulate the site to which the Cu binds (McArdle et al., 1990a). Hence, mixing albumin and Cu at pH 5.5 and then raising the pH to 7.4 results in all the Cu being associated with the specific site. In contrast, simply mixing at pH 7.4 results in a significant proportion of the Cu being bound non-specifically. Using these treatments to label human or bovine albumin, we found, somewhat surprisingly, that when the Cu was bound exclusively to the high affinity site, it was taken up more rapidly by the

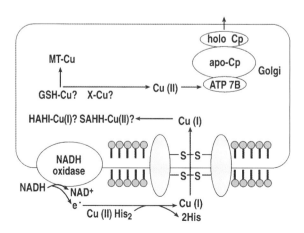

Figure 1. A model for Cu metabolism in the liver. Cu is reduced from Cu(II) to Cu(I) by NADH oxidase, transported into the cell as Cu(I). Inside the cell Cu is possibly reoxidised prior to transport across the Golgi membranes by ATP7B for incorporation into ceruloplasmin and subsequent secretion of the holo-protein. Abbreviations are SAHH-adenosyl homocysteine hydrolase; GSH-glutathione; MT-metallothionein; Cp-ceruloplasmin. For more details see text.

hepatocyte (McArdle et al., 1990a). When dog albumin, which has lost the specific binding site, was used as a substrate, there was no difference in uptake. Thus it seemed that Cu was preferentially accumulated from the complex with the highest affinity and not, as would be expected, from the complex with the lowest affinity. However, there was some capacity to transport Cu across the membrane from any of the complexes (McArdle et al., 1990a).

Cu Reduction by Plasma Membrane Metalloreductase/NADH Oxidase

Whichever is the substrate for uptake, the cell has to remove the Cu from a high affinity complex. In the 1970's Crane and colleagues suggested that a plasma membrane NADH oxidase was involved in iron uptake, reducing Fe(III) to Fe(II) (Crane et al., 1985; Sun et al., 1987). We considered the possibility that the NADH oxidase/ferrireductase could also be important in Cu uptake. Hence, if it reduced the Cu(II) to Cu(I), the affinity for the complex would be reduced and uptake could then proceed.

Accordingly, we isolated plasma membrane vesicles from rat liver and incubated them with $CuHis_2$, with and without NADH. We found a significant increase in uptake when NADH was added. This increase was dependent on NADH concentration and did not occur if NAD^+ was used instead (van den Berg and McArdle, 1994). Similarly, adding Cu increased NADH oxidation by membrane vesicles (van den Berg and McArdle, 1994). From the relative activities we calculated a stoichiometry of electron production and determined that approximately one Cu ion was transported for each molecule of NADH oxidised (see fig 1). This leaves one electron unaccounted for, but there are several possibilities which will explain the observation. Our favoured explanation is that the electron is transported with the Cu(I) or a hydrogen ion is transported the other way, to make an electroneutral exchange. This hypothesis remains to be tested.

Our data received support from experiments conducted in yeast and plants, which face a similar problem to mammals, in that Cu is presented in a complex with a variety of ligands, or indeed, as the free ion. In yeast, Cu(II) is reduced by Fre1 or Fre2, which also use NADH as the electron donor (Dancis et al., 1990; Georgatsou and Alexandraki, 1994; Shatwell et al., 1996), and in plants the reducing enzyme shows some homology to $gp91^{phox}$, an NADPH oxidase in macrophages involved in the oxidative burst (Groom et al., 1996). In these two cell types, the NADH oxidase can reduce either Cu(II) or Fe(III), which provides an elegant connection with our data and that of Crane and colleagues in mammalian liver.

Despite the fact that the plasma membrane NADH oxidase/metalloreductase has been studied for at least twenty years, it has not been isolated, nor has the gene been cloned so that as yet we have no direct evidence of its existence or regulation. However, our data certainly encourages the search for such a gene and/or gene product.

Cu Transfer across the Cell Membrane

At present, very little is known about the actual mechanism of transfer of Cu across the hepatocyte membrane. As mentioned above, it occurs through a classic carrier-mediated process, which is not dependent on metabolic energy or the Na gradient, but is dependent on the integrity of at least one and possibly two disulphide bonds. The recent identification of a mammalian protein which can reverse the mutation in ctr1- yeast mutants, i.e. a potential plasma membrane transporter for Cu (Zhou and Gitschier, 1997), will help enormously in elucidating the mechanisms involved.

Cu uptake in hepatocytes is regulated to a limited extent. Work in the early 1980s showed that uptake is increased by adrenaline (Weiner and Cousins, 1983) and our own data indicate that chronic (Bingham et al., 1995b), but not acute (McArdle et al., 1990b), Cu deficiency increases uptake, possibly due to an increase in transporter number at the plasma membrane (Bingham et al., 1995b).

Recent data from our laboratory also indicate that when iron levels in the hepatocyte are increased Cu uptake decreases, possibly by down-regulating the metalloreductase on the plasma membrane (Whitaker and McArdle, 1997) - again this is a similar mechanism to that operating in yeast (Morrissey et al., 1996).

INTRACELLULAR METABOLISM OF Cu

Once inside the cell, Cu is distributed into different pools. These can be defined as the storage, excretion and transit pools (Owen, 1980). Using a Cu chelator, diamsar, we have identified two of these pools and have characterised them extensively (Bingham et al., 1997; McArdle et al., 1989). Diamsar is a chelator which binds Cu(II) but not Cu(I). In vitro, it cannot remove Cu(I) from glutathione (Bingham et al., 1997). In the hepatocyte, however, it removes up to about 80 % of the Cu (McArdle et al., 1989). This Cu is the most recently acquired, which we demonstrated by incubating hepatocytes for increasing periods of time with $CuHis_2$ and then extracting the cells with the chelator (Bingham et al., 1997). After short incubation times, virtually all the radioactive Cu was removed from the hepatocyte. However as incubation times increased, the percentage that could be extracted decreased to a plateau. At higher ^{67}Cu concentrations, the time to plateau was shorter, showing that the extractable pool, as we termed it, was saturable in terms of both concentration and time (Bingham et al., 1997). In contrast, the Cu which diamsar was unable to remove from the cell (non-extractable) increased with incubation time and at higher ^{67}Cu concentrations suggesting that Cu moves from the extractable pool to the non-extractable pool.

In a series of kinetic experiments, we demonstrated that Cu can move back from the non-extractable pool to the extractable pool, and that the rate of reflux can be modulated by addition of Cu. Thus, if the extractable pool is reduced, Cu from the non-extractable pool will move back to fill it. In contrast, if the extractable pool is filled, Cu movement back is inhibited. It seems reasonable to deduce, therefore, that the non-extractable pool is a storage pool (Bingham et al., 1997).

The extractable pool is necessary for the synthesis of ceruloplasmin. We demonstrated this by loading cells with ^{67}Cu presented as $^{67}CuHis_2$ and then incubating with increasing concentrations of diamsar for 24 h (Bingham et al., 1997). After this time the diamsar was removed and replaced with fresh incubation media for 4 h. ^{67}Cu-labelled ceruloplasmin secreted into the media was determined by immunoprecipitation and shown to be inversely proportional to diamsar concentration. As diamsar is unable to remove ^{67}Cu directly from ceruloplasmin this implies that Cu in the extractable pool is available for incorporation into ceruloplasmin. These data will be referred to again later, since they suggest that the Cu which is used for ceruloplasmin synthesis is present as Cu(II), or goes from Cu(I) to Cu(II) at some stage prior to incorporation into the protein (Bingham et al., 1997).

From this data we can draw some conclusions about the nature of the two pools. Diamsar is a very powerful chelator of Cu(II) but does not bind Cu(I). It cannot remove Cu from glutathione and if metallothionein levels in the cell are increased, by incubating with Zn, the size of the non-extractable pool also increases. It would seem feasible, therefore, that the non-extractable (storage) pool is at least partly metallothionein. This would mitigate against, but not completely exclude metallothionein having a role as an intracellular transport protein. In addition we know that diamsar can not remove copper from enzymes such as cytochrome c oxidase and superoxide dismutase so that incorporation of Cu into enzymes could constitute movement of Cu to a "non-extractable pool". Which proteins bind Cu(II) in the transit pool however, are not certain. Those described elsewhere in this volume would seem to bind Cu as Cu(I) and also would not seem to be present at very high concentrations, since they had not been identified previously. One strong candidate though is an enzyme identified by Ettinger and colleagues - S-adenosyl homocysteine hydrolase (SAHH) (Bethin et al., 1995a; Bethin et al., 1995b).

It was first suggested that SAHH may play a role in Cu metabolism when it was shown to be under-expressed in livers from brindled mice, which have a homologous mutation to Menkes disease. It was clear that this was not the primary defect in Menkes - not least because the liver expressed no phenotype, but eventually the gene was cloned and it was demonstrated that the protein directly binds Cu and levels are regulated by intracellular Cu. Interestingly, the data also strongly indicated that SAHH binds Cu as Cu(II), and with an affinity similar to that of albumin (Bethin et al., 1995b). Why SAHH should bind Cu(II) with such a high affinity is not apparent but it has been suggested that SAHH may play a role in delivering Cu to apo SOD for synthesis of the holo-protein (Petrovic et al., 1997). Given the data presented in this volume on Cu chaperone proteins (Klomp et al., 1997), this may not be the case, but it is tempting to speculate, given our own work, that at some stage SAHH may act as an intermediary possibly mediating the transfer of Cu from the chaperones to the ATPase, as described below.

Cu TRANSPORT BY ATP7B, THE Cu-ATPase OF LIVER

The gene for Wilson disease was cloned in 1994, after the identification of the gene for Menkes disease (Bull et al., 1993; Petrukhin et al., 1994; Tanzi et al., 1993). From the structural data, they both appear to be ATPases, possibly acting as pumps for Cu in an analogous fashion to Ca-ATPases in membrane systems such as the sarcoplasmic reticulum. Given the phenotype expressed in Wilson disease - a defect in Cu excretion coupled with very low or no serum ceruloplasmin (Wilson, 1912), we hypothesised that the gene product would be found in a part of the Cu pathway common to both excretion and protein synthesis. This, we considered, would be within the microsomal fraction of the cell, rather than the bile duct, since problems directly with bile secretion would not explain the effect on ceruloplasmin production. Accordingly, we examined microsomal vesicles isolated from rat liver to determine whether we could identify ATP driven Cu uptake (Bingham et al., 1995a).

Initial data showed that accumulation was too rapid to detect using a simple incubation medium, so we adapted a method used by ourselves and other workers studying Ca uptake into sarcoplasmic reticulum vesicles. This involved the addition of oxalate to the incubation medium. Oxalate rapidly equilibrates throughout the system and as Cu is pumped into the vesicles by the ATPase, Cu-oxalate forms and is precipitated so maintaining the Cu gradient. This system worked well and we were able to characterise the ATPase (Bingham et al., 1995a). We demonstrated that the transporter was specific for ATP, and that UTP, GTP and CTP or low energy analogues of ATP, such as AMP.PCP, could not stimulate uptake. We also showed that Mg was required for uptake to reach maximal values and that uptake could be inhibited by vanadate (Bingham et al., 1995a).

All these data supported the idea that we were studying a P-type ATPase, and that this was the Wilson disease protein ATP7B. However, closer examination of the data revealed some unexpected information.

Is Cu(I) or Cu(II) Transferred by the ATPase?

The experiments described above were carried out using a Cu-glutathione complex, since it has been suggested that glutathione acts as an intracellular transporter (Freedman et al., 1989), and as such we believed Cu would present to the transporter as Cu(I). However, it became clear that uptake was maximal when Cu was present as Cu(II) (Bingham et al., 1996). A variety of factors led us to this conclusion. Firstly, increasing glutathione, which should increase the effective concentration of Cu(I), decreased uptake (Bingham et al., 1996). Secondly, other treatments which increased Cu(I) decreased ATP dependent uptake, for example, vitamin C. This effect was reversed if vitamin E was added. Finally, the use of electron spin resonance spectroscopy demonstrated directly that in the presence of oxalate, which as described above is essential for ATP-dependent uptake, Cu is present as Cu(II) (Bingham et al., 1996). This was a surprising result, not least because the predicted structure of the Cu binding sites on the ATPase suggest they should recognise Cu(I). In addition, the bacterial homologue of ATP7B, CopB has been shown to transport both Cu(I) and Ag(I) (Solioz and Odermatt, 1995). Bacterial ATP-dependent Cu transport does however require the presence of 5 mmol/L dithiothreitol which could in theory act like oxalate in our system and present Cu as Cu(II). It is therefore feasible that the Cu(I) binding sites on the ATPase are not directly involved in transport but instead act as indicators of cellular Cu status (see Chapter by Dameron in this volume).

Localisation of the Functional Cu-ATPase

Our initial hypothesis was that the ATPase would be located in the endoplasmic reticulum of the liver, since this was the simplest explanation for the defect observed in Wilson disease (Bingham et al., 1995a). However, another group suggested that it could be located in the bile canalicular membrane (Dijkstra et al., 1995). To investigate the localisation more accurately, we fractionated the microsomes and characterised the fractions using enzymes known to be associated with different membrane compartments of the cell. We chose this approach rather than immunohistochemistry because we considered it essential to correlate functional activity with the compartment. The reasoning behind this was that although immunohistochemistry would detect the localisation of the protein per se. there was a distinct possibility that the protein could be located in certain subcellular fractions and not actually be operating.

The data obtained clearly show that ATP dependent Cu(II) transporter activity is associated with enzymes of the Golgi apparatus and not, as we originally thought, with the endoplasmic reticulum (Bingham et al., 1996). The correlation is very strong (Fig 2) and is supported by observations made by Petris et al., who have shown the same localisation of the Menkes protein (Petris et al., 1996). More recently, Yang et al., have provided evidence for two forms of the Wilson protein (Yang et al., 1997). Their data indicate that the membrane-bound form is present in the Golgi of HepG2 cells (Yang et al., 1997).

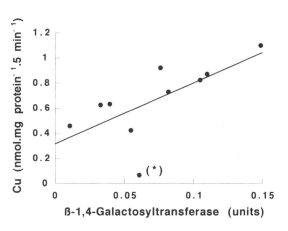

Figure 2. Scatter plot and linear regression of ATP dependent Cu(II) uptake against ß-1,4-galactosyltransferase activity. Microsomal vesicles were isolated from Wistar rats and fractionated using a continuous sucrose gradient. Ten fractions were collected from the gradient (see Bingham et al., 1996). The fractions were then assayed for the Golgi apparatus enzyme ß-1,4-galactosyltransferase and Cu(II) transport in the presence and absence of ATP. Data presented are net ATP dependent Cu(II) transport (uptake in the presence of ATP minus uptake in the absence of ATP) for each fraction plotted against the corresponding ß-1,4-galactosyltransferase activity. ß-1,4-galactosyltransferase activity was calculated using galactosyltransferase isolated from bovine milk (Sigma) as a standard. Units are defined as the change in absorbance per milligram protein per hour. $r = 0.676$, d.f. = 8, and $P < 0.05$. *Omitted: $r = 0.869$, d.f. = 7, $P < 0.01$.

The Effect of the LEC Mutation on the Physiological Function of ATP7B

The LEC rat has a mutation in a gene homologous to ATP7B in humans making it ideal as a model for Wilson disease (Wu et al., 1994). The mutation results in a 900 base pair deletion at the 3 end and a loss of 300 amino acids from the protein (Wu et al., 1994). Since the deletion was identified by sequencing of RNA from LEC rat liver, it seems reasonable to assume that the RNA is transcribed, although whether it is translated is not known. The animal does however display the same phenotype as humans with Wilson disease, hence we assumed that the LEC rat would act as an excellent control for our ATPase experiments. Our prediction for these experiments was that the deletion would result in the loss of ATP-dependent Cu transport across the Golgi and certainly this is what we found. What we did not anticipate however, was that *total* (ATP-dependent plus ATP-independent) uptake by microsomes would be the same in normal and LEC rats (Bingham et al., 1998). We considered this very surprising, and were even more intrigued when we found that, rather than localising with enzymes of the Golgi apparatus, the transporters co-localised with enzymes of the endoplasmic reticulum (Bingham et al., 1998). Our data therefore, leads us to hypothesise that the ATP binding site is important in keeping the channel closed. With the loss of this site, the channel becomes permanently open. Further, there is, we suggest, a localisation signal which is missing in the LEC protein, and without this signal the ATPase remains in the endoplasmic reticulum. Whether or not this is the case remains to be seen, but the theory could certainly be tested with the availability of site directed mutants.

SUMMARY AND CONCLUSION

This paper has, given some idea of our concepts of the processes involved in the transport of Cu across cell membranes in the liver, which we have summarised in Fig 1.

Cu(II)His$_2$ is reduced to Cu(I). This is transported across the membrane, re-oxidised, either before or after binding to glutathione (Freedman et al., 1989) or HAH1 (Klomp et al., 1997), binds to SAHH, and donates Cu(II) to the ATPase. It is very interesting that cells which are very diverse from an evolutionary point of view still use very similar methods to handle the metal. Whether regulation of transport is also the same remains to be seen. We would guess that, although there will be strong similarities, there will also be very significant differences, reflecting the different environments seen by different tissues in mammalian cells and given the different requirements of the tissues.

REFERENCES

Bethin, K.E., Cimato, T.R. and Ettinger, M.J. (1995a). Copper binding to mouse liver S-Adenosylhomocysteine hydrolase and the effects of copper on its levels. J. Biol. Chem. 270, 20703–20711.

Bethin, K.E., Petrovic, N. and Ettinger, M.J. (1995b). Identification of a major hepatic copper binding protein as S-adenosylhomocysteine hydrolase. J. Biol. Chem. 270, 20698–20702.

Bingham, M.J., Burchell, A. and McArdle, H.J. (1995a). Identification of an ATP dependent copper transport system in endoplasmic reticulum vesicles isolated from rat liver. J. Physiol. 482, 583–587.

Bingham, M.J., Ong, T.J., Ingledew, W.J. and McArdle, H.J. (1996). ATP-dependent copper transporter, in the Golgi apparatus of rat hepatocytes, transports Cu(II), not Cu(I). Am. J. Physiol. 271, G741-G746.

Bingham, M.J., Ong, T.J., Summer, K.H., Middleton, R.B. and McArdle, H.J. (1998). The physiological function of the Wilson disease gene product, ATP7B. Am. J. Clin. Invest. in press,

Bingham, M.J., Sargeson, A.M. and McArdle, H.J. (1997). Identification and characterisation of intracellular copper pools in rat hepatocytes. Am J Physiol. 272, G1400-G1407.

Bingham, M.J., van den Berg, G.J. and McArdle, H.J. (1995b). Effect of nutritional copper deficiency on copper uptake by plasma membrane vesicles isolated from rat livers. J. Physiol. 489, 125P.

Bull, P.C., Thomas, G.R., Rommens, J.M., Forbes, J.R. and Cox, D.W. (1993). The Wilson disease gene is a putative copper transporting P-type ATPase similar to the Menkes gene. Nature Genetics. 5, 327–337.

Crane, F., Sun, I., Clark, M., Grebing, C. and Low, H. (1985). Transplasma-membrane redox systems in growth and development. Biochim. Biophys. Acta. 811, 233–264.

Dancis, A., Klausner, R.D., Hinnebusch, A.G. and Barriocanal, J.G. (1990). Genetic evidence that ferric reductase is required for iron uptake in *Saccharomyces cerevisiae*. Mol Cell Biol. 10: 2294–2301.

Dijkstra, M., In 't Veld, G., van den Berg, G.J., Muller, M., Kuipers, F. and Vonk, R.J. (1995). Adenosine triphosphate-dependent copper transport in isolated rat liver plasma membrane. J. Clin. Invest. 95, 412–416.

Ettinger, M.J., Darwish, H.M. and Schmitt, R.C. (1986). Mechanism of copper transport from plasma to hepatocytes. Fed. Proc. 45, 2800–4.

Freedman, J.H., Ciriolo, M.R. and Peisach, J. (1989). The role of glutathione in copper metabolism and toxicity. J Biol Chem. 264: 5598–5605.

Georgatsou, E., and Alexandraki, D. (1994). 2 distinctly regulated genes are required for ferric reduction, the first step of iron uptake in *Saccharomyces cerevisiae*. Mol Cell Biol. 14: 3065–3073.

Groom, Q.J., Torres, M.A., Fordham-Skelton, A.P., Hammond-Kosack, K.E., Robinson, N.J. and Jones J.D.G. (1996). *rbohA*, a rice homologue of the mammalian *gp91phox* respiratory burst oxidase gene. Plant J. 10: 515–522.

Klomp, L.W.J., Lin, S., Yuan, D.S., Klausner, R.D., Culotta, V.C. and Gitlin, J.D. (1997). Identification and functional expression of *HAH1*, a novel human gene involved in copper homeostasis. J Biol Chem. 272: 9221–9226.

Lau, S.Y., Kruck, T.P.A. and Sarker, B. (1974). Peptide molecule mimicking the copper (II) transport site of human serum albumin. J Biol Chem. 246, 5878–5884.

Linder, M.C. (1991). Biochemistry of copper. Biochemistry of the Elements. New York, Elsevier.

McArdle, H.J., Gross, S.M., Creaser, I., Sargeson, A.M. and Danks, D.M. (1989). The effect of chelators on copper metabolism and copper pools in mouse hepatocytes. Am. J. Physiol. 256, G667-G672.

McArdle, H.J., Gross, S.M. and Danks, D.M. (1988). Uptake of copper by mouse hepatocytes. J. Cell. Physiol. 136, 373–378.

McArdle, H.J., Gross, S.M., Danks, D.M. and Wedd, A.G. (1990a). Role of albumin's specific copper binding site in copper uptake by mouse hepatocytes. Am J Physiol. 258G, 988–991.

McArdle, H.J., Mercer, J.F., Sargeson, A.M. and Danks, D.M. (1990b). Effects of cellular copper content on copper uptake and metallothionein and ceruloplasmin mRNA in mouse hepatocytes. J. Nutr. 120, 1370–1375.
Morrissey, J.A., Williams, P.H. and Cashmore, A.M. (1996). *Candida albicans* has a cell-associated ferric-reductase activity which is regulated in response to levels of iron and copper. Microbiol. 142: 485–492.
Owen, C.A.J. (1980). Copper and hepatic function. Ciba Found Symp. 79, 267–82.
Petris, M.J., Mercer, J.F., Culvenor, J.G., Lockhart, P., Gleeson, P.A. and Camakaris, J. (1996). Ligand-regulated transport of the Menkes copper P-type ATPase efflux pump from the Golgi apparatus to the plasma membrane: a novel mechanism of regulated trafficking. EMBO. J>. 15, 6084–6095.
Petrovic, N., Comi, A. and Ettinger, M.J. (1997). Copper incorporation into superoxide dismutase in Menkes lymphoblasts. J. Biol. Chem. 271, 28335–28340.
Petrukhin, K., Lutsenko, S., Chernov, I., Ross, B.M., Kaplan, J.H. and Gilliam, T.C. (1994). Characterization of the Wilson disease gene encoding a P-type copper transporting ATPase: genomic organization, alternative splicing and structure/function predictions. Hum. Mol. Gen. 3, 1647–1656.
Schmitt, R.C., Darwish, H.M., Cheney, J.C. and Ettinger, M.J. (1983). Copper transport kinetics by isolated rat hepatocytes. Am. J. Physiol. 244, G183-G191.
Shatwell, K.P., Dancis, A., Cross, A.R. and Klausner, R.D. (1996). The FRE1 ferric reductase of *Saccharomyces cerevisiae* is a cytochrome *b* similar to that of NADPH oxidase. J Biol Chem. 271: 14240–14244.
Solioz, M., and Odermatt, A. (1995). Copper and silver transport by CopB-ATPase in membrane vesicles of *Enterococcus hirae*. J Biol Chem. 270: 9217–9221.
Sun, I.L., Navas, P., Crane, F.L., Morro, D.J. and Low, H. (1987). NADH diferric transferrin reductase in liver plasma membrane. J Biol Chem. 262, 15915–15921.
Tanzi, R.E., Petrukhin, K., Chernov, I., Pellequer, J.L., Wasco, W., Ross, B., Romano, D.M., Parano, E., Pavone, L., Brzustowicz, L.M., Devoto, M., Peppercorn, J., Bush, A.I., Sternlieb, I., Pirastu, M., Gusella, J.F., Evgrafov, O., Penchaszadeh, G.K., Honig, B., Edelman, I.S., Soares, M.B., Scheinberg, I.H. and Gilliam, T.C. (1993). The Wilson disease gene is a copper transporting ATPase with homology to the Menkes disease gene. Nature Genetics. 5, 344–350.
van den Berg, G.J. and McArdle, H.J. (1994). A plasma membrane NADH oxidase is involved in copper uptake by plasma membrane vesicles isolated from rat liver. Biochim. Biophys. Acta. 1195, 276–280.
Weiner, A.L. and Cousins, R.J. (1980). Copper accumulation and metabolism in primary monolayer culture of rat liver parenchymal cells. Biochim. Biophys. Acta. 629, 113–125.
Weiner, A.L. and Cousins, R.J. (1983). Hormonally produced changes in caeruloplasmin synthesis and secretion in primary cultured rat hepatocytes. Relationship to hepatic copper metabolism. Biochem. J. 212, 297–304.
Weiss, K.C., and Linder, M.C. (1985). Copper transport in rats involving a new plasma protein. Am J Physiol. 249: E77-E88.
Whitaker, P. and McArdle, H.J. (1997). Iron inhibits copper uptake by down-regulating the plasma membrane NADH oxidase. T.E.M.A. 9, 237–239.
Wilson, S.A.K. (1912). Progressive lenticular degeneration: A familial nervous disease associated with cirrhosis of the liver. Brain. 34, 295–509.
Wu, J., Forbes, J.R., Chen, H.S. and Cox, D.W. (1994). The LEC rat has a deletion in the copper transporting ATPase gene homologous to the Wilson disease gene. Nature Genetics. 7, 541–545.
Yang, X.L., Miura, N., Kawarada, Y., Terada, K., Petrukhin, K., Gilliam, T.C. and Sugiyama, T. (1997). Two forms of Wilson disease protein produced by alternative splicing are localized in distinct cellular compartments. Biochem. J. 326, 897–902.
Zaitseva, I., Zaitsev, V., Card, G., Moshkov, K., Bax, B., Ralph, A. and Lindley, P. (1996). The X-ray structure of human ceruloplasmin at 3.1 A: nature of the copper centres. J. Biol. Inorg. Chem. 1, 15–23.
Zhou, B. and Gitschier, J. (1997). hCTR1: A human gene for copper uptake identified by complementation in yeast. Proc. Natl. Acad. Sci. USA. 94, 7481–7486.

4

MULTIPLE FORMS OF THE MENKES Cu-ATPase

Edward D. Harris,[1] Manchi C. M. Reddy,[1] Yongchang Qian,[1]
Evelyn Tiffany-Castiglioni,[2] Sudeep Majumdar,[1] and John Nelson[1]

[1]Department of Biochemistry and Biophysics
[2]Department of Veterinary Anatomy and Public Health
Texas A&M University
College Station, Texas, 77843

1. ABSTRACT

The 5' region of *MNK* cDNAs has a 45 bp insert terminating at the 5' end with an AGATG sequence. The ATG in the sequence is in-frame with the ATG downstream identified by Vulpe et al (1993) as a translation start site for *MNK* mRNA. Inserts of 192 bp and 45 bp have been found in the 5' region of *MNK* mRNAs from BeWo cells, Caco-2 cells and normal human fibroblasts. Extensions to the 5' end of these mRNAs could foretell a modified N-termini in certain forms of the Menkes Cu-ATPase These modified H_2N-terminal extensions are postulated to be targeting signals for post-translational processing and cellular localization. In this report, we provide evidence that the primary Menkes transcript in non-Menkes cells undergoes post-transcriptional splicing that gives rise to multiple transcripts. The data suggest that the Menkes gene is a copper locus that codes for more than one form of the Menkes Cu-ATPase and one of these forms could be a small Cu transport protein.

2. INTRODUCTION

X-linked Menkes disease is characterized by a failure to pass Cu ions completely across the intestinal mucosa. The entrapment of Cu within the intestinal cells leads to a Cu deficiency in peripheral organs, severely impairing neurological, catalytic, and connective tissue functions (Danks, 1972; Danks et al., 1988). Menkes disease is perhaps the best documented evidence for a Cu deficiency in humans. Wilson disease or hepatolenticular degeneration, is an autosomal recessive disorder that results from pathological accumulations of Cu predominantly in liver and brain tissues. The dominant symptoms relate to a failure to release Cu into the bile or to incorporate Cu into apo-ceruloplasmin, both events occurring in liver. Both diseases in addition to providing unprecedented molecular insights into genetic factors regulating Cu transport, have been found to be based on nearly identical biochemical components.

The isolation and sequencing of the Wilson (Petrukhin et al., 1993; Tanzi et al., 1993; Bull et al., 1993 Thomas et al., 1995) and Menkes (Mercer et al., 1993; Vulpe et al., 1993) genes has revealed that both code for P-type Cu-ATPases, an evolutionary family of ion-transporting ATPases akin to bacterial Cu ATPases (Pufahl et al., 1997) and related in both structure and function to membrane-bound Ca ATPases. The Wilson and the Menkes proteins transport Cu ions specifically (Solioz and Vulpe, 1996; Lutsenko et al., 1997). Moreover, the two proteins have a 57% sequence homology to one another (Vulpe and Packman, 1995). The mRNA for *MNK* protein (ATP7A), analyzed as a cDNA, is 8.3 to 8.5 kb with a single open reading frame that encompasses 23 exons and encodes a protein of exactly 1,500 amino acids (Fig. 1A) Strong expression of *MNK* mRNA is seen in muscle, kidney, lung, brain; weaker expression has been associated with placenta and pancreas, and liver shows only traces (Mercer et al., 1993; Chelly et al., 1993; Vulpe et al., 1993). The Wilson transcript is 7.5 kb and encodes a protein (ATP7B) of 1411 amino acids. In contrast to the Menkes gene, the Wilson gene is strongly expressed in liver and kidney (Yamaguchi et al., 1993; Bull et al., 1993).

While the structures may be similar, their biological functions are not. The Menkes protein regulates the release of Cu ions at the outer membrane whereas the Wilson gene is concerned with internal Cu trafficking. In our work, we have shown that sulfhydryl reagents block the release of Cu from C6 rat glioma cells presumably by inhibiting a Cu-ATPase controlling release (Qian et al., 1995). Others have shown that CHO cells selected for over expression of the Menkes Cu-ATPase are better able to tolerate toxic amounts of Cu in the extracellular environment presumably by forcing Cu extrusion from the cell (Camakaris et al., 1995) Both observations have help us understand the likely function of ATP7A in normal Cu homeostasis.

The basic structure of ATP7A has been shown in other papers in this symposium. Suffice to say the dominant features are a comparatively large 650 residue heavy metal binding (Hmb) domain with 6 tandem cysteine metal-binding clusters within the structural motif GMT/HCXSC, 8 transmembrane (Tm) regions that anchor and define the Cu channel, two flexible loops one 139 residues and the other 335 residues extending from the membrane into the cytosol. ATP7A and ATP7B are classified as type 2 membrane proteins, denoting the presence of an even number of Tm domains with the $-NH_2$ and -COOH termini on the cytosolic side of the membrane. The bacterial CopA protein contains only one GMT/HCXXC motif (Solioz et al., 1994) whereas yeast Cu-ATPase has the motif at most twice. If size is an indication of Cu transport capacity, the mammalian Cu-ATPases exceed bacteria and yeast in their ability to sequester and transport Cu. Additional important features include a Cys-Pro-Cys motif that is part of or adjacent to the channel through which the Cu ions pass (Vulpe et al., 1993).

2.1. Cellular Location of ATP7A and ATP7B

Dierick et al have localized ATP7A to the perinuclear area within the Golgi. Localization to this region was deduced from a punctated pattern that is seen when cells are stained with a fluorescent-labeled antibody to the Hmb of ATP7A (Dierick et al., 1997). Camakaris and coworkers believe that ATP7A-laden vesicles are in continuous motion transporting Cu between the Golgi and plasma membrane (Petris et al., 1996). Agents such as NH_4Cl, chloroquin, brefeldin A or bafilomycin A_1 that disrupt Golgi structure disperse the fluorescent pattern and thus indicate a major interference with localization of ATP7A. In contrast, high

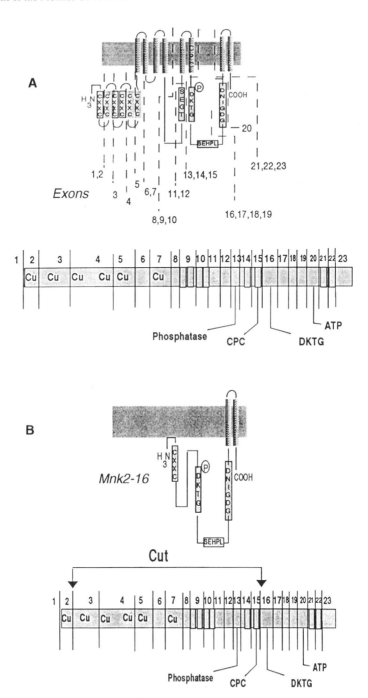

Figure 1. Structure and exon location of ATP7A. A: ATP7A mRNA contains 23 exons and codes or a protein of 1,500 amino acid residues. The first 7 exons code for 6 Cu-binding sites comprising the heavy metal domain; exons 9 and 10 code 4 or the 8 transmembrane domains, exons 13–23 code for flexible loops and 4 transmembrane (Tm) domains. A critical CPC sequence in exon 15 (Tm domain 6) and a DKGT sequence in exon 16 form part of the ion channel and phosphorylating site, respectively. B: A truncated ATP7A (ATP7A2–16) representing the product encoded by a 1.9 kb transcript that lacks exons 3–15 and fuses exons 2 with 16.

concentrations of Cu in the exterior of the cell induce a movement of the fluorescent marker to the cell boundary (Petris et al., 1996). The Wilson protein is believed to reside in two locations: an internal cellular compartment (either ER or Golgi) where it transports Cu to apoceruloplasmin and the apical surface of parenchymal cells in association with the biliary canaliculi where it forces the release of Cu into the bile (Dijkstra et al., 1995; Yang et al., 1997). A truncated homologue of ATP7B lacking four of the eight membrane-spanning domains is unable to fix to the membrane and resides in the cytosol (Yang et al., 1997).

A specific membrane location for ATP7A and ATP7B implies the existence of targeting signals that specify organelle location. As integral membrane proteins, The Menkes and Wilson ATPases undergo synthesis and maturation that follows other eukaryotic membrane proteins. Secretory proteins and plasma membrane proteins are thought to travel from the ER to the Golgi, and from the Golgi to organelles in small vesicles that pinch off from one larger vesicle and fuse with other membranes. Biosynthesis requires the nascent polypeptide chain to become resident in the endoplasmic reticulum (ER) and, once correctly folded and assembled, transported out of the ER to the excretory pathway via the *trans*-Golgi. Within the ER environment individual segments fold in accordance with residues positioned in internal and external domains of the protein. Targeting to the lysosome, Golgi, and plasma membrane is largely dependent on sorting signals in the form of small stretches of amino acid residues that are strategically placed at the NH_2- or -COOH termini of the protein, the former being more prominent (Parks, 1996). These signal/anchor sequences (S/A) are a necessary component of the trafficking mechanism. As is typical of type 2 membrane proteins, S/A signals position the protein and direct its vectorial movement to a post-Golgi compartment or anchor it to the ER. S/A signals also intercede in the retrieval of a protein from the Golgi back to the ER. The latter blocks the movement of the protein into the secretory pathway. Considerable progress has been made in defining S/A signals for specific proteins. As yet, S/A sequences have not been reported for the Menkes or Wilson ATPases. That information is absolutely essential if we are to understand and explain cellular localization of the Menkes and Wilson ATPases.

The discovery of two forms of the Wilson protein that arise through alternative splicing have fortified the concept of distinct cellular compartments for the Cu-ATPases (Yang et al., 1997) and our recent data (see below) on alternative spliced forms of ATP7A suggest Menkes transcripts also undergo a similar type of post-transcriptional processing.

3. CELLS EXPRESS MULTIPLE MENKES TRANSCRIPTS

In our work we have used pure cell lines to study Cu transport in human cells. BeWo cells, a human choriocarcinoma placental cell line, and Caco-2 cells, a coloncarcinoma cell line comprise the bulk of the studies. With the goal of obtaining a cDNA with the complete ORF of ATP7A, we designed primers based on a full length cDNA reported by Vulpe et. al. (1993) that would generate a 5.8 kb cDNA fragment. These were:

Forward [13–39]   5'- GCTACTGTGACTTCTCCGATTGTGTGA-3'
Reverse [5844–5815] 5'- ACCCCGTCTCTACTGAAAAATATGAAAAT-3'.

The cDNA fragments were excised from agarose gel and cloned into the eukaryotic TA cloning vector pCR3.1. Orientation was determined by screening the colonies with oligonucleotides covering 4391–4882 and confirmed by restriction endonuclease digestion and electrophoresis. The sequence data showed zero frequency of mutated bases.

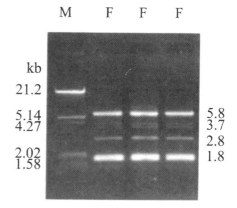

Figure 2. Multiple cDNA products generated with menkes-specific primers. cDNA products were obtained by RT-PCR amplification of *MNK* mRNA. The primers were designed to generate a 5.8 kb fragment. A prominent 1.9 band was a second major product. (M): Lambda HindIII/EcoR1 molecular weight markers. (F): filter-grown BeWo cells, the source of the RNA.

In addition to the expected 5.8 kb fragment, we discovered four smaller cDNA fragments ranging in size from 1.9 to 3.8 kb (Fig. 2). All PCR products were subcloned into a pCR3.1 vector for further analysis. Of the 60 plasmids screened, 9 had inserts with ATP7A sequences. One clone with Menkes sequences was modified at the 5' end by a 45 bp insert. The insert interrupted the traditional sequences at nt. position -121 and extended into the down stream sequences at nt position -166. The insert also terminated with a AGATG sequence at its 5' end (Fig. 3). Moreover, the ATG in the sequence was in-frame with the down stream ATG and hence appeared to represent a second start codon site further upstream from the start codon reported by Vulpe et al. (1993). If this was true, the intervening base sequence between the two ATGs may code for a 22 amino-acid extension at the N-terminal of ATP7A as shown in Fig. 3.

Screening RNAs from other cells showed that transcripts from Caco-2 cells also had a 45 bp insert on the 5' end in the exact same position as the BeWo cell insert. Caco-2 cells and fibroblasts, however, had a second 5' insert of 192 bp at that was contiguous with the 3' end of the 45 bp insert (see Fig 4) and extended into the traditional ATP7A sequences at nt. position -166.

The presence of two inserts on the 5' end of the Menkes mRNA has now been confirmed by sequencing RT-PCR cDNAs in at least four human cell lines (Table 1). Those that expressed cDNAs with a modified 5' end displayed the parent ATP7A mRNA, i.e., a

Table 1. Status of alternative transcripts in human cell lines

Cell line	Organ	2-16	45 bp	45 + 192 bp
Bewo[b]	Placenta	+	+	−
Caco-2	Intestine	+	+	+
Normal Fibroblasts	Skin	+	+	+
Neuroblastoma	Brain	+	+	+
HepG2	Liver	+	ND	ND

[a]2-16 refers to a transcript missing exons 3-15, but retaining one GMTCXSC binding motif; 45 bp refers to a transcript with a 45 bp insert 5' to the ATG start site; 45+192 bp refers to a transcript with both a 45 bp and 192 bp insert 5' to the ATG start site. (+) detected, (-) not detected, ND, not determined. [b]BeWo cells were induced to express ATP7A mRNA by culturing on filter surfaces (Qian et al., 1996)

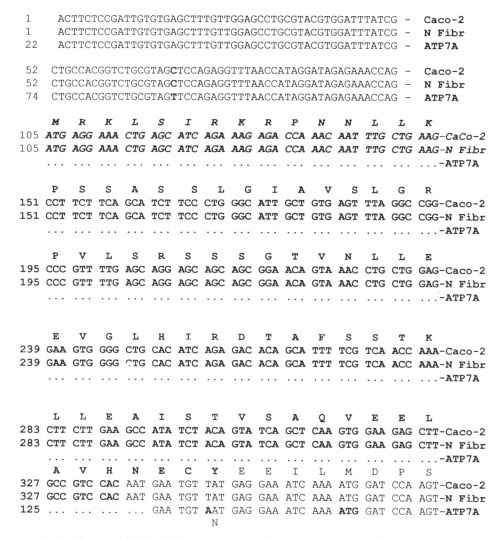

Figure 3. Modification of ATP7A cDNA at the 5' terminus. The above data are from RNA extracted from Caco-2 cells and normal fibroblasts. The 45 bp insert is shown in italics. The 192 insert is shown at the 3' end of the 45 (nt. position 151) and extends into the ATP7A sequences at nt. position 138 (Vulpe et al., 1993). A single base substitution in the untranslated region (T-C) was noticed at position 92 of ATP7A and T-A substitution was seen at nt. position 143. The ATG start site reported by Vulpe et al. is in bold on the lower right. Dotted lines indicate no base in this position.

transcript with no modifications. Caco-2 cells expressed one with 45 bp and both 45 and 192. They did not express 192 bp alone. BeWo cells gave no evidence for a 192 bp insert.

3.1. A 192 bp Insert on ATP7A Codes for a Golgi Targeting Protein

The polypeptide encoded by the intervening bases of the 192 bp insert has been examined by a BLAST search. Two sequences were brought up. One corresponded to p115, a bovine protein that has been shown to be essential for binding vesicles to the interphase Golgi

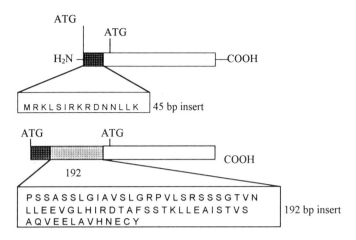

Figure 4. Analysis of sequences at the 5' end of ATP7A cDNA.

membranes and may be required for fusion with target membranes (Waters et al., 1992). The second match was to TAP (transcytosis accessory protein), a rat homologue of p115 that also has been identified with membrane protein trafficking (Barroso et al., 1995). Sequence comparisons between p115, TAP, and the 192 bp peptide product are shown in Fig. 5.

Table 1 summarizes the status of alternative transcripts found by RT-PCR amplified RNAs using primers directed specifically for ATP7A sequences. The 1.9 kb designates a transcript that is missing exons 3–15 while maintaining an in-frame link between exons 2 and 16 to 23. This transcript is found in every cell we have examined including HepG2 cells, a human hepatoma cell line which would not be expected to express ATP7A. Although the 1.9 kb transcript has stop and start codons flanking ATP7A sequences, it would

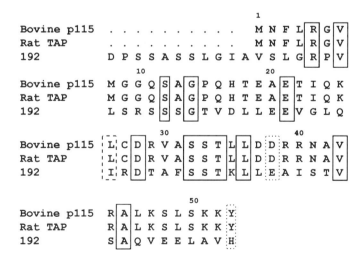

Figure 5. Sequence comparison of 192 bp insert with golgi locating and anchoring proteins. Solid boxes indicate identical residues. Boxes with dotted lines indicate conservative substitutions. The sequence alignment overall was 28 percent.

Figure 6. Analysis of ATP7A sequences in genomic DNA. Shown are the relative positions of the 45 bp and 192 bp insertion sequences and the 540 bp linking region between the two.

not code for an ATPase with membrane transporting properties. Further discussion of this unique transcript is given below.

3.2. Genomic Analysis of Menkes DNA Is Consistent with Alternative Splicing

The genomic localization of the 45 and 192 nucleotide sequences has been confirmed by PCR analysis with DNA extracted from HepG2 cells. A forward primer complementary to the 45 bp and a reverse primer complementary to the 192 bp sequences generated a DNA fragment of about 540 bp. A link of the 45 or 192 sequences with either exon 1 or exon 2 sequences using a Taq polymerase and PCR, however, has not been successful, suggesting the two lie within a region that is outside the range of PCR generated fragments and within a genomic DNA segment identified with intron 1.

3.3. All Cells Express an ATP7A Variant that Lacks Exons 3–15

Oligonucleotide primers designed to generate a 5.8 kb RT-PCR fragment generated several cDNAs with ATP7A sequences. A 1.9 kb was one of the major cDNAs (Fig. 2). Sequence data revealed that this cDNA had an ORF of 1512 bp which encoded a protein of 504 amino acid residues (theoretical molecular mass = 55,400 Da). The truncated ORF resulted from an unusual in-frame fusion between exons 2 and 16 (Fig. 7). The polypeptide encoded by the sequences (designated ATP7A2–16) would lack the phosphatase domain, six of the eight transmembrane domains, and 5 of the six Cu binding sites of ATP7A (see Fig. 1B). The absence of a CPC group in the transmembrane channel makes its unlikely

```
              M   D   P   S   M   G   V   N   S  V/A  T   I   S   V   E
146           ATG GAT CCA AGT ATG GGT GTG AAT TCT GTT ACC ATT TCT GTT GAG-ATP7A
159           ATG GAT CCA AGT ATG GGT GTG AAT TCT GCT ACC ATT TCT GTT GAG-BeWo 1.9

              G   M   T   C   N   S   C   V   W   T   I   E   Q   Q   I
191           GGT ATG ACT TGC AAT TCC TGT GTT TGG ACC ATT GAG CAG CAG ATT-ATP7A
214           GGT ATG ACT TGC AAT TCC TGT GTT TGG ACC ATT GAG CAG CAG ATT-BeWo 1.9
                                                (Exon 2-3)
              G   K   V   N   G   V   H   H   I   K   V   ( S   L   E   E
235           GGA AAA GTG AAT GGT GTG CAT CAC ATT AAG GTA TCA CTG GAA GAA-ATP7A
248           GGA AAA GTG AAT GGT GTG CAT CAC ATT AAG GTA AAG GTA GTG GTA-BeWo 1.9
                                                        ( K   V   V   V
                                                (Exon 2-(-16))
```

Figure 7. Sequence of ATP7A2–16: Evidence for a 2–16 and 2–3 Link. ATP7A2–16 codes for a protein of 503 amino acid residues. Nine bases in the modified transcript did not match with ATP7A sequences. Shown above is the sequence of bases at the exon 2–3 and 2–16 junctions. The splice site obeys Chambon's basic GT-AG rule (Breathnach et al., 1978) with an uninterrupted valine at the 2–16 exon junction.

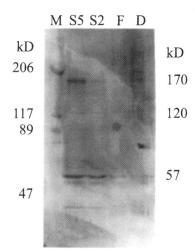

Figure 8. Western blot analysis of immunoreactive proteins. The antibody was directed against the heavy metal binding domain of ATP7A. (S5) and (S2) indicate total proteins extracted from BeWo cells that had been treated with 5.0 mM and 2.5 mM sodium butyrate, respectively, for 24–36 hr to induce the appearance ATP7A. (F) and (D) indicate BeWo cells grown on filter surfaces and dish surfaces, respectively. (M): molecular weight markers.

ATP7A2–16 could export Cu. However, the presence of a single GMTCXSC sequence on the N-terminal is reason to believe that ATP7A2–16 protein could binds and transports Cu.

3.4.1. Western Blots Provide Evidence for a Smaller Transcript. An important question is whether the ATP7A2–16 protein exists. The protein would lack a membrane anchor and hence could be soluble in the cytosol. Several points become germane to the existence of this protein. First, probes that had sequences that complemented the 3 kb core of ATP7A mRNA would be handicapped in detecting *MNK* transcripts by Northern analysis because the binding stringency would be weakened by the missing segment. Such probes would have a stronger affinity to bind ATP7A over ATP7A2–16 mRNA. Chelly et al. (1993) observed a moderately strong band by Northern blotting that had a size of about 5.5 kb. They dismissed the smaller transcript as a degradation product and further observed that a 5.5 kb mRNA was not a common occurrence for all tissues. The 5.5kb would be the predicted size for a Menkes transcript that lacked exons 3–15 (total size of exon 3–15 = 2.99 kb (Dierick et al., 1995). Second, we have observed on numerous occasions the presence of an immunoreactive protein 57–59 kDa on Western Analysis (Fig. 8). The protein reacts positively with polyclonal antibodies to the Hmb region of ATP7A. We have seen this protein in soluble extracts from C6 rat glioma cells (Qian et al., unpublished). Although the tendency is to consider lower molecular weight proteins as degradation fragments of full length ATP7A, it still remains to determine by sequence data if the 57–59 kDa protein is ATP7A2–16 protein.

4. SUMMARY AND CONCLUSIONS

Numerous pure cell lines express ATP7A. From studies of these cell lines has come evidence that non-Menkes cells, as part of a normal processing event, perform alternative splicing of the primary Menkes transcript that gives rise to isoforms of ATP7A. It is too soon to tell, but conceivable, that alternative splicing creates isoforms of ATP7A for localization to specific subcellular compartments or entrance into the secretory pathway. What appears to be a second start site has been found 66 bases upstream from the published start site. The second site occurs within a 45 bp insert at the 5' end of some transcripts and features an ATG in-frame with the downstream ATG site. A 45 bp insert has been seen in cDNAs from BeWo cells and an insert that combines the 45 with 192 bps has been observed in Caco-2 cells and human fibroblasts. The data suggests the presence of additional sequences in numerous cells is not a PCR artifact. Moreover, the sequences in the 192 insert correspond to a short run of amino acids that have been identified with the N-terminal region of Golgi anchoring proteins. This observation supports the concept that the 192 and 45 inserts are coding sequences and perhaps take part in imparting Golgi locating signals to the N-terminal region of ATP7A proteins. At this time, we cannot exclude the possibility these inserts arise from a second Menkes gene. Evidence against this, however, comes from genomic DNA analysis showing the 45 bp and 192 bp sequences are in the intron 1 region of genomic DNA and appear to be flanked by consensus sequences that define splice sites. The data support only one *MNK* gene expressed in cells and fortify the occurrence of alternative splicing of a single primary transcript.

One of the novel transcripts we observed was a truncated form of the *MNK* mRNA. The sequence data suggested this transcript encoded a peptide that was missing most of the transmembrane domains and the ion channel sequences. Because it lacked exons 3–15, the protein made by the transcript would not be expected to bind firmly to membranes and would likely appear in the soluble phase. The presence of a single GMTCXSC motif leaves open the possibility that the protein we have called ATP7A2–16 can exchange Cu ions with ATP7A or ATP7B, providing Cu to the corresponding binding motifs on these proteins. A small Cu-binding protein, HAH1, an human analogue of a yeast Cu-binding protein Atx1, has been reported at this meeting. HAH1 has a single MTCXGC binding motif similar to the Cu-ATPases. The 8 kDa HAH1 protein and similar variants are postulated to donate Cu to enzymes including the Cu-ATPases by a simple ligand exchange reaction (Klomp et al., 1997). As reported elsewhere at this symposium, the discovery has provided the first rationale for soluble proteins acting as Cu chaperons that move Cu into enzymes or exclude Cu from the cell. It is important, therefore, to note that an ATP7A2–16 transcript has be found in every cell we examined and because of its ubiquity and predicted ATP7A-like structure, it is conceivable that ATP7A2–16 protein could share transport functions with HAH1.

Fig. 9 summarizes the ATP7A transcripts discovered thus far. A splicing scheme that would give rise to the alternative transcripts of Menkes mRNA is shown in the lower part of the figure. In the scheme, we predict a series of introns positioned between exons one and two that surround the newly discovered 45 and 192 bp sequences. What is referred to as ATP7A (the parent structure) arises through a simple union between exon 1 and exon 2. An alternative splice site at the boundary between exon 1 and the 45 bp insert could give rise to the 45 bp insert on the 5' end of ATP7A (referred to as ATP7A-1). The removal of the 540 bp sequences between the 45 bp and 192 bp gives an additional splice variant that has a union between the 45 and 192 which joins with exon 1 on the 5' end and exon 2 on the 3' end giving rise to an ATP7A transcript with both extensions and with the alternative ATG start site (ATP7A-2). Finally, the small soluble protein arises by a fusion between exon 2 and 16 which maintains the in-frame reference to exon 23. Because exons 3–7 code for 5 of the 6 Cu binding sites, only one site is left intact. As yet, we have not discovered an ATP7A2–16 transcript with 45 bp or 192 bp inserts on the 5' end.

4.1. Future Work

A hypothesis guiding this work is that ATP7A (Menkes) and ATP7B (Wilson) owe their unique cellular locations to signal anchoring sequences on the 5' end of the respective proteins. If this hypothesis is correct, there is reason to believe that what is referred to as the

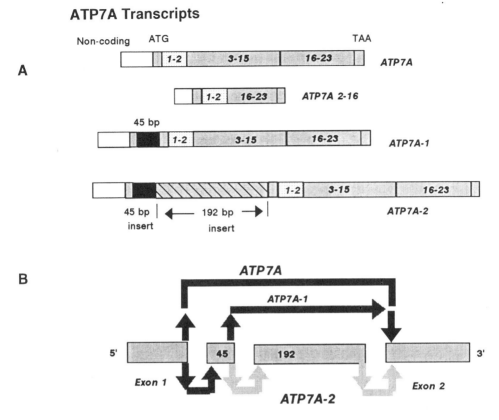

Figure 9. Summary of alternative transcripts of ATP7A. A: Transcripts containing inserts at 5' end. B: Hypothetical splicing mechanism that leads to their Formation.

Menkes gene could in reality be a *copper locus* with the capacity to synthesize numerous transcripts that code for proteins involved in Cu transport. Alternative splicing governs the relative quantities of each of the transcripts and dictates the cellular location and function of the protein products. Research is needed to determine if ATP7A proteins with N-terminal extensions are translation products as the cDNA evidence would predict. Indeed, thus far we have only cDNA data for their existence. The reality of a soluble Cu transporter with ATP7A sequences must also be confirmed. That alternative splicing can lead to other proteins bearing the GMTCXSC structure suggest a single genetic locus also takes part in the synthesis of internal Cu binding and transport proteins designed to deliver Cu to ATP7A and ATP7B. Additional research will clarify if splicing factors that act on ATP7A and ATP7B primary transcripts are themselves regulated by the Cu status of the cell and are thus part of the mechanism controlling Cu homeostasis.

ACKNOWLEDGMENTS

This research was supported in part by USDA Grant 94-37200-0375 and NIEHS Grant ES-05781.

REFERENCES

Barroso, M., Nelson, D.S., and Sztul, E. (1995). Transcytosis-associated protein (TAP)/p115 is a general fusion factor required for binding of vesicles to acceptor membranes. Proc. Nat. Acad .Sci. (USA) 92, 527–531.

Breathnach, R., Benoish, C., O'Hare, K., Gannon, F., and Chambon, P. (1978). Ovalbumin gene: evidence for a leader sequence in mRNA and DNA sequences at the intron-exon boundaries. Proc. Nat. Acad. Sci. (USA) 75, 4853–4857.

Bull, P.C., Thomas, G.R., Rommens, J.M., Forbes, J.R., and Cox, D.W. (1993). The Wilson disease gene is a putative copper transporting P-type ATPase similar to the Menkes gene. Nature Genet. 5, 327–336.

Camakaris, J., Petris, M.J., Bailey, L., Shen, P.Y., Lockhart, P., Glover, T.W., Barcroft, C.L., Patton, J., and Mercer, J.F.B. (1995). Gene amplification of the Menkes (MNK; ATP7A) P-type ATPase gene of CHO cells is associated with copper resistance and enhanced copper efflux. Hum. Mol. Genet. 4, 2117–2123.

Chelly, J., Tümer, Z., Tonnesen, T., Petterson, A., Ishikawa-Brush, Y., Tommerup, N., Horn, N., and Monaco, A.P. (1993). Isolation of a candidate gene for Menkes disease that encodes a potential heavy metal binding protein. Nature Genet. 3, 14–19.

Danks, D.M., Campbell, P.E., Walker-Smith, J., Stevens, B.J., Gillespie, J.M., Blomfield, J., and Turner, B. (1972). Menkes' kinky-hair syndrome. Lancet 1, 1100–1103.

Danks, D.M. (1988). Copper Deficiency in Humans. Ann.Rev.Nutr. 8, 235–257.

Dierick, H., Ambrosini, L., Spencer, J., Glover, T.W., and Mercer, J.F.B. (1995). Molecular structure of the Menkes disease gene (*ATP7A*). Genomics 28, 462–469.

Dierick, H.A., Adam, A.N., Escara-Wilke, J.F., and Glover, T.W. (1997). Immunocytochemical localization of the Menkes copper transport protein (ATP7A) to the *trans*-Golgi network. Hum. Molec. Genet. 6, 409–416.

Dijkstra, M., Veld, G.I., van den Berg, G.J., Muller, M., Kuipers, F., and Vonk, R.J. (1995). *In vitro* modeling of liver membrane copper transport. J. Clin. Invest. 95, 412–416.

Klomp, L.W., Lin, S.J., Yuan, D.S., Klausner, R.D., Culotta, V.C., and Gitlin, J.D. (1997). Identification and functional expression of HAH1, a novel human gene involved in copper homeostasis. J. Biol. Chem. 272, 9221–9226.

Lutsenko, S., Petrukhin, K., Cooper, M.J., Gilliam, C.T., and Kaplan, J.H. (1997). N-terminal domains of human copper-transporting adenosine triphosphatases (the Wilson's and Menkes disease proteins) bind copper selectively in vivo and in vitro and with stoichiometry of one copper per metal-binding repeat. J. Biol. Chem. 272, 18939–18944.

Mercer, J.F.B., Livingston, J., Hall, B., Paynter, J.A., Begy, C., Chandrasekharappa, S., Lockhart, P., Grimes, A., Bhave, M., Siemieniak, D., and Glover, T.W. (1993). Isolation of a partial candidate gene for Menkes disease by positional cloning. Nature Genet. 3, 20–25.

Parks, G.D. (1996). Differential effects of changes in the length of a signal/anchor domain on membrane insertion, subunit assembly, and intracellular transport of a type II integral membrane protein. J. Biol. Chem. 271, 7187–7195.

Petris, M.J., Mercer, J.F.B., Culvenor, J.G., Gleeson, P.A., and Camakaris, J. (1996). Ligand-regulated transport of the Menkes copper P-type ATPase efflux pump from the Golgi apparatus to the plasma membrane: a novel mechanism of regulated trafficking. EMBO J. 15, 6084–6095.

Petrukhin, K., Fischer, S.G., Pirastu, M., Tanzi, R.E., Chernov, I., Devoto, M., Brzustowicz, L.M., Cayanis, E., Vitale, E., Russo, J.J., Matseoane, D., Boukhgalter, B., Wasco, W., Figus, A.L., Loudianos, J., Cao, A., Sternlieb, I., Evgrafov, O., Parano, E., Pavone, L., Warburton, D., Ott, J., Penchaszadeh, G.K., Scheinberg, I.H., and Gilliam, T.C. (1993). Mapping, cloning and genetic characterization of the region containing the Wilson disease gene. Nature Genet. 5, 338–343.

Pufahl, R.A., Singer, C.P., Peariso, K.L., Lin, S.-J., Schmidt, P.J., Fahrni, C.J., Culotta, V.C., Penner-Hahn, J.E., and O'Halloran, T.V. (1997). Metal ion chaperone function of the soluble Cu(I) receptor Atx1. Science 278, 853–856.

Qian, Y.C., Tiffany-Castiglioni, E., and Harris, E.D. (1995). Copper transport and kinetics in cultured C6 rat glioma cells. Am. J. Physiol. 269, C892-C898.

Qian, Y.C., Majumdar, S., Reddy, M.C.M., and Harris, E.D. (1996). Coincident expression of Menkes gene with copper efflux in human placental cells. Am. J. Physiol. 270, C1880-C1884.

Solioz, M., Odermatt, A., and Krapf, R. (1994). Copper pumping ATPases: common concepts in bacteria and man. FEBS Lett. 346, 44–47.

Solioz, M. and Vulpe, C. (1996). Cpx-type ATPases: a class of P-type ATPases that pump heavy metals. Trends Biochem. Sci. 21, 237–241.

Tanzi, R.E., Petrukhin, K., Chernov, I., Pellequer, J.L., Wasco, W., Ross, B., Romano, D.M., Parano, E., Pavone, L., Brzustowicz, L.M., Devoto, M., Peppercorn, J., Bush, A.I., Sternlieb, I., Pirastu, M., Gusella, J.F.,

Evgrafov, O., Penchaszadeh, G.K., Honig, B., Edelman, I.S., Soares, M.B., Scheinberg, I.H., and Gilliam, T.C. (1993). The Wilson disease gene is a copper transporting ATPase with homology to the Menkes disease gene. Nature Genet. 5, 344–350.

Thomas, G.R., Forbes, J.R., Roberts, E.A., Walshe, J.M., and Cox, D.W. (1995). The Wilson disease gene: spectrum of mutations and their consequences. Nat. Genet. 9, 210–217.

Vulpe, C., Levinson, B., Whitney, S., Packman, S., and Gitschier, J. (1993). Isolation of a candidate gene for Menkes disease and evidence that it encodes a copper-transporting ATPase. Nature Genet. 3, 7–13.

Vulpe, C.D. and Packman, S. (1995). Cellular copper transport. Ann.Rev.Nutr. 15, 293–322.

Waters, M.G., Clary, D.O. and Rothman, J.E. (1992). A novel 115-kD peripheral membrane protein is required for intercisternal transport in the Golgi stack. J. Cell Biol. 118, 1015–1026

Yamaguchi, Y., Heiny, M.E., and Gitlin, J.D. (1993). Isolation and characterization of a human liver cDNA as a candidate gene for Wilson's disease. Biochem. Biophys. Res. Commun. 197, 271–277.

Yang, X., Miura, N., Kawarda, Y., Terada, K., Petrukhin, K., Gilliam, C., and Sugiyama, T. (1997). Two forms of Wilson disease protein produced by alternative splicing are localized in distinct cellular compartments. Biochem. J. 325, 897–902.

5

THE CELL BIOLOGY OF THE MENKES DISEASE PROTEIN

Michael J. Petris,[1,2] Julian F. B. Mercer,[2] and James Camakaris[1]

[1]Genetics Department, University of Melbourne
Victoria, Australia
[2]Murdoch Institute, Royal Children's Hospital
Victoria, Australia

1. INTRODUCTION

1.1. Copper

Copper is a trace element which is readily converted between cuprous and cupric forms under physiological conditions. This redox property has been harnessed in biological systems where copper forms an integral component of enzymes whose catalytic function involves electron exchange. However, this same property of copper also makes it toxic when present within the cell at elevated levels. Consequently, intracellular copper levels must be carefully controlled, presumably by regulated transport mechanisms. Disruptions of some of these mechanisms can have a genetic basis as illustrated in the case of both Menkes disease and Wilson disease (Danks, 1995).

1.2. Menkes Disease

Menkes disease is an X-linked disorder of copper metabolism which was first described by John Menkes and co-workers in 1962, and later discovered to be caused by a systemic copper deficiency (Danks et al., 1972). Menkes disease is characterised by coarse and brittle depigmented hair, neonatal problems including hypothermia and defective temperature control, severe mental retardation, skeletal abnormalities and connective tissue defects (Danks, 1995). The symptoms of Menkes disease can mostly be understood as a consequence of reduced activities of a several copper-dependent enzymes. Lysyl oxidase deficiency causes arterial disease due to reduced cross-linking of collagen and elastin, necessary in the strengthening of arterial walls. Pigmentation defects are the result of reduced tyrosinase activity, and cytochrome c oxidase deficiencies probably account for reduced body temperatures and neurological defects observed in Menkes patients.

The copper deficiency in Menkes patients is due, in part, to malabsorption of dietary copper, resulting in reduced copper levels in brain, liver, and serum of affected individuals. Interestingly, copper accumulates in the cells lining both the small intestine and the proximal tubules of the kidney (Danks et al, 1995). Both these tissues contain polarised cells necessary for the vectorial transport of copper across the basolateral membrane into serum. Defects in the polarised transport of copper also occurs in the cells lining the blood brain barrier, leading to an accumulation of copper in these cells (Kodama, 1993). This accumulation of copper in polarised tissues of Menkes patients initially suggested a defect in polarised copper transport. However, defects in copper transport are present in non-polarised cell types cultured from Menkes patients including fibroblasts (Beratis et al., 1978; Camakaris et al., 1980; Chan et al., 1978; Goka et al., 1976), amniotic cells (Camakaris et al., 1980; Horn, 1976) and continuous lymphoid cell lines (Camakaris et al., 1980; Riordan and Jolicoeur-Paquet, 1982). Later studies investigating rates of copper uptake and efflux in cultured Menkes fibroblasts suggested that the accumulation of copper is a consequence of defective intracellular copper transport (Herd et al., 1987).

1.3. Wilson Disease

Wilson disease is an autosomally inherited disorder of copper toxicity caused by defective excretion of copper from the liver into bile. This results in an accumulation of hepatic copper which can eventually lead to excessive copper in the brain as a secondary effect. The aberrant copper levels in Wilson patients most often result in hepatic failure, but can also manifest as neurological problems, acute haemolytic crisis, renal stones, renal tubular damage and copper deposits on the cornea, termed Kayser-Fleischer rings. In addition to biliary excretion, the loading of copper onto apocaeruloplasmin for copper transport to extrahepatic tissues is also reduced in Wilson disease patients and suggests the Wilson gene product may function in both processes.

1.4. Menkes and Wilson Gene Products: Putative Copper Transporting P-Type ATPases

Menkes disease is caused by mutations in the *MNK* (*ATP7A*) gene (Chelly et al., 1993; Mercer et al., 1993; Vulpe et al., 1993), whilst the gene defective in Wilson disease is *WND* (*ATP7B*) (Bull et al., 1993; Tanzi et al., 1993; Yamaguchi et al., 1993). These genes are closely related and encode proteins which are members of the P-type ATPase family. Members of this family are involved in the ATP-dependent transport of cations across membranes in both prokaryotes and eukaryotes (Pederson and Carafoli, 1987). P-type ATPases use the energy obtained from the hydrolysis of ATP to drive the transport of cation across membranes, and become transiently phosphorylated at an aspartic acid residue as part of the reaction cycle. The formation of this aspartyl-phosphate bond is characteristic of all P-type ATPases, as is the ability of sodium orthovanadate to inhibit the formation of this phospho-intermediate. On completion of cation transfer across the membrane, an autophosphatase function cleaves the aspartyl-phosphate bond returning the enzyme to its original dephosphorylated state for second cycle of cation transport.

The primary sequences of both Menkes and Wilson proteins are highly conserved and share 59% identity. Both proteins are predicted to have eight transmembrane domains as well as regions common to all members of the P-type ATPase family. These include a phosphorylation domain, an ATP-binding sequence, a phosphatase domain and a transduction domain (Figure 4). An interesting feature of both Menkes and Wilson proteins is the presence of six repeating motifs minimally, GMXCXXC (where X represents any amino acid), close to the amino-terminal region of both molecules. Each motif bears a strong resemblance to the cadmium-binding sequence of the cadmium transporting ATPase, cadA, of *Staphyloccoccus aureus*. Given the propensity of cysteines to bind copper, and the association of Menkes and Wilson diseases with copper homeostasis, the six repeating motifs were proposed to bind copper (Chelly et al., 1993; Mercer et al., 1993; Vulpe et al., 1993). Indeed, recent experiments have demonstrated the NH_2-terminal domains of both MNK and WND bind copper *in vitro* and *in vivo*, in a stoichiometry of 5–6 nmol of copper/nmol protein, suggesting the binding of one copper atom per motif (Lutsenko et al., 1997).

Since the cloning of the Menkes gene, several putative copper-transporting P-type ATPases have been identified in a range of lower organisms. These organisms include *Enterococcus hirae* (Odermatt et al., 1993), *Synchococcus* species PCC7942 (Kanamaru et al., 1994), *Saccharomyces cerevisiae* (Rad et al.,1994), *Helicobacter pylori* (Ge et al., 1995), and *Caenorhabditis elegans* (Sambongi et al., 1997). The existence of putative copper ATPases across this broad range of organisms suggests ATP-driven transmembrane copper transport is evolutionarily well conserved, with probable origins amongst early life forms. Interestingly, the ATPases from bacteria and yeast contain only one or two of the putative copper-binding GMXCXXC motifs. The more complex multicellular nematode, *C.elegans*, has three copies of this motif within its WND/MNK homologue. In humans, the MNK and WND proteins have six repeating motifs, however, it is notable that rodent homologues of WND have only five complete GMXCXXC motifs, the fourth being altered to a presumably non-functional form (Wu et al., 1994; Theophilos et al., 1996). The remarkable conservation of these motifs suggests they are functionally important, perhaps in regulating the catalytic activity of these copper pumps. However, the significance of the progressive increases in the number of motifs which occurs from unicellular bacteria and yeast, to multicellular organisms such as the nematode and mammals, is not understood.

Another notable feature of the primary sequences conserved between Menkes and Wilson proteins is the presence of a Cys-Pro-Cys motif within the membrane transduction domain. It has been proposed that the cysteines (and histidine where present) of this motif bind copper atoms, and facilitate transfer through the membrane (Solioz and Vulpe, 1996). Indeed, the presence of a Cys-Pro-X (where X is either Cys or His) within the transduction domain has been used as a signature sequence for the further classification of ATPases into a subgroup known as CPX-ATPases (Solioz et al, 1994; Solioz and Vulpe, 1996).

1.5. Menkes P-Type ATPase: Evidence for Copper Efflux Function from Studies of Copper Resistant CHO Cells

The copper accumulation phenotype of cultured Menkes cells, together with the primary sequence of the MNK gene product, suggest the MNK protein is involved in the efflux of copper from cells. Evidence supporting this hypothesis was obtained from studies of three copper-resistant Chinese hamster ovary (CHO) cell lines. These copper resistant lines, termed CUR1, CUR2 and CUR3 (CUR= Cu resistance) were isolated as spontaneous variants resistant to elevated copper in the growth medium. The same mode of copper resistance occurred in each of the CUR cell lines and involved an increased ability to maintain homeostasis by an elevated copper efflux when cultured in the presence of high copper in the growth medium (Camakaris et al., 1995). Evidence for a role in copper efflux for the hamster homologue of the MNK protein (haMNK) was obtained when the levels of elevated copper efflux in the

CUR cell lines correlated with ha*MNK* gene amplification and over-expression of ha*MNK* mRNA and protein levels (Camakaris et al., 1995). The ha*MNK* gene was found to be over-expressed 2-, 10-, and 70- fold in CUR1, CUR2 and CUR3 cell lines respectively, above the parental cell line CHO-K1. This suggests the haMNK protein has an important role in copper homeostasis by functioning in copper efflux and this proposal is consistent with the elevated intracellular copper observed when the MNK protein is deficient or non-functional, as in the case of cultured cells from Menkes disease (Camakaris et al., 1980; Goka et al., 1976; Herd et al., 1987; Horn, 1976). Indeed, the finding that the predicted hamster MNK amino acid sequence (Genebank accession no. U29946) shares 95% similarity with the human sequence (across 1476 amino acids of available sequence) suggests both proteins are likely to function equivalently, and provides a degree of confidence in extrapolating the observations in CHO cells to predict the function of the human protein.

2. THE SUBCELLULAR LOCALISATION OF THE haMNK PROTEIN

One of the major questions to be addressed following the evidence for a causal link between over-expression of haMNK and elevated copper efflux in the CUR cell lines, was the subcellular localisation of the haMNK protein: is the protein localised in the plasma membrane or in an intracellular membrane? This was an important aim as it was expected to shed light on the function of the protein in copper efflux, and possibly the cellular phenotype of Menkes disease. The fortuitous discovery of haMNK over-expression in the CUR cell lines provided an advantageous system for investigating the subcellular localisation of the haMNK protein.

2.1. Immunofluorescence Microscopy

2.1.1. Localisation of haMNK to the Trans-Golgi Network. Confocal immunofluorescence microscopy using the affinity purified MNK-antibodies was used to investigate the subcellular localisation of haMNK in the copper-resistant CUR cell lines. Within each CUR line, a perinuclear fluorescent signal was observed when cells were cultured in basal medium (Petris et al., 1996). Figure 1A shows this perinuclear localisation of haMNK in CUR3 cells. The intensity of this signal across the CUR cells together with the absence of staining when preimmune serum was used indicated the source of the perinuclear staining was the haMNK protein (Petris et al., 1996). The perinuclear location of the haMNK protein in the CUR cells suggested haMNK is localised to the Golgi apparatus. This was confirmed in CUR3 cells when the Golgi-disrupting agent, nocodazole, dispersed the perinuclear labelling of haMNK (Petris et al., 1996). Using the drug, brefeldin A, evidence was obtained suggesting haMNK resides within the distal compartment of the Golgi apparatus, known as the *trans*-Golgi network (TGN). This was further supported when double labelling experiments demonstrated co-localisation of haMNK with the marker for the TGN, p230 in CUR3 cells (Petris et al., 1996).

Given the predicted role in copper efflux for the haMNK protein, it was initially expected that haMNK would reside in the plasma membrane, since this was the location where other ATPases, such as the P-glycoprotein (mdr), efflux toxic substances, and like haMNK, are amplified under appropriate selective conditions. This led to experiments

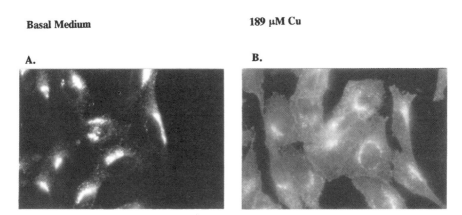

Figure 1. Immunfluorescence localisation of the haMNK protein in CUR3 cells. CUR3 cells were grown for 24 h in basal media (A) or media supplemented with 189 µM copper (CuCl$_2$) (B) and then stained by indirect immunofluorescence with affinity purified antibodies to MNK (1:50). After primary antibody reactions cells were reacted with FITC-conjugated sheep antibodies to rabbit IgG (Silenus, Australia) (1:300).

which investigated the localisation of haMNK in the CUR cells grown in elevated copper where their elevated efflux phenotype is apparent (Camakaris et al., 1995).

2.1.2. Copper-Induced Redistribution of haMNK from the TGN to the Plasma Membrane. In each of the CUR cells cultured in medium supplemented with 189 µM copper, the distribution of haMNK shifted from being predominantly perinuclear to a punctate staining pattern which extended throughout the cytoplasm to the peripheries of the cells (Petris et al., 1996). This striking effect of copper on haMNK distribution is shown for CUR3 cells in figure 1B. Using confocal immunofluorescence microscopy of isolated plasma membrane fragments, a shift of 5.5 fold to the plasma membrane for haMNK was observed in CUR3 cells cultured in elevated copper (Petris et al., 1996). This shift in haMNK distribution from the TGN to the plasma membrane occurs within 15 minutes following the addition of 189 µM to the medium and is independent of *de novo* protein synthesis. Interestingly, when the CUR cells are returned from elevated copper to basal medium, haMNK rapidly returns to a steady state TGN localisation, indicating the haMNK is able to recycle from the plasma membrane to the TGN (Petris et al., 1996). Further analysis using inhibitors of endosomal recycling suggested haMNK constitutively recycles between the TGN and plasma membrane in cells cultured in basal medium.

2.1.3. The Redistribution of haMNK Occurs at Low Copper Levels in the Growth Medium. The level of copper supplemented to the growth medium where haMNK has been shown to relocate from the TGN to the plasma membrane (189 µM (12 µg/ml)) is far greater than those found physiologically (Petris et al., 1996). Initially this copper level was chosen for immunofluorescence experiments because there is a significant increase in detectable copper efflux when cells are exposed to 189 µM Cu and it is at this concentration of copper where differences in efflux activities amongst each of the CUR cell lines is readily distinguishable (unpublished data). To determine whether the copper-induced relocalisation of haMNK from the TGN occurs at copper concentrations close to physiological

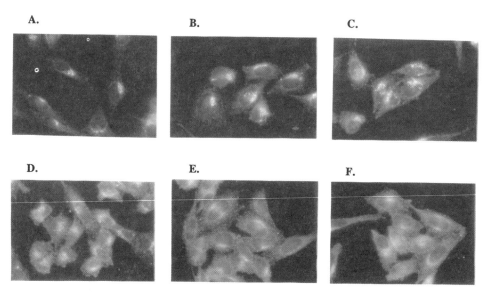

Figure 2. Localisation of haMNK in CUR2 cells in a range of media copper concentrations. CUR2 cells were cultured for 24 h in basal medium (0.8 μM Cu) (A) or in the presence of 10, 20, 50, 100, and 189 μM media copper (B, C, D, E and F respectively) then processed for immunofluorescence using the affinity purified MNK-antibodies (1:50). Primary antibodies were detected using FITC-conjugated sheep antibodies to rabbit IgG (1:300).

levels, the localisation of haMNK was investigated in a range of media copper concentrations in CUR2 cells. Figure 2 shows that the extent of the dispersal of haMNK from the TGN in CUR2 cells was dependent on the level of copper in the growth medium. Relocalisation to a punctate cytoplasmic distribution became apparent between 10- and 20- μM copper (Figure 2), suggesting this process may occur at copper concentrations encountered by cells *in vivo*. This is an important observation as it suggests efflux via haMNK may regulate copper homeostasis not only as a response to excessively high copper concentrations, but under physiological conditions by small fluctuations in serum- and intracellular copper concentrations. This proposal will be tested in future studies by investigating the range of intracellular copper levels at which haMNK relocalisation occurs from the TGN and whether this occurs simultaneously with a switching on of elevated copper efflux.

3. COPPER INDUCED RELOCALISATION OF MNK: IS IT A GENERAL PHENOMENON?

The copper-induced redistribution of haMNK is a novel observation which increases our understanding of cellular copper homeostasis in the CUR cell lines. However, it was important to establish whether this phenomenon occurs in cells where the MNK protein is not over-expressed and indeed if it occurs in human cells. To date the subcellular localisation of the MNK protein to the TGN has been demonstrated within wild-type CHO cells (Dierick et al., 1997) and the human epithelial cell line, HeLa (Yamaguchi et al., 1996). These observations suggest the localisation of haMNK in the CUR cells in basal medium is not affected by over-expression of haMNK. Using confocal immunofluorescence microcopy the effect of copper on the subcellular localisation of the MNK protein in both CHO-K1 and HeLa cells was investigated.

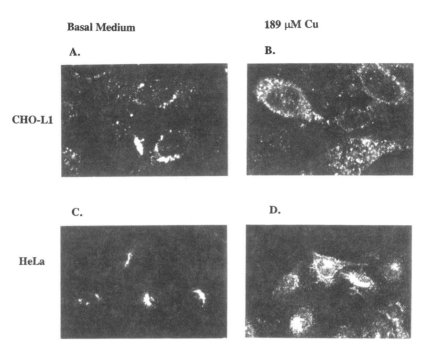

Figure 3. The effect of copper on the localisation of the Menkes protein in CHO-K1 and HeLa cells. CHO-K1 and HeLa cells were cultured for 2 h in basal medium (A and C, respectively) and media supplemented with 189 μM copper (B and D, respectively), prior to fixing cells. Cells were incubated with the affinity purified anti-MNK antibodies (1:30) before detection using FITC-conjugated sheep antibodies to rabbit IgG (1:300). Cells were imaged using confocal microscopy.

3.1. Copper-Induced Redistribution of MNK in CHO-K1 and HeLa Cells

As shown in figure 3 (A and B), a perinuclear distribution of haMNK in CHO-K1 and HeLa cells was observed when cultured in basal medium. When cultured for 2 h in medium containing 189 μM copper there was no longer a predominantly perinuclear distribution of MNK, rather a punctate staining pattern was observed which extended throughout the cytoplasm to the peripheries of the cells (Figure 3B and D). This copper induced distribution of MNK is clearly very similar to that which occurs in CUR cells where the haMNK is over-expressed (Figure 1) (Petris et al., 1996). The copper-induced redistribution of MNK in HeLa cells occurs within 15 minutes of the addition of 100 μM copper and occurs at copper concentrations as low as 20 μM (data not shown). Hence, both the time taken, as well as the minimum media copper concentration required to induce this redistribution, is approximately the same when comparing CUR2 cells with HeLa cells.

4. A ROLE FOR COPPER ATPases IN COPPER DELIVERY INTO THE SECRETORY PATHWAY

Given the evidence from the CUR cells that the MNK homologue is involved in copper efflux, and is redistributed from the TGN to the plasma membrane by elevated copper, a reasonable hypothesis is that both processes are related. That is, efflux of copper via MNK involves a relocalisation from the TGN to the plasma membrane. The MNK protein may function in copper efflux only upon arrival at the plasma membrane, or alternatively, MNK may pump copper into vesicles *en route* to the plasma membrane, where the eventual fusion of the vesicle with the plasma membrane releases copper to the extracellular environment.

The steady-state localisation of MNK in the TGN in cells cultured in basal medium is consistent with a second role for the enzyme. In cultured fibroblasts from Menkes disease patients (MNK loss of function) the activity of the secreted copper-dependent enzyme, lysyl oxidase, is greatly reduced (Royce et al., 1980). Hence, in the TGN, MNK may be required to deliver copper to newly synthesised membrane-bound or secreted copper-dependent apoenzymes as they move through the exocytic pathway. Interestingly, the closely-related copper ATPase, WND, has also been shown to localise to the TGN of hepatoma cell lines, and therefore, may also deliver copper to copper-dependent enzymes in this compartment (Hung et al., 1997). Consistent with this proposed role of WND was the finding of defective copper loading onto caeruloplasmin in the Golgi of hepatocytes cultured from the Long-Evans Cinnamon rat, an animal model of Wilson disease (Murata et al., 1995; Terada et al., 1995). It is also notable that the MNK/WND homologue of *S.cerevisiae*, Ccc2p, is localised to the late-Golgi or a post-Golgi secretory compartment (Yuan et al., 1997). Ccc2p is required for the delivery of copper into the lumen of a post-Golgi secretory compartment to a caeruloplasmin-like oxidase required for iron uptake, Fet3p (Yuan et al., 1997). Significantly, the expression of human WND and the Menkes/Wilson homologue of *C.elegans,* can complement a *ccc2* mutant by delivering copper to Fet3p in the secretory pathway (Hung et al., 1997; Sambongi et al., 1997). Hence, the overall evidence suggests that one function of eukaryotic copper ATPases is to pump copper into the lumen of late Golgi compartments, thereby providing essential copper to copper-dependent enzymes as they migrate through the exocytic pathway.

4.1. In Search of the TGN-Localisation Signal of the MNK Protein

4.1.1. Internalisation Signals Can Function as Determinants of TGN Localisation.
Proteins which localise to the TGN and other compartments of the secretory pathway contain signals which confer this targeting and exclude the protein from continuing the default route, known as bulk flow, to the plasma membrane. Indeed there are many examples of proteins which function at more than one location within the secretory pathway and possess signals which specify sorting between different intracellular compartments. In a range of recycling proteins, the information for internalisation from the cell surface and targeting to the endosomal pathway resides within short, linear stretches of amino acids within the cytoplasmic domains of the proteins. Most of these targeting signals fall into two groups of consensus sequences or 'motifs', namely the tyrosine-based motifs and the di-leucine-based motifs. The tyrosine-based signal has a consensus Y-X-X-hydrophobic amino acid (Collawn et al., 1990), and participates in the sorting of proteins from the plasma membrane to endosomes via clathrin coated pits (Bos et al., 1993; Humphrey et al., 1993; Wilcox et al., 1992). Di-leucine-motifs are involved in the sorting of proteins to lysosomal compartments from the TGN (Trowbridge et al., 1993) but, like the tyrosine-based motifs, function in the internalisation of proteins from the cell surface to various intracellular compartments. Interestingly, these endocytic targeting motifs are responsible for maintaining the steady state TGN-localisation of proteins which constitutively recycle between this compartment and the cell surface. For example the glucose transporter, GLUT4, is maintained in the TGN and associated tubulovesicular structures, in part, by

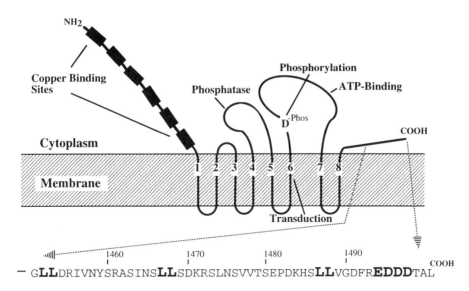

Figure 4. A schematic representation of the MNK protein highlighting candidate endocytic targeting sequences of the carboxyl terminus. The terminal 50 amino acids are shown with three di-leucine signals and the acidic sequence shown in bold.

the endocytic activity of a di-leucine motif in the carboxyl tail of the protein (Corvera et al., 1994). TGN38 is a protein which constitutively recycles between the TGN and the cell surface, and maintains steady state localisation in the TGN by constant retrieval of surface protein via a tyrosine motif in the carboxyl tail (Bos et al., 1993). More recently, stretches of acidic amino acids have been shown to confer TGN localisation by functioning as internalisation signals from the plasma membrane to the TGN (Alconada et al., 1996; Schäfer et al., 1995; Voorhees et al., 1995).

4.1.2. The Carboxyl Tail of MNK Possesses Several Candidate Internalisation Motifs. The intracellular localisation of MNK in the TGN, together with the evidence for constitutive recycling (Petris et al., 1996), suggests the MNK protein may be targeted to the TGN via similar motifs to the above TGN-localised proteins. Again, by analogy with these proteins, the mutation of such a motif within MNK may result in a shift in localisation to the plasma membrane. Given that endocytic motifs commonly occur in the cytoplasmic carboxyl tails of proteins, this region of MNK was analysed for potential endocytic targeting motifs. This resulted in the identification of several candidate endocytic motifs within the last 50 amino acids of the carboxyl-terminus of MNK, including three di-leucines and a continuous stretch of four acidic amino acids (Figure 4). In order to determine whether the last 50 amino acids contain information necessary in conferring TGN localisation, the effect of removing this region on subcellular distribution was investigated.

4.1.3. Deletion of the Carboxyl Tail of the Human MNK Protein. Using a plasmid mammalian expression construct containing the coding region of the human MNK cDNA (kindly provided by S. La Fontaine), *in vitro* mutagenesis was used to introduce a stop codon at position 1451 resulting in the deletion of the last 50 amino acids of the MNK pro-

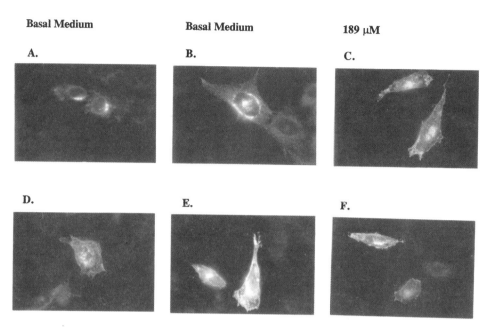

Figure 5. Localisation of the wildtype- and carboxyl-terminal truncated forms of the Menkes protein transiently expressed in CHO cells cultured in basal and elevated copper. The localisation of the human wildtype protein (A and B) and the truncated form of MNK (D and E) is shown in CHO-K1 cells cultured in basal medium for 24 h. The effect of a 2 h exposure to 189 μM copper in the growth medium is shown for the wildtype MNK protein (C) and truncated MNK protein (F).

tein. This was achieved using the Transformer mutagenesis kit (Clonetech) on a 445 bp subclone of the MNK cDNA containing the last 332 bp of coding sequence. Fidelity of the *in vitro* mutagenesis reaction was determine by automated sequencing.

4.1.4. Localisation of MNK to the TGN Requires the Carboxyl Terminal 50 Amino Acids. The importance of the carboxyl-terminal 50 amino acids of MNK in maintaining basal TGN localisation was investigated in CHO-K1 cells transiently expressing the truncated MNK protein. Immunofluorescence fixation conditions and antibodies were chosen so that low levels of endogenous haMNK within CHO-K1 cells would not produced a signal. The localisation of the MNK (wildtype) protein transiently expressed in CHO-K1 cells grown in both basal- and elevated copper medium is shown in figure 5. Two levels of protein expression are shown for cells in basal medium (Figure 5A and B). At both levels of expression, the predominant localisation of the MNK (wildtype) protein was perinuclear and staining of the cell periphery was not observed. The effect of elevated copper in the growth medium induced the relocalisation of the MNK(wildtype) protein from the perinuclear region to the plasma membrane (Figure 5C). Clearly, these observations indicate the basal localisation and copper-induced targeting of the transiently-expressed MNK protein appears identical to both that of stably-expressed MNK (La Fontaine et al., 1997; La Fontaine et al., 1998) and the endogenous haMNK protein of the CUR cells (Figure 1) (Petris et al., 1996). Thus, the method of transient expression is a suitable system of investigating the localisation of the carboxyl-terminal MNK deletion.

The distribution of the deleted form of MNK expressed in CHO-K1 cells grown in basal medium is shown at two levels of expression (Figure 5D and E). The localisation was not predominantly perinuclear as for MNK(wildtype) (Figure 5A and B), but rather mislocalised throughout the cytoplasm with strong labelling of the cell periphery, although there was some staining within the perinuclear region. This is a significant finding as it suggests a region or signal contained within the last 50 amino acids of MNK contains information essential for maintaining TGN localisation. The distribution of the deleted form of MNK was also investigated in CHO-K1 cells grown in elevated copper concentrations (Figure 5F). Under these conditions the localisation was indistinguishable from the mislocalised distribution observed in basal medium described above, and the copper relocalisation effect on the wildtype MNK protein (Figure 5C).

5. DISCUSSION

The data presented herein indicate the TGN localisation and copper-induced redistribution of the MNK protein that was originally identified in haMNK over-expressing CUR cells also occurs in both parental CHO cells and HeLa cells. The minimal concentration of copper in the growth medium required to produce this effect in the CHO cell lines and HeLa cells occurred well below toxic levels, and was close to the range of copper predicted to be encountered under physiological conditions. Thus, the data are consistent with the hypothesis that the role of the MNK protein in cells is in the maintenance of copper homeostasis under normal conditions rather than as a response to toxic concentrations of copper. The second role of MNK in the delivery of copper to copper-dependent enzymes as they migrate through the secretory pathway has been proposed (Petris et al., 1996; Yamaguchi et al., 1996) as well as for the Wilson ATPase (Hung et al., 1997). The data presented identify the carboxyl terminal 50 amino acids as essential in conferring TGN localisation of the MNK protein, suggesting that the three di-leucine motifs and a sequence of four acidic amino acids within this region may determine this targeting (Figure 4).

When considering the overall sequence similarity between both Menkes and Wilson proteins as well as the roles both proteins are predicted to have in copper delivery into the secretory pathway, it is reasonable to speculate that both proteins localise to the TGN via the same process and signals. Interestingly, the carboxyl terminal region of the Wilson protein also possesses some of the same candidate TGN-targeting motifs identified in MNK. These include a tri-leucine sequence at a position corresponding to the di-leucine (1487–1488) of MNK, as well as a stretch of acidic amino acids located downstream. *In vitro* mutagenesis of this region of both Menkes and Wilson proteins is currently being pursued to determine the amino acids essential in mediating TGN localisation.

The mechanism by which C-terminal region of MNK functions to mediate TGN localisation remains to be shown. The simplest hypothesis, and one which is consistent with the function of TGN-targeting determinants of other recycling proteins, is that this region is an endocytic signal which functions in the retrieval of MNK from the plasma membrane into vesicles destined for the TGN. Given the evidence that MNK constitutively recycles, the accumulation of MNK at the cell surface observed when the carboxyl tail is deleted could be explained by a continuous low-level supply of MNK from the TGN. An alternative possibility is that this region forms part of a TGN-retention signal. This may explain the partial localisation of MNK observed in the TGN, as well as the strong plasma membrane labelling of protein which has escaped to the cell surface.

It is noteworthy that the deletion of the MNK carboxyl tail produced an effect on MNK localisation which appeared very similar to that of elevated copper on the MNK wildtype protein. This observation suggests a model whereby increased intracellular copper levels may confer relocalisation to the plasma membrane by regulating the activity of the carboxyl tail in conferring TGN localisation of MNK. Two models account for how this may occur. The first proposes that the binding of copper to the amino-terminal copper binding motifs may stimulate increased exocytic trafficking of MNK from the TGN to the plasma membrane. If the di-leucines are involved in retention in the TGN, the possibility exists that the binding of copper to the copper-binding site of MNK may result in a conformational change which reduces affinity of the di-leucine signal with proteins involved in this retention. This is then predicted to result in the sorting of MNK into vesicles for bulk flow to the plasma membrane. Alternatively, if the di-leucine is involved in internalisation, the binding of copper to the amino terminus of MNK may induce a conformation whereby the di-leucines are not recognised by the internalisation machinery. The net effect of this would be to shift the steady-state localisation of MNK toward the plasma membrane, maintained by a constant low-level supply from the exocytic pathway.

6. CONCLUSIONS

In summary, the data suggest that the intracellular localisation of the Menkes protein within the TGN and its redistribution in response to copper is a general phenomenon not restricted to the Menkes homologue of CHO cells with elevated levels of protein. The copper-induced relocalisation of MNK is the first known mechanism to explain cellular copper homeostasis and, in addition, is a novel form of regulated protein trafficking. The data suggest the last 50 amino acids of the carboxyl terminus of the Menkes protein contain information essential in maintaining basal localisation in the TGN. The results reveal the possibility that, in addition to mutations which reduce or abolish the catalytic activity of MNK, Menkes disease may be caused by mutations which affect the basal localisation and / or copper-induced trafficking to the plasma membrane. Indeed, if aberrant localisations of MNK are discovered in cultured Menkes patient cells along with the causative mutations, this may contribute to an understanding of the regions of the molecule which determine localisation and copper-induced relocalisation. An evaluation of the clinical phenotype of these patients, may then reveal the effects of mislocalisation of MNK *in vivo*.

REFERENCES

Alconada, A., Bauer, U., and Hoflack, B. (1996). A tyrosine-based motif and a casein kinase II phosphorylation site regulate the intracellular trafficking of the varicella-zoster virus glycoprotein I, a protein localized in the trans-Golgi network. EMBO.J. 15, 6096–110.

Beratis, N.G; Price, P., Labadie, G., and Hirschhorn, K. (1978). ^{64}Cu metabolism in Menkes and normal cultured skin fibroblasts. Pediatr.Res. 12, 699–702.

Bos, K., Wraight, C., and Stanley, K.K. (1993). TGN38 is maintained in the trans-Golgi network by a tyrosine-containing motif in the cytoplasmic domain. EMBO.J. 12, 2219–2228.

Bull, P.C., Thomas, J.R., Rommens, J.M., Forbes, J.R., and Cox, D.C. (1993). The Wilson disease gene is a putative copper transporting P-type ATPase similar to the Menkes gene. Nature Genet. 5, 327–337.

Camakaris, J., Danks, D.M., Ackland, L., Cartright, E., Borger, P., and Cotton, R.G.H. (1980). Altered copper metabolism in cultured cells from human Menkes' syndrome and mottled mouse mutants. Biochemical Genetics. 18, 117–131.

Camakaris, J., Petris, M.J., Bailey, L., Shen, P., Lockhart, P., Glover, T.W., Barcroft, C.L., Patton, J., and Mercer, J.F.B. (1995). Gene amplification of the Menkes (MNK; ATP7A) P-type ATPase gene of CHO cells is associated with copper resistance and enhanced copper efflux. Hum. Mol. Genet. 4, 2117–2123.

Chan, W.Y., Garnica, A.D., and Rennert, O.M. (1978). Metal-binding studies of metallothioneins in Menkes kinky hair disease. Clin.Chim.Acta. 88, 221–228.

Chelly, J., Tumer, Z., Tonneson, T., Petterson, A., Ishikawa-Brush, Y., Horn, N., and Monaco, A.P. (1993). Isolation of a candidate gene for Menkes disease that encodes a potential heavy metal binding protein. Nature Genet. 3, 14–19.

Collawn, J.F., Stangel, M., Kuhn, L.A., Esekowu, V., Jing, S.Q., Trowbridge, I.S., and Tainer, J.A. (1990). Transferrin receptor internalization sequence YXRF implicates a tight turn as the structural recognition motif for endocytosis. Cell. 63, 1061–1072.

Corvera, S., Chawla, A., Chakrabarti, R., Koly, M., Buxton, J., and Czech, M. P. (1994). A double leucine within the GLUT4 glucose transporter COOH-terminal domain functions as an endocytosis signal. J.Cell.Biol. 126, 979–989.

Danks, D.M. (1995). Disorders of copper transport. In The Metabolic and Molecular Basis of Inherited Disease, C.R. Scriver, A.L. Beaudet, W.M. Sly and D. Valle, eds. (New York: McGraw-Hill), pp. 2211–2235.

Danks, D.M., Campbell, P.E., Walker, Smith, J., Stevens, B.J., Gillespie, J.M., Blomfield, J., and Turner, B. (1972). Menkes' kinky-hair syndrome. Lancet. 1, 1100–1102.

Dierick, H.A., Adam, A.N., Escara-Wilke, J.F., and Glover, T.W. (1997). Immunocytochemical localization of the Menkes copper transport protein (ATP7A) to the trans-Golgi network. Hum.Mol.Genet. 6, 409–416.

Fu, D., Beeler, T. J., and Dunn, T. M. (1995). Sequence, mapping and disruption of CCC2, and gene that crosscomplements the Ca-2+- sensitive phenotype of csg1 mutants and encodes a P-type ATPase belonging to the Cu-2+-ATPase subfamily. Yeast. 11, 283–292.

Ge, Z., Hiratsuka, K., and Taylor, D.E. (1995). Nucleotide sequence and mutational analysis indicate that two Helicobacter pylori genes encode a P-type ATPase and a cation-binding protein associated with copper transport. Mol.Microbiol. 15, 97–106.

Goka, T.J., Stevenson, R.E., Hefferan, P.M., and Howell, R.R. (1976). Menkes disease: a biochemical abnormality in cultured human fibroblasts. Proc. Natl. Acad. Sci. USA. 73, 604–606.

Herd, S.M., Camakaris, J., Christofferson, R., Wookey, P., and Danks, D.M. (1987). Uptake and efflux of copper-64 in Menkes'-disease and normal continuous lymphoid cell lines. Biochem.J. 247, 341–347.

Horn, N. (1976). Copper incorporation studies on cultured cells for prenatal diagnosis of Menkes' disease. Lancet. 1, 1156–1158.

Humphrey, J.S., Peters, P.J., Yuan, L.C., and Bonifacino, J.S. (1993). Localization of TGN38 to the trans-Golgi network: involvement of a cytoplasmic tyrosine-containing sequence. J.Cell.Biol. 120, 1123–1135.

Hung, I.H., Suzuki, M., Yamaguchi, Y., Yuan, D.S., Klausner, R.D., and Gitlin, J.D. (1997). Biochemical characterization of the Wilson disease protein and functional expression in the yeast Saccharomyces cerevisiae. J.Biol.Chem. 272, 21461–21466.

Kanamaru, K., Kashiwagi, S., and Mizuno, S. (1994). A copper-transporting P-type ATPase found in the thylakoid membrane of the cyanobacterium Synechococcus species PCC7942. Mol.Microbiol. 13, 369–377.

Kodama, H. (1993). Recent developments in Menkes disease. J.Inherit.Metab.Dis. 16, 791–799.

La Fontaine, S., Firth, S.D., Lockhart, P.J., Brooks, H., Camakaris, J., and Mercer, J.F.B. (1998). Functional Analysis of the Menkes protein (MNK) expressed from a cDNA construct. In Copper Transport and Its Disorders: Molecular and Cellular Aspects. A. Leone and J.F.B. Mercer, eds. (New York: Plenum Publishing Corporation), pp.(This Volume).

La Fontaine, S., Firth, S.D., Lockhart, P.J., Paynter, J.A., and Mercer, J.F.B. (1997). Low copy number plasmid vectors for eukaryotic gene expression: transient expression of the Menkes protein. Plasmid. (Accepted for publication).

Lutsenko, S., Petrukhin, K., Cooper, M.J., Gilliam, C.T., and Kaplan, J.H. (1997). N-terminal domains of human copper-transporting adenosine triphosphatases (the Wilson's and Menkes disease proteins) bind copper selectively in vivo and in vitro with stoichiometry of one copper per metal-binding repeat. J.Biol.Chem. 272, 18939–18944.

Mercer, J.F.B., Livingston, J., Hall, B., Paynter, J.A., Begy, C., Chandrasekharappa, S., Lockhart, P., Grimes, A., Bhave, M., Siemieniak, D., and Glover, T.W. (1993). Isolation of a partial candidate gene for Menkes disease by positional cloning. Nature Genet. 3, 20–25.

Menkes, J. H., Alter, M., Steigleder, G.K., Weakley, D.R., and Sung, J. H. (1962). A sex-linked recessive disorder with retardation of growth, peculiar hair and focal cerebral degeneration. Pediatrics. 29, 764–779.

Murata, Y., Yamakawa, E., Iizuka, T., Kodama, H., Abe, T., Seki, Y., and Kodama, M. (1995). Failure of copper incorporation into ceruloplasmin in the Golgi apparatus of LEC rat hepatocytes. Biochem.Biophys.Res.Comm. 209, 349–355.

Odermatt, A., Suter, H., Krapt, R., and Solioz, M. (1993). Primary structure of two P-type ATPases involved in copper homeostasis in Enterococcus hirae. J.Biol.Chem. 268, 12775–12779.

Pederson, P.L., and Carafoli, E. (1987). Ion motive ATPases. I Ubiquity, properties and significance to cell function. Trends.Biochem.Sci. 12, 146–150.

Petris, M.J., Mercer, J.F., Culvenor, J.G., Lockhart, P., Gleeson, P.A., and Camakaris, J. (1996). Ligand-regulated transport of the Menkes copper P-type ATPase efflux pump from the Golgi apparatus to the plasma membrane: a novel mechanism of regulated trafficking. EMBO.J. 15, 6084–6095.

Rad, M.R., Kirchrath, L., and Hollenberg, C.P. (1994). A putative P-type Cu-2+-transporting ATPase gene on chromosome II of Saccharomyces cerevisiae. Yeast. 10, 1217–1225.

Riordan, J.R., and Jolicoeur-Paquet, L. (1982). Metallothionein accumulation may account for intracellular copper retention in Menke's disease. J.Biol.Chem. 257, 4639–4645.

Royce, P.M., Camakaris, J., and Danks, D.M. (1980). Reduced lysyl oxidase activity in skin fibroblasts from patients with Menkes' syndrome. Biochem.J. 192, 579–586.

Sambongi, Y., Wakabayashi, T., Yoshimizu, T., Omote, H., Oka, T., and Futai, M. (1997). Caenorhabditis elegans cDNA for a Menkes/Wilson disease gene homologue and its function in a yeast CCC2 gene deletion mutant. J.Biochem. 121, 1169–1175.

Schäfer, W., Stroh, A., Berghofer, S., Seiler, J., Vey, M., Kruse, M.L., Kern, H.F., Klenk, H.D., and Garten, W. (1995). Two independent targeting signals in the cytoplasmic domain determine trans-Golgi network localization and endosomal trafficking of the proprotein convertase furin. EMBO.J. 14, 2424–2435.

Solioz, M., Odermatt, A., and Krapf, R. (1994). Copper pumping ATPases: common concepts in bacteria and man. FEBS.Lett. 346, 44–47.

Solioz, M., and Vulpe, C. (1996). CPx-type ATPases: a class of P-type ATPases that pump heavy metals. Trends.Biochem.Sci. 21, 237–241.

Tanzi, R.E., Petrukhin, K., Chernov, I., Pellequer, J.L., Wasco, W., Ross, B., Romano, D.M., Parano, E., Pavone, L., Brzustowicz, L.M., Devoto, M., Peppercorn, J., Bush, A.I., Sternlieb, I., Piratsu, M., Gusella, J.F., Evgrafov, O., Penchaszadeh, G.K., Honig, B., Edelman, I.S., Soares, M.B., Scheinberg, I.H., and Gilliam, T.C. (1993). The Wilson disease gene is a copper transporting ATPase with homology to the Menkes disease gene. Nature Genet. 5, 344–350.

Terada, K., Kawarada, Y., Miura, N., Yasiu, O., Koyama, K., and Sugiyama, T. (1995). Copper incorporation into ceruloplasmin in rat livers. Biochim.Biophys.Acta. 1270, 58–62.

Theophilos, M.B., Cox, D.W., and Mercer, J.F. (1996). The toxic milk mouse is a murine model of Wilson disease. Hum.Mol.Genet. 5, 1619–1624.

Trowbridge, I.S., Collawn, J.F., and Hopkins, C.R. (1993). Signal-dependent membrane protein trafficking in the endocytic pathway. Annu. Rev. Cell. Biol. 9, 129–161.

Voorhees, P., Deignan, E., van Donselaar, E., Humphrey, J., Marks, M.S., Peters, P.J., and Bonifacino, J.S. (1995). An acidic sequence within the cytoplasmic domain of furin functions as a determinant of trans-Golgi network localization and internalization from the cell surface. EMBO.J. 14, 4961–4975.

Vulpe, C., Levinson, B., Whitney, S., Packman, S., and Gitschier, J. (1993). Isolation of a candidate gene for Menkes disease and evidence that it encodes a copper-transporting ATPase. Nature Genet. 3, 7–13.

Wilcox, C.A., Redding, K., Wright, R., and Fuller, R.S. (1992). Mutation of a tyrosine localization signal in the cytosolic tail of yeast Kex2 protease disrupts Golgi retention and results in default transport to the vacuole. Mol.Biol.Cell. 3, 1353–1371.

Wu, J., Forbes, J.R., Chen, H.S., and Cox, D.W. (1994). The LEC rat has a deletion in the copper transporting ATPase gene homologous to the Wilson disease gene. Nature Genet. 7, 541–545.

Yamaguchi, Y., Heiny, M.E., and Gitlin, J. (1993). Isolation and characterization of a human liver cDNA as a candidate gene for Wilson disease. Biochem.Res.Comm. 197, 271–277.

Yamaguchi, Y., Heiny, M.E., Suzuki, M., and Gitlin, J.D. (1996). Biochemical characterization and intracellular localization of the Menkes disease protein. Proc.Natl.Acad.Sci.U.S.A. 93, 14030–14035.

Yuan, D.S., Dancis, A., and Klausner, R.D. (1997). Restriction of copper export in Saccharomyces cerevisiae to a late Golgi or post-Golgi compartment in the secretory pathway. J.Biol.Chem. 272, 25787–25793.

Yuan, D.S., Stearman, R., Dancis, A., Dunn, T., Beeler, T., and Klausner, R.D. (1995). The Menkes/Wilson disease gene homologue in yeast provides copper to a ceruloplasmin-like oxidase required for iron uptake. Proc. Natl. Acad. Sci. USA. 92, 2632–2636.

FUNCTIONAL ANALYSIS OF THE MENKES PROTEIN (MNK) EXPRESSED FROM A cDNA CONSTRUCT

Sharon La Fontaine,[1] Stephen D. Firth,[1] Paul J. Lockhart,[1] Hilary Brooks,[2] James Camakaris,[2] and Julian F. B. Mercer[1]

[1]The Murdoch Institute for Research into Birth Defects
Royal Children's Hospital
Parkville 3052, Australia
[2]Department of Genetics, University of Melbourne
Parkville 3052, Australia

INTRODUCTION

A detailed structure/function analysis of the Menkes protein (MNK) is required to elucidate its role in maintaining cellular copper homeostasis. It was recently demonstrated that over-expression of the MNK protein confers a copper-resistance phenotype upon CHO-K1 cells (Camakaris et al., 1995), and that MNK in these cells is located primarily in the *trans*-Golgi network (Petris et al., 1996), but traffics to the plasma membrane in elevated copper levels. To investigate the molecular mechanism of the copper-induced trafficking of MNK, and copper translocation across cellular membranes, stable expression of the full length protein in mammalian cells is necessary, followed by the generation of cell lines that contain *in vitro* mutated MNK constructs.

The first step towards achieving MNK expression involved the generation of a cDNA construct which comprised the entire coding region of MNK. However, this cDNA has proven difficult to manipulate in all of several standard high copy number plasmid vectors and *E. coli* strains used (our unpublished observations; (Francis et al., 1996)). The vectors included pUC18 (Vieira and Messing, 1982), pBluescript (Stratagene) and pSP65 (Promega). Plasmid instability associated with the full length cDNA fragment was evident from gross rearrangements detected by agarose gel electrophoresis, while plasmids of the expected size were found to carry several point mutations, some of which led to the introduction of stop codons.

An appropriate vector system is crucial to successful gene expression, and there are many commercially available mammalian expression vectors that are suitable for most applications. However, many of these vectors are based on plasmids that replicate to intermediate or high copy number in *Escherichia coli*, and there are some circumstances in which they preclude the successful cloning and manipulation of genes in *E. coli*, which is, in general, the first step towards obtaining mammalian gene expression. For example, high copy number plasmids often are unsuitable for the cloning of genes whose over-expression is detrimental to the cell, such as those which encode bacterial regulatory or membrane proteins. The cloning and manipulation of such genes has been achieved with the use of low copy number plasmids (Sambrook et al., 1989).

Using high copy number vectors, the manipulation of DNA encoding mammalian transmembrane proteins, such as the cystic fibrosis transmembrane regulatory protein (CFTR) (Gregory et al., 1990; Krauss et al., 1992), also has proven difficult in *E. coli*. In this case it was postulated that the observed plasmid instability was due to the combination of the full length cDNA existing at a high gene dosage, and inappropriate expression of the cDNA from a cryptic bacterial promoter within the mammalian sequence, resulting in a product that was toxic to the bacterial host. The successful propagation of plasmids carrying the Ca^{2+}-ATPase cDNA, and the ensuing functional expression of the transmembranous protein may have been possible because the vector employed contained the pMB1 origin of replication which yields only 15 to 20 plasmid copies per cell (Maruyama and MacLennan, 1988).

2. GENERATION OF VECTORS FOR MNK EXPRESSION

The bacterial low copy number expression vectors pWSK29 and pWKS30 have a pSC101 origin of replication which yields six to eight plasmid copies per cell (Wang and Kushner, 1991). These vectors allow the stable propagation in *E. coli* of constructs containing inserts that are problematic in high copy number vectors. Therefore, a cDNA fragment comprising the entire coding region of the MNK gene, 10 bp of the 5' untranslated region (UTR), and 120 bp of the 3' UTR was generated as a 4.6 kb *Bam*HI fragment by RT-PCR, and was cloned into pWSK29 to produce a plasmid construct designated pJFM19. Consequently, the plasmid instability problems that were previously encountered with this cDNA fragment were alleviated as determined by the lack of gross plasmid rearrangements or point mutations within the sequence.

For expression of MNK in mammalian cells the pWSK29 and pWKS30 vectors were modified to create low copy number mammalian gene expression vectors, designated pJFM43 and pJFM48 (Fig.1) (La Fontaine et al., 1997). In addition to the ampicillin resistance marker for selection in bacteria that was already present, the modified vectors contained a neomycin resistance marker for the selection of stable transformants in the presence of G418. They also contained the CMV promoter (Boshart et al., 1985; Thomsen et al., 1984) for high level constitutive expression of cloned genes (pJFM43), or the sheep metallothionein 1a (sMT-1a) promoter which may be induced with heavy metals such as copper and zinc (Peterson et al., 1988) (pJFM48). A poly(A) signal for polyadenylation and transcription termination, and a chimaeric intron located 5' of any introduced coding sequence for the purpose of enhancing gene expression in transgenic mice (Brinster et al., 1988) also are present. An SV40 origin of replication was included to enable transient expression studies to be carried out in cells that expressed the SV40 large T antigen (Gluzman, 1981; Kriegler, 1990).

When using conventional methods for preparing low copy number plasmid DNA, although the yields of plasmid DNA are reduced compared with high copy number vectors, quantities of DNA that were sufficient for cloning reactions and transfections were obtained using standard alkaline lysis procedures (Morelle, 1989; Sambrook et al., 1989), with or without the use of plasmid purification columns (QIAGEN). For mammalian cell

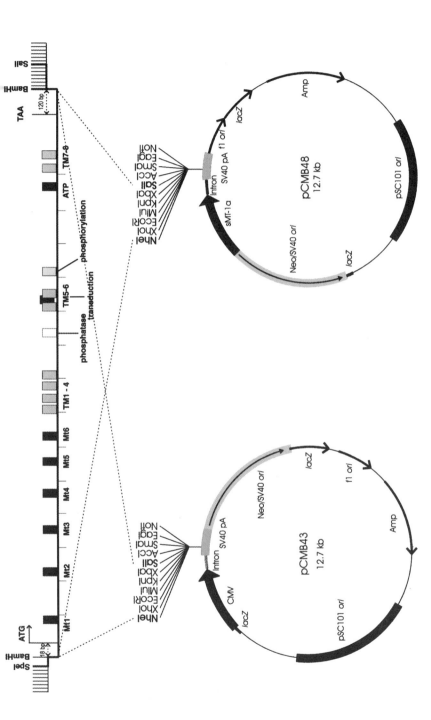

Figure 1. MNK cDNA construct and expression vectors. Schematic diagram of the MNK (ATP7A) cDNA showing the locations of the various functional domains within the encoded protein. Mt1–6 represents the metal binding sites, TM1–8, the transmembrane domains, and ATP, the ATP binding site. The relative positions of the phosphatase, transduction and phosphorylation domains are also indicated, as are the ATG start and TAA stop codons. The relevant 5' and 3' restriction sites within the multiple cloning site (MCS) of pJFM19 are indicated. Shown below the cDNA are the expression vectors, pJFM43 and pJFM48 containing the human cytomegalovirus (CMV) or sheep metallothionein promoters, respectively, chimaeric intron, pCI MCS (Promega), SV40 polyadenylation signal, and neomycin resistance cassette and SV40 origin of replication (Neo/SV40 *ori*). The MNK cDNA was cloned as a *SpeI/SalI* fragment into the *NheI/SalI* site of pJFM43 and pJFM48 to generate pJFM50 and pJFM51, respectively.

transfections that involved the electroporation of DNA into cells, further purification of DNA on caesium chloride density gradients was required. Using these methods approximately 80 - 100 μg of DNA was consistently obtained from 500 ml bacterial overnight cultures grown in standard Luria-Bertaini (LB) media (Sambrook et al., 1989).

3. TRANSIENT EXPRESSION OF MNK IN COS-7 CELLS

The level and inducibility of gene expression from the new low copy number expression vectors was assessed by cloning the MNK cDNA fragment from pJFM19 as a SpeI/SalI fragment into the NheI/SalI site of the expression vectors pJFM43 and pJFM48 to generate the plasmids, pJFM50 and pJFM51, respectively (Fig. 1). These plasmids were subsequently used in transient expression studies using COS-7 cells.

Both constitutive and zinc-inducible expression of MNK from the plasmid constructs was detected by Northern blot analysis of RNA isolated from transfected COS-7 cells. These expression data were confirmed by Western blot analysis of transfected COS-7cells, using an antibody directed against the MNK N-terminus (Camakaris et al., 1995; Petris et al., 1996). A protein band close to the expected size for MNK, previously reported at 178 kDa (Camakaris et al., 1995), was detected at levels above that of untransfected cells. It was concluded that MNK expression was possible with the new low copy number expression vectors.

4. STABLE EXPRESSION OF MNK IN CHO-K1 CELLS

4.1. Isolation of Stable MNK-Expressing CHO Clones

The plasmid pJFM50 (Fig. 1) was transfected into CHO-K1 cells, and four clones designated 900-5 #4, 600-5 #3, 600-6 #3, and 600-6 #7 were isolated. Southern hybridization analysis showed that the MNK cDNA construct was present in multiple copies in these clones, and that there were significant differences in its copy number between the transfected cell lines (Fig.2A). Northern blot analysis of the clones confirmed the production of a specific 4.8 kb plasmid-derived transcript, which included 0.2 kb of 5'untranslated vector-derived sequence in addition to the 4.6 kb MNK coding plus 3' non-coding sequence (Fig.2B). Clone 900-5#4 produced the least amount of the cDNA-derived transcript.

Using an antibody directed against the MNK N-terminal metal binding region (Camakaris et al., 1995; Petris et al., 1996), Western blot analysis of whole cell protein extracts from the clones was carried out. For comparison of protein profiles, extracts derived from the parental CHO-K1 cell line and the copper-resistant CUR cell lines were included. Expected amounts of the 178 kDa MNK protein were detected in CHO-K1, CUR2 and CUR3 cell extracts (Fig.2C). The levels of MNK expressed in the transfected clones varied, with clone 600-5 #3 expressing MNK at a level similar to CUR3, while clones 900-5 #4, 600-6 3# and 600-6 #7 expressed MNK at approximately CUR2 levels, as determined over several experiments. Among the four clones selected for analysis, the MNK cDNA copy number did not correlate with the level of MNK expression. This result was probably due to random integration of the construct into genomic regions in which gene expression is influenced by either inhibitory or enhancing factors.

The MNK product of the transfected clones was approximately 5 kDa smaller than the MNK band in CHO-K1, CUR2 and CUR3 extracts, but 5 kDa larger than the MNK product

Functional Analysis of the Menkes Protein (MNK) Expressed from a cDNA Construct

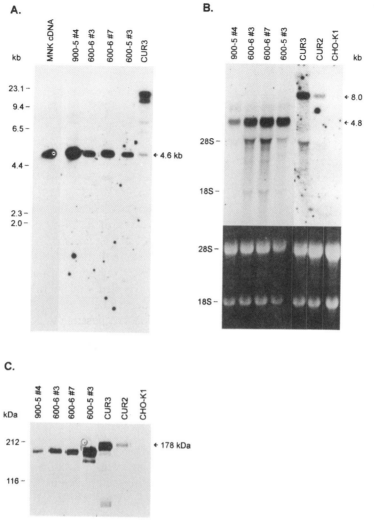

Figure 2. Southern, Northern and Western blot analysis of the transfectant clones. **A)** Southern blot analysis of genomic DNA from the clones. Approximately 10 μg of genomic DNA from the four clones (900-5 #4, 600-6 #3, 600-6 #7 and 600-5 #3) and 20 μg of CUR3 genomic DNA was digested with *Bam*HI, fractionated by gel electrophoresis and transferred to a Hybond-N+ membrane. The pJFM19 *SpeI/SalI* fragment containing the MNK cDNA also was included (far left lane) and was used as the probe for hybridization. The position of the integrated 4.6 kb *Bam*HI fragment containing the MNK cDNA is indicated. The positions of the λcI857 *sam7* HindIII fragments are shown in kilobases (kb). **B)** Northern blot analysis of transfectant clones. Top panel: Total RNA was isolated from the four clones, CUR3, CUR2 and CHO-K1 cells, fractionated in a 1% formaldehyde gel and transferred to Hybond-N+ membranes. The filter was hybridized with a combination of three fragments derived from the 5' end of the cDNA and which encoded the first five metal binding sites of MNK. The position of the 28S and 18S rRNAs are indicated as are the position of the endogenous 8.0 kb MNK transcript, and the plasmid-derived 4.8 kb transcript. Bottom panel: Gel photo showing amount and integrity of RNA in each lane. **C)** Western blot analysis of transfectant clones. Whole cell protein extracts were prepared from the four clones, CUR3, CUR2 and CHO-K1 cells, fractionated by SDS-PAGE (7% gel), and transferred to nitrocellulose filters. The filters were probed with an affinity purified preparation of anti-MNK antibodies directed against the MNK N-terminus (Camakaris et al., 1995; Petris et al., 1996), while the secondary antibody consisted of horseradish peroxidase-conjugated sheep anti-rabbit IgG (Silenus). Protein detection was carried out using the Chemiluminescent POD substrate (Boehringer Mannheim). The positions of the molecular weight markers (Bio-Rad) are indicated on the left in kilodaltons (kDa). The position of the endogenous MNK protein is also indicated with an arrow.

Table 1. Copper resistance of MNK-expressing clones and comparison with CUR cells

[Cu] (µg/ml)	% Survival						
	CHO-K1	CUR2	CUR3	900-5 #4	600-6 #3	600-6 #7	600-5 #3
0	100	100	100	100	100	100	100
6	93	94	98	100	99	100	98
10	86	91	97	100	100	100	100
12	79	91	97	96	97	100	98
14	72	87	92	96	97	100	98
16	62	88	92	92	96	100	100
18	56	80	93	89	89	100	97
20	49	82	92	87	97	100	100
22	38	78	92	83	86	97	100
24	25	65	75	77	82	88	100
26	13	61	70	64	53	56	78
28	11	54	69	60	47	53	72
30	12	42	71	55	50	46	67
32	12	28	68	49	49	44	69
34	12	24	67	41	44	43	67
36	12	21	69	35	39	37	63
38	11	15	65	23	36	30	61
40	10	14	67	17	28	27	51
42	8	11	66	11	31	22	47

The percentage of cells surviving in increasing concentrations of copper was assessed using an MTT cytotoxicity assay (La Fontaine et al., 1997). The survival of CHO-K1, CUR2, and CUR3 at each copper concentration is shown for comparison. Each data point represents the average of duplicate measurements.

that was transiently expressed from the same plasmid construct in COS-7 cells. It has been previously reported that the processing of proteins expressed from the same gene in different cells may be affected by differences in those host cells (Walls et al., 1989). Recently, it was shown that MNK was glycosylated, and that the MNK band on a Western blot was reduced from 178 to 170 kDa after deglycosylation (our unpublished results; (Yamaguchi et al., 1997)). It is possible that the human MNK product expressed from the cDNA in both CHO-K1 and COS-7 cells, was only partially glycosylated due to sequence differences and the recognition of different glycosylation signals by different cell types. However, the size and possible glycosylation differences between the endogenous and cDNA-derived MNK did not appear to affect the function of MNK produced from the cDNA construct in CHO cells.

4.2. Copper Resistance Phenotype of MNK-Expressing Clones

Using an MTT (3-(4,5-dimethylthiazol-2-yl)-2,5-diphenyl tetrazolium bromide) cytotoxic assay (La Fontaine et al., 1997) the copper-resistance characteristics of the clones were determined by assessing cell survival in increasing amounts of copper, and then compared with the copper-resistance phenotype of CHO-K1, CUR2 and CUR3 (Table 1). This assay is a colorimetric assay where live cells reduce the MTT dye in the media to a purple colour, and the absorbance at 600 nm (A_{600}) is read using an EIA Plate Reader. At copper concentrations of up to 18 to 26 µg/ml, the survival of the transfected clones was approximately equivalent to that of CUR3. Thereafter, the survival of 600-5 #3 resembled CUR3 survival, while the copper resistance of 900-5 #4, 600-6 #3 and 600-6 #7 was similar to that of CUR2. In contrast, CHO-K1 survival within this range of copper concentrations was significantly reduced and was negligible by 26 µg/ml.

Table 2. Accumulation and efflux of copper by the transfectant clones

Cell line	Cu accumulation (pmole Cu/µg DNA)	Cu efflux (%)
CHO-K1	26.0	<1%
CUR2	3.0	47
CUR3	6.2	47
600-6 #3	3.9	44
600-5 #3	8.4	54

Cells of clones 600-6 #3 and 600-5 #3, as well as CHO-K1, CUR2 and CUR3 for comparison, were incubated for 2 hr in 22 µg/ml copper medium in the presence of Cu^{64}. Accumulation was determined from the radioactivity in the cell pellets. Efflux was for 15 sec following the accumulation period, and % efflux was determined from the radioactivity remaining in the cell pellet.

The copper-resistance data corresponded with copper accumulation and efflux data obtained with clones 600-6 #3 and 600-5 #3 (Table 2). That is, both clones accumulated significantly less, and effluxed significantly more copper than the parental CHO-K1 cell line. In addition, as with the copper-resistance results, these data for 600-6 #3 and 600-5 #3 closely paralleled that obtained for CUR2 and CUR3, respectively. Therefore, the levels of MNK protein expressed by the clones correlated with their degree of copper resistance, and this copper-resistance phenotype reflected their enhanced ability to efflux copper which in turn led to reduced intracellular accumulation of copper. A similar phenomenon was described for the CHO-K1-derived copper-resistant CUR cells (Camakaris et al., 1995). It was concluded that the over-produced MNK product in the transfectant clones was functional, and that over-expression of MNK was the major contributor to the copper resistance phenotype in both the clones and CUR cells.

4.3. Intracellular Localization of MNK

Immunofluorescence microscopy with the MNK antibody directed against the N-terminus, was used to determine the intracellular location of the plasmid-derived human MNK protein in the transfectant clones as well as in the parental CHO-K1 cells. Under conditions of low media copper, all cell lines showed fluorescent staining in the perinuclear region (Fig. 3) which, in the CUR cell lines, was shown to represent TGN localization (Petris et al., 1996). The perinuclear staining was clearly evident in the parental CHO-K1 cell line, and the fluorescence intensity corresponded with the level of MNK protein expression in each cell line. There was also punctate staining within the cytoplasm indicative of vesicle-like structures. When the media copper levels were increased, the intensity of the perinuclear signal was decreased in all cell lines, including the parental CHO-K1 cell line, and a more dispersed, punctate staining pattern within the cytoplasm was observed. Staining at the periphery of cells was most apparent in the transfected clones and indicated the presence of MNK at the plasma membrane. These results were consistent with those previously reported for the CUR cells (Petris et al., 1996), and demonstrated that the cDNA-derived MNK product was able to traffic in response to copper.

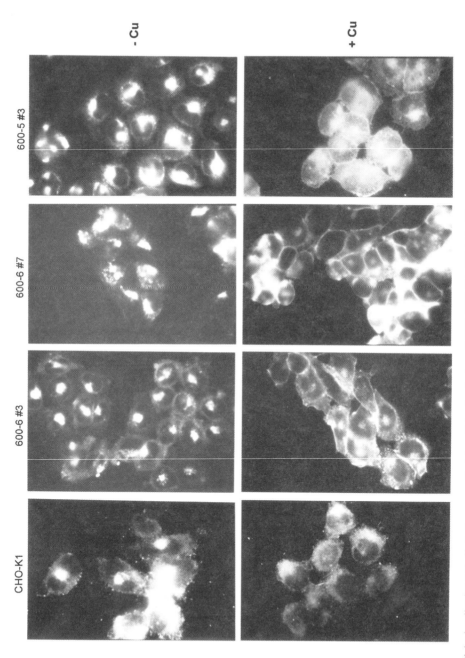

Figure 3. Subcellular localization and effect of copper on the intracellular distribution of MNK. CHO-K1, 600-6 #3, 600-6 #7, and 600-5 #3 were cultured for 48 h in basal medium and then incubated in the absence or in the presence of 189 μM copper ($CuCl_2$) for 2 to 5 h at 37°C. The cells were fixed and stained with a sodium sulphate precipitated preparation of anti-MNK antibodies, followed by FITC-conjugated sheep anti-rabbit IgG antibodies.

The localization of MNK to the TGN in CHO-K1 cells, human fibroblasts, and HeLa cells was also previously demonstrated by other groups (Dierick et al., 1997; Yamaguchi et al., 1997). However, the fact that the copper-induced relocalization of MNK also occurred in the parental CHO-K1 cells demonstrated for the first time that this phenomenon is not a specific property that is confined to copper-resistant cells.

MNK localization to the TGN was supported by immunofluorescence experiments using the fungal metabolite Brefeldin A (BFA), which caused the fluorescent signal to contract to form a more compact, juxtanuclear signal (data not shown). This effect of BFA also was observed when an antibody to a TGN protein, TGN38, was used in immunofluorescence experiments on the clones, and is typically observed with TGN proteins due to initial fusion of the TGN with the endosomal pathway, and the eventual collapse of the TGN and endosomal compartments around the microtubule organizing centre (MTOC) at the centre of the cell (Lippincott-Schwartz et al., 1991; Pelham, 1991; Reaves and Banting, 1992). Confirmation of MNK localization to the TGN under basal conditions, and its trafficking to the plasma membrane under conditions of elevated copper was provided by ultrastructural analysis of clone 900-5 #4 (data not shown; (La Fontaine et al., 1997)).

A time course experiment using clone 900-5 #4 showed that within 30 min following the transfer of cells to media with elevated copper (189 μM), the intensity of the perinuclear signal was significantly reduced while cytoplasmic staining was increased and was punctate in appearance (Fig.4). Staining at the cell periphery also was evident. Similar results were obtained when this experiment was repeated with CUR2 and the other clones (data not shown). MNK relocation to the TGN was evident within 15 min of transferring the cells back to basal media, (Fig. 4), and by 60 min the perinuclear signal was completely reconstituted (Fig. 4). Therefore, the kinetics of the copper response closely resembled that of CUR2 cells (data not shown and (Petris et al., 1996)).

Since the media copper concentrations (189 μM) used in the experiments described above were considerably higher than normal physiological copper levels (adult plasma copper concentrations range from 16 - 27 μM (Versieck and Cornelis, 1980)), a copper titration experiment was carried out using clone 600-6 #3. Cells were incubated in media containing copper at concentrations ranging from 0 to 150 μM, prior to immunofluorescence analysis. The copper-induced movement of MNK from the TGN into the cytosol, and towards the plasma membrane, was first evident between 20 and 50 μM copper, and was detectable at the plasma membrane at 150 μM (Fig. 5).

5. CONCLUSIONS AND IMPLICATIONS

The immunofluorescence data together with ultrastructural analysis of the MNK-expressing transfectant clones (La Fontaine et al., 1997) confirmed the previous results obtained with the CUR cells (Petris et al., 1996). These data also supported the hypothesis which stemmed from the CUR cell observations, that the copper-induced, vesicular movement of MNK forms the cellular basis for copper homeostasis in allowing the cell to regulate the intracellular concentrations of this essential but potentially toxic cation. At low to normal intracellular copper levels, the protein is primarily located in the TGN where it is thought to provide copper to secreted, copper-dependent enzymes such as lysyl oxidase. When intracellular copper levels increase due to elevated extracellular copper, there is a net increase in MNK at the plasma membrane resulting in increased copper efflux. This movement of MNK is reversible, and MNK appears to continually recycle between the TGN and plasma membrane in basal media (Petris et al., 1996). In addition, since MNK movement is detectable at low extracellu-

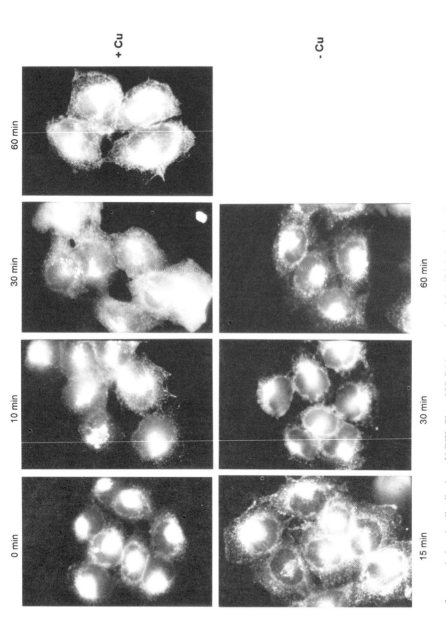

Figure 4. Time course of copper-induced redistribution of MNK. Clone 900-5 #4 was cultured for 48 h in basal medium and then incubated in the absence or in the presence of 189 µM copper (CuCl$_2$) at 37°C for 10 min, 30 min or 60 min. After 60 min in copper-containing medium, cells were returned to basal medium for 15 min, 30 min or 60 min. The cells were fixed and stained with a sodium sulphate precipitated preparation of anti-MNK antibodies, followed by FITC-conjugated sheep anti-rabbit IgG antibodies.

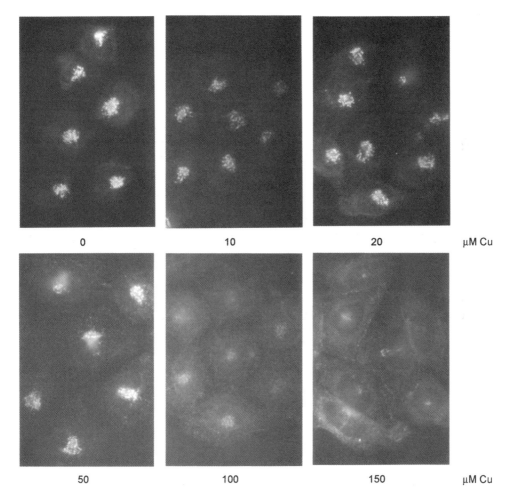

Figure 5. Effect of extracellular copper concentration on the copper-induced redistribution of MNK. Clone 600-6 #3 was cultured for 48 h in basal medium and then incubated for 3 hr in media containing 0, 10, 20, 50, 100, or 150 μM copper ($CuCl_2$). The cells then were fixed and stained with a sodium sulphate precipitated preparation of anti-MNK antibodies, followed by FITC-conjugated sheep anti-rabbit IgG antibodies.

lar copper concentrations, it is likely to represent a phenomenon that is responsive to physiological fluctuations in intra- and extracellular copper levels.

The elucidation of the molecular defects that result in Menkes and Wilson diseases has provided valuable insights into components of the mammalian copper transport pathway which form a vital part of the homeostatic mechanism that maintains a precise cellular balance of copper. The importance of these components which include intracellular copper transport for the delivery of copper to copper-requiring enzymes within the cell, and the export of excess copper from the cell, is reflected in the consequences of their breakdown as occurs in the disease states. The localization of MNK to the TGN was consistent with previous suggestions of a role for MNK in intracellular copper transport (Danks, 1995; Herd et al., 1987; Vulpe and Packman, 1995). The TGN is located at the *trans* face of the Golgi apparatus, and is the compartment in which secretory proteins are sorted into different types of vesicles for transport to different locations within the cell (Griffiths and Simons, 1986). Therefore, the absence of MNK, or a defective MNK at the TGN would preclude the delivery of copper into the TGN lumen for incorporation into secreted cuproenzymes such as lysyl oxidase, even in the presence of adequate copper. In support of this hypothesis, the activity of lysyl oxidase is reduced in patients with Occipital Horn Syndrome (OHS) and in the mottled mouse mutant Mo^{blo} (blotchy) (Das et al., 1995), and cannot be restored by copper supplementation (Kuivaniemi et al., 1985; Royce et al., 1982). Both OHS patients and the blotchy mouse have similar splice site mutations in the *MNK* (*ATP7A*) gene and its murine homologue, respectively, which are predicted to result in the production of a small amount of normal protein. In yeast, the homologue of MNK and WND, CCC2p, and the WND protein itself, recently were reported to mediate copper delivery to an intracellular, late or post-Golgi compartment for incorporation into a yeast iron oxidase Fet3p (Hung et al., 1997; Stearman et al., 1996; Yuan et al., 1997; Yuan et al., 1995). The location of MNK at the plasma membrane, the reduced intracellular copper accumulation by the cell lines that over-express MNK (Camakaris et al., 1995; La Fontaine et al., 1997; Petris et al., 1996), and the copper accumulation phenotype of cells from Menkes patients and the mottled mouse mutants (Danks, 1995), are consistent with a further role for MNK in the export of excess copper from the cell via an active efflux mechanism.

Therefore, the mechanism of MNK-mediated copper transport is a novel and efficient mechanism whereby the transport and efflux of a ligand is regulated by the ligand itself, and raises questions pertaining to the molecular interactions which occur to maintain a tightly regulated system of copper transport within and out of the cell. Within MNK, putative C-terminal amino acid motifs and transmembrane domains that may be involved in retention in the TGN, targeting from the TGN to the plasma membrane, and/or internalization from the plasma membrane were previously identified (Dierick et al., 1997; Petris et al., 1996). However, the most striking difference between the MNK and WND proteins, and other heavy metal-transporting P-type ATPases from bacteria and yeast is in the number of metal binding motifs at the N-terminus (Bull and Cox, 1994; Solioz and Vulpe, 1996), which in the case of MNK and WND were recently shown to bind copper (Lutsenko et al., 1997). Mammalian MNK and WND have five to six (rodent WND has five), whereas bacteria and yeast have one to two such motifs. A plausible role for this metal binding region is in the delivery of copper ions to the channel for the translocation/efflux of copper through the cell membane.

Based on the increased number of putative copper binding motifs in MNK and the effect of copper in regulating MNK trafficking, we have proposed a model which also incorporates a role for the extended MNK N-terminal domain in copper-induced MNK trafficking (Fig. 6). It is unlikely that MNK and WND require all six copper binding sites for copper transport activity, but they may serve to specify the intracellular copper level required for the activation of MNK. That is, the MBR serves as a copper-sensing domain

Functional Analysis of the Menkes Protein (MNK) Expressed from a cDNA Construct

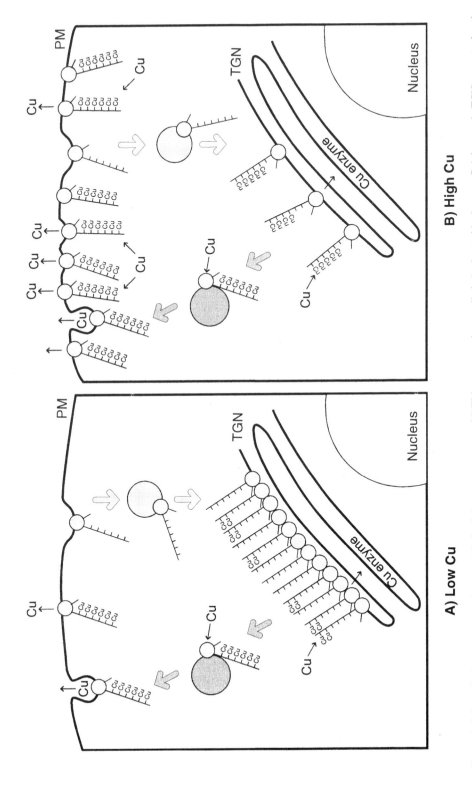

Figure 1. Schematic diagram representing a model of cellular copper transport. MNK is represented as small circles located in the *trans* Golgi network (TGN), on cytoplasmic vesicles (large circles), or at the plasma membrane (PM). The arrows indicate the direction of vesicle trafficking. The steady state distribution of MNK in both low (A) and high (B) copper states is shown.

such that saturation of the metal binding sites when intracellular copper levels increase, triggers a conformational change in MNK, which disrupts the interaction between MNK and TGN adaptor proteins, and MNK enters the exocytic pathway, as previously suggested (Petris et al., 1996). Alternatively, under high copper conditions, copper remains bound to MNK and facilitates or inhibits protein interactions that maintain MNK at the cell surface. In both cases the exocytic and endocytic rate constants would be altered, and the steady-state distribution of MNK shifted so that there is a net increase in MNK at the cell surface. With the MBR on the cytoplasmic side of the vesicle membrane MNK-containing vesicles may move towards and then fuse with the plasma membrane, leaving MNK integrated within the membrane through which copper is translocated from the cytoplasmic to the extracellular side. Alternatively, MNK may transport copper into vesicles *en route* to the plasma membrane, which is then released from the cell on fusion of vesicles with the plasma membrane. The punctate, albeit low level, cytoplasmic staining observed under basal conditions is consistent with MNK movement *via* vesicles. In this model, when the intracellular copper concentration decreases, copper ions are released from MNK, allowing its return to the TGN *via* vesicles.

The system for the functional expression of MNK in mammalian cells described in this chapter will enable future studies using *in vitro* mutagenesis, to evaluate the validity of the proposed model. In addition, the identification of other functional motifs involved in the intracellular location, trafficking, and copper transport activity of MNK will be possible, and will provide further insight into the role of MNK, and its interactions with other cellular components in maintaining cellular copper homeostasis. This system will also permit an assessment of the effects of Menkes patient mutations on MNK function. Together, these data may eventually contribute to the development of improved therapeutic strategies for patients with Menkes disease.

REFERENCES

Boshart, M., Weber, F., Jahn, G., Dorsch-Hasler, K., Fleckenstein, B., and Schaffner, W. (1985). A very strong enhancer is located upstream of an immediate early gene of human cytomegalovirus. Cell *41*, 521–530.

Brinster, R. L., Allen, J. M., Behringer, R. R., Gelinas, R. E., and Palmiter, R. D. (1988). Introns increase transcriptional effciency in transgenic mice. Proc. Natl. Acad. Sci. USA *85*, 836–840.

Bull, P. C., and Cox, D. W. (1994). Wilson disease and Menkes disease: new handles on heavy-metal transport. Trends Genet. *10*, 246–252.

Camakaris, J., Petris, M. J., Bailey, L., Shen, P., Lockhart, P., Glover, T. W., Barcroft, C. L., Patton, J., and Mercer, J. F. B. (1995). Gene amplification of the Menkes (MNK; ATP7A) P-type ATPase gene of CHO cells is associated with copper resistance and enhanced copper efflux. Hum. Mol. Genet. *4*, 2117–2123.

Danks, D. M. (1995). Disorders of copper transport. In The Metabolic and Molecular Basis of Inherited Disease, C.R. Scriver, A.L. Beaudet, W.M. Sly and D. Valle, eds. (New York: McGraw-Hill), pp. 2211–2235.

Das, S., Levinson, B., Vulpe, C., Whitney, S., Gitschier, J., and Packman, S. (1995). Similar splicing mutations of the Menkes/mottled copper-tranporting ATPase gene in occipital horn syndrome and the blotchy mouse. Am. J. Hum. Genet. *56*, 570–579.

Dierick, H. A., Adam, A. N., Escara-Wilke, J. F., and Glover, T. W. (1997). Immunocytochemical localization of the Menkes copper transport protein (ATP7A) to the *trans* Golgi network. Hum. Mol. Genet. *6*, 409–416.

Francis, M. J., Jones, E., Levy, E. R., Chelly, J., and Monaco, A. P. (1996). Functional analysis of the Menkes disease gene. Am. J. Hum. Genet. *59 (suppl.)*, A149.

Gluzman, Y. (1981). SV40-transformed Simian cells support the replication of early SV40 mutants. Cell *23*, 175–182.

Gregory, R. G., Cheng, S. H., Rich, D. P., Marshall, J., Paul, S., Hehir, K., Ostegaard, L., Klinger, K. W., Welsh, M. J., and Smith, A. E. (1990). Expression and characterization of the cystic fibrosis transmembrane conductance regulator. Nature *347*, 382–386.

Griffiths, G., and Simons, K. (1986). The trans Golgi network: sorting at the exit site of of the Golgi complex. Science *234*, 438–443.

Herd, S. M., Camakaris, J., Christofferson, R., Wookey, P., and Danks, D. M. (1987). Uptake and efflux of copper-64 in Menkes'-disease and normal continuous lymphoid cell lines. Biochem. J. *247*, 341–7.

Hung, I. H., Suzuki, M., Yamaguchi, Y., Yuan, D. S., Klausner, R. D., and Gitlin, J. D. (1997). Biochemical characterization of the Wilson disease protein and functional expression in the yeast Saccharomyces cerevisiae. J. Biol. Chem. *272*, 21461–21466.

Krauss, R. D., Bubien, J. K., Peiper, S. C., Collins, F. S., Kirk, K. L., Frizzell, R. A., and Rado, T. A. (1992). Transfection of wild-type CFTR into cystic fibrosis lymphocytes restores chloride conductance at G_1 of the cell cycle. Embo J. *11*, 875–883.

Kriegler, M. P. (1990). Gene transfer and expression. A laboratory manual. (New York: Freeman).

Kuivaniemi, H., Peltonen, L., and Kivirikko, K. I. (1985). Type IX Ehlers-Danlos syndrome and Menkes syndrome: the decrease in lysyl oxidase activity is associated with a corresponding deficiency in the enzyme protein. Am. J. Hum. Genet. *37*, 798–808.

La Fontaine, S., Firth, S. D., Lockhart, P. J., Brooks, H., Parton, R. G., Camakaris, J., and Mercer, J. F. B. (1997). Functional analysis and intracellular localization of the human Menkes Disease protein (MNK) stably expressed from a cDNA construct in Chinese Hamster Ovary cells (CHO-K1). Submitted for publication.

La Fontaine, S., Firth, S. D., Lockhart, P. J., Paynter, J. A., and Mercer, J. F. B. (1997). Low copy number plasmid vectors for eukaryotic gene expression: transient expression of the Menkes protein. Plasmid. (Accepted for publication).

Lippincott-Schwartz, J., Yuan, L., Tipper, C., Amherdt, M., Orci, L., and Klausner, R. D. (1991). Brefeldin A's effects on endosomes, lysosomes, and the TGN suggest a general mechanism for regulating organelle structure and membrane traffic. Cell *67*, 601–616.

Lutsenko, S., Petrukhin, K., Cooper, M. J., Gilliam, C. T., and Kaplan, J. H. (1997). N-terminal domains of human copper-transporting adenosine triphosphatases (the Wilson's and Menkes disease proteins) bind copper selectively *in vivo* and *in vitro* with stoichiometry of one copper per metal-binding repeat. J. Biol. Chem. *272*, 18939–18944.

Maruyama, K., and MacLennan, D. H. (1988). Mutation of aspartic acid-351, lysine-352, and lysine-515 alters the Ca^{2+} transport activity of the Ca^{2+}-ATPase expressed in COS-1 cells. Proc. Natl. Acad. Sci. USA *85*, 3314–3318.

Morelle, G. (1989). A plasmid extraction procedure on a miniprep scale. Focus *11*, 7–8.

Pelham, H. R. B. (1991). Multiple targets for Brefeldin A. Cell *67*, 449–451.

Peterson, M. G., Hannan, F., and Mercer, J. F. B. (1988). The sheep metallothionein gene family. Eur. J. Biochem. *174*, 417–424.

Petris, M. J., Mercer, J. F. B., Culvenor, J. G., Lockhart, P., Gleeson, P. A., and Camakaris, J. (1996). Ligand-regulated transport of the Menkes copper P-type ATPase efflux pump from the Golgi apparatus to the plasma membrane: a novel mechanism of regulated trafficking. EMBO J. *15*, 6084–6095.

Reaves, B., and Banting, G. (1992). Perturbation of the morphology of the *trans*-Golgi network following Brefeldin A treatment: redistribution of a TGN-specific integral membrane protein, TGN38. J. Cell Biol. *116*, 85–94.

Royce, P. M., Camakaris, J., Mann, J. R., and Danks, D. M. (1982). Copper metabolism in mottled mouse mutants. Biochem. J. *202*, 369–371.

Sambrook, J., Fritsch, E. F., and Maniatis, T. (1989). Molecular cloning. A laboratory manual., 2nd Edition (New York: Cold Spring Harbor Laboratory Press).

Solioz, M., and Vulpe, C. (1996). CPX-type ATPases: a class of P-type ATPases that pump heavy metals. Trends Biochem. Sci. *21*, 237–241.

Stearman, R., Yuan, D. S., Yamaguchi-Iwai, Y., Klausner, R. D., and Dancis, A. (1996). A permease-oxidase complex involved in high-affinity iron uptake in yeast. Science *271*, 1552–1557.

Thomsen, D. R., Stenberg, R. M., Goins, W. F., and Stinski, M. F. (1984). Promoter-regulatory region of the major immediate early gene of human cytomegalovirus. Proc. Natl. Acad. Sci. USA *81*, 659–663.

Versieck, J., and Cornelis, R. (1980). Normal levels of trace elements in human blood plasma or serum. Anal. Chim., Acta *116*, 217–254.

Vieira, J., and Messing, J. (1982). The pUC plasmids, an M13mp7-derived system for insertion mutagenesis and sequencing with synthetic universal primers. Gene *19*, 259–268.

Vulpe, C. D., and Packman, S. (1995). Cellular copper transport. Annu. Rev. Nutr. *15*, 293–322.

Walls, J. D., Berg, D. T., Yan, S. B., and Grinnell, B. W. (1989). Amplification of multicistronic plasmids in the human 293 cell line and secretion of correctly processed recombinant human protein C. Gene *81*, 139–149.

Wang, R. F., and Kushner, S. R. (1991). Construction of versatile low-copy-number vectors for cloning, sequencing and gene expression in *Escherichia coli*. Gene *100*, 195–199.

Yamaguchi, Y., Heiny, M. E., Suzuki, M., and Gitlin, J. D. (1997). Biochemical characterization and intracellular localization of the Menkes disease protein. Proc. Natl. Acad. Sci. USA 93, 14030–14035.

Yuan, D. S., Dancis, A., and Klausner, R. D. (1997). Restriction of copper export in *Saccharomyces cerevisiae* to a late Golgi or post-Golgi compartment in the secretory pathway. J. Biol. Chem. 272, 25787–25793.

Yuan, D. S., Stearman, R., Dancis, A., Dunn, T., Beeler, T., and Klausner, R. D. (1995). The Menkes/Wilson disease gene homologue in yeast provides copper to a ceruloplasmin-like oxidase required for iron uptake. Proc. Natl. Acad. Sci. USA 92, 2632–2636.

MUTATION SPECTRUM OF *ATP7A*, THE GENE DEFECTIVE IN MENKES DISEASE

Zeynep Tümer,[1,2] Lisbeth Birk Møller,[2] and Nina Horn[2]

[1] Department of Medical Genetics, Panum Institute, University of Copenhagen
Copenhagen, Denmark
[2] The John F. Kennedy Institute
Glostrup, Denmark

1. ABSTRACT

Our knowledge about Menkes disease (MD) has expanded greatly since its description in 1962 as a new X-linked recessive neurodegenerative disorder of early infancy. Ten years later a defect in copper metabolism was established as the underlying biochemical deficiency. In the beginning of 1990s efforts were concentrated on the molecular genetic aspects. The disease locus was mapped to Xq13.3 and the gene has been isolated by means of positional cloning. This was the beginning of a series of new findings which have greatly enhanced our understanding of copper metabolism not only in human, but also in other species. This review will focus on the molecular genetic aspects of Menkes disease and its allelic form occipital horn syndrome. The mutations will be compared briefly with those described in the animal model mottled mouse, and in Wilson disease, the autosomal recessive disorder of copper metabolism.

2. BACKGROUND

Copper is an essential trace element in biological systems being an integral component of several important enzymes, such as cytochrome c oxidase (COX), superoxide dismutase (SOD), lysyl oxidase (LOX) or tyrosinase. These metalloenzymes utilize the oxidative potential of copper, a property which can easily convert this micronutrient into a highly toxic element, causing cellular damage by enhancing free radical generation, oxidizing proteins and lipids in membranes. Consequently, efficient mechanisms must exist to regulate cellular copper homeostasis, responsive to the requirements at trace levels without allowing toxic accumulation of the free ion. Three hereditary disorders of intracellular copper transport reflect the contradictory manifestations of an unbalanced cellular copper homeostasis: The X-linked recessive Menkes disease (MD) and the occipital horn syndrome (OHS) represent the vital importance of copper, and the autosomal recessive Wilson disease (WD) the adverse effects.

Menkes disease (Menkes et al., 1962) is a multisystemic lethal disorder of copper metabolism (Danks et al., 1972) dominated by neurodegenerative symptoms and connective tissue abnormalities. Most of the clinical features of Menkes disease are attributable to malfunctioning of one or more copper requiring enzymes. Though 90–95% of MD patients present a severe clinical course, milder forms can also be distinguished. OHS has been suggested to be a mild allelic form of MD based on the biochemical and clinical resemblances, and this hypothesis has been confirmed subsequent to the cloning of the Menkes disease gene (*ATP7A*) (Kaler et al., 1994; Das et al., 1995). *ATP7A* was isolated in 1993 (Vulpe et al., 1993; Chelly et al., 1993; Mercer et al., 1993) and the gene was predicted to encode a 1500 amino acid copper translocating P-type ATPase (ATP7A) (Vulpe et al., 1993). The six repetitive domains, with the consensus GMXCXSC motif, at the amino terminal of this protein have recently been shown to bind copper (CuI) selectively (one copper per repeat) (Lutsenko et al., 1997). Further studies suggested that the ~165 kDa ATP7A was predominantly localized to the *trans*-golgi network at normal copper concentrations (Petris et al., 1996; Dierick et al., 1997). At elevated copper levels the protein was likely to be trafficked to the plasma membrane (Petris et al., 1997). These findings were in line with the suggested role of ATP7A in providing copper to metalloenzymes and in extrusion of the metal from the cell.

2.1. Mottled Mouse

In mouse, more than 20 spontaneous or induced mutations at the X-linked mottled locus *(Mo)* lead to a mottled coat pigmentation in the female heterozygotes and the affected males show neurological and connective tissue abnormalities, differing greatly in severity (Green, 1989). Biochemical and phenotypic similarities between MD patients and the mottled alleles, along with their conserved positions on the X chromosome, have long suggested that both defects were caused by mutations at homologous loci. Following the isolation of *ATP7A*, the mouse homologue *(atp7a)* was cloned using human sequences (Mercer et al., 1994; Levinson et al., 1994). The predicted protein product was also a copper binding P-type ATPase, showing 89% sequence homology to ATP7A. Later, identification of an *atp7a* mutation in one of the alleles confirmed the status of mottled mouse as the animal model for MD (Das et al., 1995). This allele was mottled blotchy, the suggested animal model for OHS (Rowe et al., 1977).

2.2. Wilson Disease

Wilson disease is mainly characterized by different degrees of liver disease, neurologic or psychiatric symptoms (Danks, 1995). The clinical features of WD are attributable to the toxic accumulation of copper in the liver, and subsequent overflow to different tissues, as kidney and brain. This accumulation is the result of deficient incorporation of copper into ceruloplasmin and impaired biliary copper excretion, a cellular physiopathology similar to that underlying MD. Following the identification of *ATP7A*, the gene defective in Wilson disease (designated *ATP7B*) was isolated and the predicted protein product was also a copper binding P-type ATPase (ATP7B) (Bull et al., 1993; Tanzi et al., 1993; Yamaguchi et al., 1993). Soonafter, the rat and mouse homologues of *ATP7B* have been isolated, and mutations were established in the respective genes of LEC rat and toxic milk mouse, making them confirmed animal models of WD (Wu et al., 1994;

Theophilos et al., 1996). In LEC rat a 900 bp of the 3'-coding region is deleted removing the ATP binding domain and downstream of the gene, and the toxic milk mouse has a missense mutation in the last transmembrane domain.

2.3. Copper Binding ATPases

ATP7A, ATP7B and the predicted protein products of their murine homologues are closely related to bacterial heavy metal binding P-type ATPases, in *Enterococcus hirae* (Odermatt et al., 1993) and *Synechococcus* 7942 (Phung et al., 1994), and *Saccharomyces cerevisiae* (CCC2) (Fu et al., 1995). These proteins belong to a large family of P-type ATPases specialized in energy dependent translocation of various cations in prokaryotes and eukaryotes. Independent of the cations they translocate, P-type ATPases have highly conserved motifs in the ATPase core such as a phosphorylation domain with a conserved aspartate residue and a phosphatase domain. The predicted amino acid sequences of the copper-translocating ATPases show all the features common for P-type ATPases. In addition Cu-ATPases contain a characteristic transduction motif (CPC) and a highly conserved histidine containing motif (SEHPL) common for other heavy metal translocating ATPases (Tanzi et al., 1993). Another feature of this subfamily is the copper binding domain(s) located at the amino terminal. The bacterial Cu-ATPases have only one copper binding domain; yeast two; mouse and rat atp7b five; mouse atp7a, human ATP7A and ATP7B have six domains. Unlike the functional domains located at the ATPase core, the copper binding domains allow more sequence diversity among species. An exception to this is the consensus GXXCXXC motif, which is involved in specific binding of copper (Lutsenko et al., 1997).

2.4. Genomic Organization of *ATP7A* in Comparison with *atp7a* and *ATP7B*

The 8.5 kb coding sequence of *ATP7A* is organized in 23 exons spanning a genomic region of about 150 kb (Tümer et al., 1995; Dierick et al., 1995). The first exon is a leader exon containing only untranslated sequences and the ATG translation start codon is in the second exon. The last exon contains a 274 bp translated sequence, the TAA translation termination site, the 3.8 kb 3'-untranslated region, and a polyadenylation site. The exon structure of the mouse gene *atp7a* is almost identical to its human counterpart. The 8.0 kb transcribed sequence consists of 23 exons spanning about 120 kb and, all the introns interrupt the aligned coding regions of *ATP7A* and *atp7a* at the same positions (Cecchi et al., 1996). The sizes of the exons are thus identical, except exons 1, 2 and 5, which are shorter in the mouse corresponding to the 9 amino acid difference in the respective proteins. Most interestingly in both genes the splice donor site of intron 9 (GT_GC) does not conform to the AG/GT rule indicative of the high conservation of these genes through evolution.

The 8 kb *ATP7B* gene has 22 exons (Petrukhin et al., 1994; Thomas et al., 1995) and its genomic organization shows remarkable similarity to *ATP7A* and *atp7a* (Tümer et al., 1995; Cecchi et al., 1996). Starting from the exons coding for the 5th copper binding domain (exon 5 in *ATP7A* and exon 3 in *ATP7B)* the coding regions of both genes are organized into 19 exons, showing an almost identical structure (Tümer et al., 1995).

3. MUTATIONS DETECTED IN *ATP7A*

3.1. Mutation Analysis at the John F. Kennedy Institute

Since 1973 more than 300 unrelated male patients or affected fetuses (n=309) from 27 different countries have been referred to the John F. Kennedy Institute for biochemical diag-

Table 1. Review of the mutations affecting *ATP7A*

Type of mutation	Total	JFK[a]	Others[b]
Chromosome mutations	7	1	6
Partial gene deletions	35	35	-
Point mutations			
Deletion/insertion	39	34	5
Nonsense	35	32	2
Splice site	43[c]	35	8[d]
Missense	32	30	3
	191	167	23

[a] Published or unpublished mutations datected at the John F. Kennedy Institute
[b] Gene mutations (Das et al., 1994,1995; Kaler et al., 1994, 1995, 1996; Ronce et al., 1997; Levinson et al., 1997) or chromosome abnormalities (Barton et al., 1983; Kapur et al., 1987; Beck et al., 1994; M. Tsukahara, D. Wattana, S. Mohammed, personal communication) identified by other groups
[c] Includes 2 deletion/insertion, 1 nonsense and 4 missense mutations occurring at the consensus splice site sequences;
[d] This value may represent a bias as atypical patients were investigated with priority.

nosis. Approximately 90–95% of these patients presented the classical severe phenotype and 5–10% was affected with one of the atypical forms, including the occipital horn syndrome.

The patients were investigated cytogenetically and in case of an abnormality the chromosome breakpoints were analysed with fluorescence in situ hybridization (FISH) using yeast artificial chromosomes (YAC) containing the Menkes disease gene (Tümer et al., 1992a). To identify partial deletions 230 patients were screened with Southern blot hybridisation using various genomic or cDNA probes (Chelly et al., 1993; manuscript in preparation) and 55 patients were investigated with exon-PCR using primers described previously (Tümer et al., 1996, 1997). For small nucleotide changes SSCP (single strand conformation polymorphism) or ddF (dideoxy fingerprinting) analyses were performed (Tümer et al., 1997). The promoter region upstream to exon 1 and the 3'-untranslated region have not been investigated for point mutations yet.

At the John F. Kennedy Institute a total of 143 different *ATP7A* mutations were identified in 167 unrelated families (Tümer et al., 1992b; 1994a,b; 1996; 1997; unpublished data) (table 1). Only one of these mutations was a chromosome aberration detected through a systematic screening of approximately 300 patients for a cytogenetical abnormality (Tümer et al., 1992b). Partial deletion of *ATP7A* was detected in 35 of the 285 investigated patients and a total of 107 different point mutations (the small base pair changes) were identified in 131 unrelated families (table 1, figure 1).

3.2. Mutation Spectrum of *ATP7A*

To date a total of 191 mutations affecting *ATP7A* has been identified in unrelated MD patients with the classical severe form or with one of the atypical phenotypes (table 1).

Chromosome abnormalities affecting *ATP7A* were detected in seven patients. One of these patients was a male with a unique chromosome abnormality, where the segment Xq13.3-q21.1 was inserted into the short arm of the X chromosome (Tümer et al., 1992b). One of the female patients was mosaic for the Turner karyotype (Barton et al., 1983) and the rest had X;autosome translocations (Barton et al., 1983; Kapur et al., 1987;; Beck et al., 1994; M. Tsukahara, D. Wattana, and S. Mohammed, personal communication).

Of the 184 gene mutations identified, 35 were partial gene deletions (manuscript in preparation). The largest defect observed, was the deletion of the whole gene except for

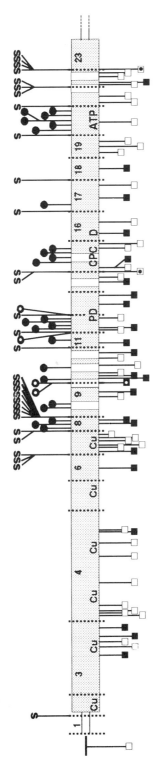

Figure 1. Localization of the 118 different point mutations identified within *ATP7A*. The predicted copper binding domains are indicated with Cu1–6 and the transmembrane domains with vertical white bars. PD, phosphatase domain; CC, cation channel with the consensus CPC motif; D, phosphorylation domain; ATP, ATP-binding domain. The coding region of *ATP7A* is represented by a gray horizontal box. The vertical dotted lines indicate the positions of the introns and the exons are indicated by numbers. The 5'- and 3'-untranslated regions are shown by a horizontal empty box and the flanking genomic sequences of the 5'-untranslated region by a horizontal line. Empty squares, insertion/deletion mutations; filled squares, nonsense mutations; filled circles missense mutations; S, splice site mutations. Mutations at the coding sequences which are predicted to affect normal splicing are marked with a small dot.

the first two exons, and the smallest one detectable with Southern blot hybridisation was the deletion of the first exon.

The total number of point mutations identified within *ATP7A* is 149, comprising 74% of all the mutations. Eighteen of these mutations were published by other groups (Das et al., 1994, 1995; Kaler et al., 1994, 1995, 1996; Ronce et al., 1997; Levinson et al., 1997) and 44 by ours (Tümer et al., 1994, 1996, 1997) (table 1). Seventeen point mutations were observed more than once in unrelated families and six of these have been published previously (Das et al., 1994; Kaler et al., 1994). One of these mutations was a missense mutation which was detected in 9 unrelated families (Das et al., 1994; manuscript in preparation). Furthermore, a couple of codons were targets of different mutations.

The four types of point mutations (deletions/insertions, missense and nonsense mutations, splice site mutations) are represented almost equally in *ATP7A* (table 1). Half of the point mutations are frameshift (22%) and nonsense mutations (14%), which are predicted to result in a non-functional truncated protein. These truncating mutations are distributed almost equally throughout the gene, though none has been observed in exons 1, 2, 5, or 23 (figure 1). In these four exons, except for the splice site mutation observed in IVS1, other point mutations were not detected either. A clustering of mutations could be observed in the exons which represent the functionally important and highly conserved regions of the protein. These exons are 12, 15, and 20, coding for the phosphatase, transduction and ATP binding domains of the protein, respetively. An exception to this is exon 8, where a total of 15 point mutations (including the acceptor and donor site mutations) have been identified (Das et al., 1994; Tümer et al., 1997; Ronce et al., 1997). Exon 8 encodes a region between the last metal binding domain and the first transmembrane domain of ATP7A (figure 2). Though a specific function has not been attributed to this region, it may play an important role in the folding of the protein and serve as a 'stalk' joining the metal binding domains and the ATPase core.

In the copper binding domains of ATP7A encoded by exons 2–7, no missense mutations were observed. This was also valid for exons 2–5 of *ATP7B* coding for these domains. The lack of missense mutations in this region is in line with the amino acid diversity allowed among species (figure 1). Furthermore the homologues of *ATP7B* both in rat and mouse contain only 5 copper binding domains and this finding is also suggestive for the flexibility of the region. In contrast to the amino terminal copper binding region, missense mutations were frequent in the ATPase core, especially in the conserved motifs, underscoring the importance of these domains for the normal function of the protein. *ATP7A* mutations occurring at the conserved motifs within the functionally important domains of the protein are shown in figure 2.

3.3. Brief Comparison of *ATP7A* and *ATP7B* Mutations

In the mutation spectrum of *ATP7A* all kinds of mutations are represented, while chromosome mutations or partial gene deletions have not been reported for *ATP7B* (table 2). The lack of chromosome aberrations disrupting *ATP7B* is not unexpected, as WD is inherited as an autosomal recessive trait and chromosome abnormalities are very rare in such disorders. The largest deletion observed in *ATP7B* was a 24 bp deletion (Figus et al., 1997; Orru et al., 1997), while in *ATP7A* an approximately 100 kb deletion removed the whole gene except for the first two exons, the leader exon and exon 2 with the translation start site.

In *ATP7B* all the mutations leading to the WD phenotype are point mutations. Distribution of different kinds of point mutations also differ from that observed in *ATP7A*. In

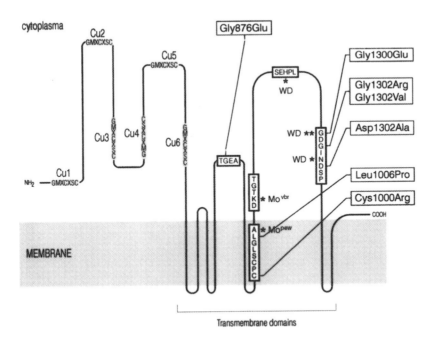

Figure 2. The missense mutations identified in the conserved motifs of the functionally important domains of ATP7A: GMXCXSC, copper binding domain (Cu1–6); TGEA, phosphatase domain; ALGLSCPC, transduction domain; TGTKD, phosphorylation domain; SEHLP, loop region; GDGINDSP, ATP binding domain. The missense mutations affecting the conserved motifs of *ATP7A* are shown in boxes. *, indicate the mutations identified in *ATP7B* and *atp7a*. Mo^{pew}, mottled pewter; Mo^{vbr}, mottled viable brindled. Mutations occurring at the conserved residiues of the transmembrane domains are not included in the figure.

Table 2. Comparison of the mutation profiles of *ATP7A* and *ATP7B*

Type of mutation	*ATP7A		ATP7B	
Chromosome mutations	7	<4%	-	
Partial gene deletions	35	22%	-	
Point mutations				
Deletion/insertion	38	24%	30	35%
Nonsense	22	14%	7	8%
Splice	36	22%	7	8%
Missense	23	14%	43	49%
Total	161		87	

Only the novel mutations are included and the percentages of different kinds of mutations are presented in italic. * Mutations published by Bull et al., 1993; Tanzi et al., 1993; Thomas et al., 1995a,b,c; Figus et al., 1995; Shimizu et al., 1995; Chuang et al., 1996; Loudianos et al., 1996; Nanji et al., 1997; Orru et al., 1997; Shah et al., 1997; Waldenström et al., 1997. Due to nomenclature differences used for *ATP7B*, the numbers may not represent the true values.

Table 3. *atp7a* mutations detected in mottled alleles

Localization	Mutation	Allele	Phenotype	Reference
promoter+exon 1	1.8 deletion	Dappled	Intrauterine death	Levinson et al., 1997
exon 11	6bp deletion	Brindled	Classical MNK	Grimes et al., 1997; Reed et al., 1997
IVS11	DS a_c +3	Blotchy	Mild phenotype	Das et al., 1995
IVS14	DS g_c +1	1Pub	Intrauterine death	Cechhi et al., 1997
exon 15	Ala998Thr	Pewter	Mild phenotype	Levinson et al., 1997
exon 16	Lys1036Thr	Viable brindled	Mild phenotype	Cechhi et al., 1997; Reed et al., 1997
exon 21	Ala1364Asp	11H	Intrauterine death	Cechhi et al., 1997
exon 22	Ser1381Pro	Macular	Classical MNK	Ohta et al., 1997; Mori et al., 1997; Murata et al., 1997

ATP7B the frequency of frameshift (35%) and missense mutations (43%) are significantly higher than nonsense (8%) and splice site mutations (8%).

As for *ATP7A*, some codons of *ATP7B* are also targeted more than once and there is a common missense mutation affecting the histidine residue of the highly conserved SEHPL motif (figure 2) (Bull et al., 1993; Tanzi et al., 1993; Thomas et al., 1995a,b,c; Figus et al., 1995; Houwen et al., 1995; Shimizu et al., 1995; Chuang et al., 1996; Loudianos et al., 1996; Nanji et al., 1997; Orru et al., 1997; Shah et al., 1997; Waldenström et al., 1997; Czlonkowska et al., 1997). Interestingly, a mutation affecting this motif has not been detected in *ATP7A*.

3.4. Mutations in the Mouse Model

In mouse 20–30 mutations lead to the mottled phenotype and *atp7a* mutations have been defined in 8 of them (table 3). These mutations include, a partial gene deletion, a 6 bp deletion, four missense and two splicing mutations. Small insertions or nonsense mutations have not been identified, but the present data is not large enough to make a conclusion for the mutation spectrum of *atp7a*.

Two of the mottled alleles have mutations similar to those found in MD patients. Mottled blotchy has a splice site mutation affecting RNA splicing, similar to that observed in several OHS patients (Kaler et al., 1994; Das et al., 1995). Interestingly, Kaler et al. (1994) have described a mildly affected patient (now classified as OHS) with a mutation occurring at the same splice site (IVS 11). Mottled 1Pub, where the affected males die in utero, the mutation is at the consensus donor splice site of IVS 14 (Cecchi et al., 1997). In three unrelated patients we have investigated, the same basepair was mutated and all three patients suffered from the classical severe form of MD (unpublished data).

Mutations of the mottled viable brindled (Cecchi et al., 1997; Reed et al., 1997) and mottled pewter (Levinson et al., 1997) affect the residues belonging to the conserved motifs of atp7a, but mutations affecting these residues have not been described for ATP7A.

Partial deletion of *atp7a* was detected only in mottled dappled mouse (Levinson et al., 1997). George et al. (1994) have previously screened 10 alleles, including mottled dappled, for partial gene deletions, without observing any. In this study human cDNA clones were used as hybridisation probes. However, these clones did not include exon 1 and the failure in detecting the deletion is likely to be due to this, rather than employment of human cDNA against mouse DNA.

4. GENOTYPE-PHENOTYPE CORRELATION

There is no obvious correlation between the mutations and the clinical course of Menkes disease, though patients with a milder phenotype has a higher proportion of mutations predicted to result in a normal protein with residual activity.

Until now 6 point mutations have been reported in patients with non-classical phenotypes (Kaler et al., 1994; Das et al., 1995; Levinson et al., 19967; Ronce et al., 1997). One patient had a milder phenotype (Kaler et al., 1994) and five had OHS (Kaler et al., 1994; Das et al., 1995; Levinson et al., 1996; Ronce et al., 1997). In all these patients, except for the one described by Levinson et al., (1996) the DNA mutation was leading to RNA splicing defects. In the patient with the mild phenotype (Kaler et al., 1994) and in three OHS patients (Kaler et al., 1994; Das et al., 1995) the mutations were at the consensus splice site sequences of affecting the normal RNA splicing. Besides the abnormal size transcripts, a normal size transcript was always present in low amounts in each of these cases. In one OHS patient (Ronce et al., 1997) a missense mutation within exon 8 resulted in a normal size transcript bearing the mutation besides abnormal size transcripts. In this case the missense mutation could have a milder effect on the function of the otherwise normal size protein. Exon skipping due to point mutations within the coding sequences is a phenomenon observed in *ATP7A* previously (Das et al., 1994) and we are currently investigating the mRNA transcripts of the patients with atypical phenotypes in attempt to make a genotype-phenotype correlation.

The patient described by Levinson et al. (1996) had a mutation different than the others. One of the three repeated elements in the regulatory region of *ATP7A* was deleted (Levinson et al., 1996), but the transcription level was normal. In this patient no other DNA mutations could be detected in *ATP7A* and the relation of this deletion to OHS remained obscure.

5. CONCLUSION

The most important impact of the identification of *ATP7A* mutations in Menkes disease patients is on the diagnosis of the disease. For MD a definitive biochemical diagnosis exists and it is based on the intracellular accumulation of copper in cultured cells due to impaired efflux. Prenatal diagnosis is carried out by measuring radioactive copper accumulation in cultured amniotic fluid cells in the second trimester and by determining the total copper content in chorionic villi in the first trimester, a test very susceptible to exogenous copper contamination. These analyses demand expertise and are carried out only in a few centers in the world. Demonstration of a defect in *ATP7A* will be the ultimate diagnostic proof and may eventually decentralize the diagnosis of MD. However, mutation detection in Menkes disease is a formidable task: the 8.5 kb *ATP7A* transcript is organized in 23 exons and the genetic defect shows a great variety. A prenatal molecular genetic diagnosis will thus be possible, only if the mutation of the family has already been identified.

In MD carrier determination is also possible by measuring ^{64}Cu-uptake in cultured fibroblasts. However, carrier identification of X-linked disorders by biochemical means are not reliable in case of negative results, due to random inactivation of the X chromosomes. Mutation analyses will therefore provide the ultimate proof for heterozygosity. This will also have an impact on the amount of prenatal diagnoses referred to The John F. Kennedy Institute, as about 80% of the male fetuses tested until now were found to be unaffected. This indicated that a substantial number of the mothers were indeed not carriers. Until now we have performed carrier diagnosis in 61 families and in 17 of these the mother of the index patient was not carrier (Tümer et al., 1994a,b; unpublished data). Although germ line mosaicism cannot be excluded in these cases, the results are crucial for genetic counselling of other female members of such families. In families where the mutation is unknown, intragenic polymorphic markers may also enable carrier diagnosis. The Val767Leu polymorphism identified in exon 10 (Das et al., 1994), and the two polymorphic CA repeats within intron 2 and intron 5 (Begy et al., 1995) are currently being used for carrier diagnosis in the families where the mutation is yet unknown (unpublished data).

REFERENCES

Barton, N.W., Dambrosia, J.M., and Barranger, J.A. (1983). Menkes' Kinky-Hair Syndrome: Report of a case in a female infant. Neurology 33 [Suppl 2], 154.

Beck, J., Enders, H., Schliephacke, M., Buchwald-Saal, M., and Tümer, Z. (1994). X;1 translocation in afemale Menkes patient: characterization by fluorescence in situ hybridization. Clin Genet 46, 295–298.

Begy, C.R., Dierick, H.A., Innis, J.W., Glover, T.W. (1995). Two highly polymorphic CA repeats in the Menkes gene (ATP7A). Hum Genet 96, 355–356.

Bull, P.C., Thomas, G.R., Rommens, J.M., Forbes, J.R., and Cox, D.W. (1993). The Wilson disease gene is a putative copper transporting P-type ATPase similar to the Menkes gene. Nat Genet 5, 327–337.

Cecchi, C. and Avner, P. (1996). Genomic organization of the mottled gene, the mouse homologue of the human menkes disease gene. Genomics 37, 96–104.

Cecchi, C., Biasotto, M., Tosi, M., and Avner, P. (1997). The mottled mouse as a model for human Menkes disease: identification of mutations in the Atp7a gene. Hum. Mol. Genet. 6, 425–433.

Chelly, J., Tümer, Z., Tønnesen, T., Petterson, A., Ishikawa-Brush, Y., Tommerup, N., Horn, N., and Monaco, A.P. (1993). Isolation of a candidate gene for Menkes disease that encodes a potential heavy metal binding protein. Nat Genet 3, 14–19.

Chuang, L.-M., Wu, H.-P., Jang, M.-H., Wang, T.-R., Sue, W.-C., Lin, B.J., Cox, D.W., and Tai, T.-W. (1996). High frequency of two mutations in codon 778 in exon 8 of the ATP7B gene in Taiwanese families with Wilson disease. J Med Genet 33, 521–523.

Czlonkowska, A., Rodo, M., Gajda, J., Ploos van Amstel, H.K., Juyn, J., and Houwen, R.H. (1997). Very high frequency of the His1069Gln mutation in Polish Wilson disease patients [letter] [In Process Citation]. J. Neurol. 244, 591–592.

Danks, D.M., Stevens, B.J., Campbell, P.E., Gillespie, J.M., Walker-Smith, J., Blomfield, J., and Turner, B. (1972). Menkes' kinky-hair syndrome. Lancet 1, 1100–1102.

Danks, D.M. (1995). Disorders of Copper Transport. In The Metabolic Basis of Inherited Disease. J.R. Scriver, A.L. Beaudet, W.S. Sly, and D. Valle, eds. (New York: McGraw-Hill), pp. 2211–2235.

Das, S., Levinson, B., Whitney, S., Vulpe, C., Packman, S., and Gitschier, J. (1994). Diverse mutations in patients with Menkes disease often lead to exon skipping. Am J Hum Genet 55, 883–889.

Das, S., Levinson, B., Vulpe, C., Whitney, S., Gitschier, J., and Packman, S. (1995). Similar splicing mutations of the Menkes/Mottled copper transporting ATPase gene in occipital horn syndrome and the blotchy mouse. Am J Hum Genet 56, 570–576.

Dierick, H.A., Ambrosini, L., Spencer, J., Glover, T.W., and Mercer, J.F.B. (1995). Molecular structure of the Menkes disease gene (ATP7A). Genomics 28, 462–469.

Dierick, H.A., Adam, A.N., Escara-Wilke, J.F., and Glover, T.W. (1997). Immunocytochemical localization of the Menkes copper transport protein (ATP7A) to the trans-Golgi network. Hum. Mol. Genet. 6, 409–416.

Figus, A., Angius, A., Loudianos, G., Bertini, C., Dessi, V., Loi, A., Deliana, M., Lovicu, M., Olla, N., Sole, G., De Virgiliis, S., Lilliu, F., Farci, A.M.G., Nurchi, A., Giacchino, R., Barabino, A., Marazzi, M., Zancan, L., Greggio, N.A., Marcellini, M., Solinas, A., Deplano, A., Barbera, C., Devoto, M., Ozsoylu, S., Kocak, N., Akar, N., Karayalcin, S., Mokini, V., Cullufi, P., Balestrieri, A., Cao, A., and Pirastu, M. (1995). Molecular pathology and haplotype analysis of Wilson disease in Mediterranean populations. Am J Hum Genet 57, 1318–1324.

Fu, D., Beeler, T.J., and Dunn, T.M. (1995). Sequence, mapping and disruption of CCC2, a gene that cross-complements the Ca2+-sensitive phenotype of csg1 mutants and encodes a P-type ATPase belonging to the Cu2+-ATPase subfamily. Yeast *11*, 283–292.

George, A.M., Reed, V., Glenister, P., Chelly, J., Tümer, Z., Horn, N., Monaco, A.P., and Boyd, Y. (1994). Analysis of Mnk, the murine homologue of the locus for Menkes' disease, in normal and mottled (Mo) mice. Genomics *22*, 27–35.

Green, M.C. (1989). Catalog of mutant genes and polymorphic loci. In Genetic variants and strains of the laboratory mouse. M.F. Lyon and A.G. Searle, eds. (Oxford: Oxford University Press), pp. 241–244.

Grimes, A., Hearn, C.J., Lockhart, P., Newgreen, D.F., and Mercer, J.F. (1997). Molecular basis of the brindled mouse mutant (Mo(br)): a murine model of Menkes disease. Hum. Mol. Genet. *6*, 1037–1042.

Horn, N., Tønnesen, T., and Tümer, Z. (1995). Variability in clinical expression of an X-linked copper disturbance, Menkes disease. In Genetic response to metals. B. Sarkar, ed. (New York: Marcel and Dekker), pp. 285–303.

Houwen, R.H.J., Juyn, J., Hoogenraad, T.U., Ploos van Amstel, J.K., and Berger, R. (1995). H714Q mutation in Wilson disease is associated with late, neurological presentation. J Med Genet *32*, 180–182.

Kaler, S.G., Gallo, L.K., Proud, V.K., Percy, A.K., Mark, Y., Segal, N.A., Goldstein, D.S., Holmes, C.S., and Gahl, W.A. (1994). Occipital horn syndrome and a mild Menkes phenotype associated with splice site mutations at the MNK locus. Nat Genet *8*, 195–202.

Kaler, S.G., Buist, N.R.M., Holmes, C.S., Goldstein, D.S., Miller, R.C., and Gahl, W.A. (1995). Early copper therapy in classic Menkes disease patients with a novel splicing mutation. Ann Neurol *38*, 921–928.

Kaler, S.G., Das, S., Levinson, B., Goldstein, D.S., Holmes, C.S., Patronas, N.J., Packman, S., and Gahl, W.A. (1996). Successful early copper therapy in Menkes disease associated with a mutant transcript containing a small in-frame deletion. Biochemical and Molecular Medicine *57*, 37–46.

Kapur, S., Higgins, J.V., Delp, K., and Rogers, B. (1987). Menkes syndrome in a girl with X-autosome translocation. Am J Med Genet *26*, 503–510.

Levinson, B., Vulpe, C., Elder, B., Martin, C., Verley, F., Packman, S., and Gitschier, J. (1994). The mottled gene is the mouse homologue of the Menkes disease gene. Nat Genet *6*, 369–373.

Levinson, B., Conant, R., Schnur, R., Das, S., Packman, S., and Gitschier, J. (1996). A repeated element in the regulatory region of the *MNK* gene and its deletion in a patient with occipital horn syndrome. Human Molecular Genetics *5*, 1737–1742.

Levinson B., Packman, S., and Gitschier, J (1997). Mutation analysis of mottled pewter. Mouse Genome *95*, 163–165.

Levinson, B., Packman, S., and Gitschier, J. (1997). Deletion of the promoter region of the *Atp7a* gene of the *mottled dappled* mouse. Nature Genetics *16*, 223–224.

Loudianos, G., Dessi, V., Angius, A., Lovicu, M., Loi, A., Deiana, M., Akar, N., Vajro, P., Figus, A., Cao, A., and Pirastu, M. (1996). Wilson disease mutations associated with uncommon haplotypes in Mediterranean patients. Human Genetics *98*, 640–642.

Lutsenko, S., Petrukhin, K., Cooper, M.J., Gilliam, C.T., and Kaplan, J.H. (1997). N-terminal domains of human copper-transporting adenosine triphosphatases (the Wilson's and Menkes disease proteins) bind copper selectively in vivo and in vitro with stoichiometry of one copper per metal-binding repeat. J. Biol. Chem. *272*, 18939–18944.

Menkes, J.H., Alter, M., Steigleder, G., Weakley, D.R., and Sung, J.H. (1962). A sex-linked recessive disorder with retardation of growth, peculiar hair and focal cerebral and cerebellar degeneration. Pediatrics *29*, 764–779.

Mercer, J.F.B., Livingston, J., Hall, B., Paynter, J.A., Begy, C., Chandrasekharappa, S., Lockhart, P., Grimes, A., Bhave, M., Siemieniak, D., and Glover, T.W. (1993). Isolation of a partial candidate gene for Menkes disease by positional cloning. Nat Genet *3*, 20–25.

Mercer, J.F.B., Grimes, A., Ambrosini, L., Lockhart, P., Paynter, J.A., Dierick, H.A., and Glover, T.W. (1994). Mutations in the murine homologue of the Menkes gene in dappled and blotchy mice. Nat Genet *6*, 374–378.

Mori, M. and Nishimura, M. (1997). A serine-to-proline mutation in the copper-transporting P-type ATPase gene of the macular mouse. Mamm. Genome *8*, 407–410.

Muramatsu, Y., Yamada, T., Moralejo, D.H., Cai, Y., Xin, X., Miwa, Y., Izumi, K., and Matsumoto, K. (1995). The rat homologue of the Wilson's disease gene was partially deleted at the 3' end of its protein-coding region in long-evans cinnamon mutant rats. Res Commun Mol Pathol Pharmacol *89*, 421–424.

Murata, Y., Kodama, H., Abe, T., Ishida, N., Nishimura, M., Levinson, B., Gitschier, J., and Packman, S. (1997). Mutation analysis and expression of the mottled gene in the macular mouse model of Menkes disease [In Process Citation]. Pediatr. Res. *42*, 436–442.

Nanji, M.S., Nguyen, V.T., Kawasoe, J.H., Inui, K., Endo, F., Nakajima, T., Anezaki, T., and Cox, D.W. (1997). Haplotype and mutation analysis in Japanese patients with Wilson disease. Am. J. Hum. Genet. *60*, 1423–1429.

Odermatt, A., Suter, H., Krapf, R., and Solioz, M. (1993). Primary structure of two P-type ATPases involved in copper homeostasis in Enterococcus hirae. J Biol Chem 268, 12775–12779.

Ohta, Y., Shiraishi, N., and Nishikimi, M. (1997). Occurrence of two missense mutations in Cu-ATPase of the macular mouse, a Menkes disease model [In Process Citation]. Biochem. Mol. Biol. Int. 43, 913–918.

Ono, T., Fukumoto, R., Kondoh, Y., and Yoshida, M.C. (1995). Deletion of the Wilson's disease gene in hereditary hepatitis LEC rats. Jpn J Genet 70, 25–33.

Orru, S., Thomas, G., Cox, D.W., and Contu, L. (1997). 24 bp deletion and Ala_{1278} to Val mutation of the ATP7B gene in a Sardinian family with Wilson disease. Am J Hum Genet 10, 84–85.

Petris, M.J., Mercer, J.F.B., Culvenor, J.G., Lockhart, P., Gleeson, P.A., and Camakaris, J. (1996). Ligand-regulated transport of the Menkes copper P-type ATPase efflux pump from the Golgi apparatus to the plasma membrane: a novel mechanism of regulated trafficking. EMBO J 15, 6084–6095.

Petrukhin, K., Lutsenko, S., Chernov, I., Ross, B.M., Kaplan, J.H., and Gilliam, T.C. (1994). Characterization of the Wilson disease gene encoding a P-type copper transporting ATPase: genomic organization, altrenative splicing, and structure/function predictions. Hum Molec Genet 3, 1647–1656.

Phung, L.T., Ajlani, G., and Haselkorn, R. (1994). P-type ATPase from the cyanobacterium Synechococcus 7942 related to the human Menkes and Wilson disease gene products. Proc Natl Acad Sci,USA 91, 9651–9654.

Reed, V. and Boyd, Y. (1997). Mutation analysis provides additional proof that mottled is the mouse homologue of Menkes' disease. Hum. Mol. Genet. 6, 417–423.

Ronce, N., Moizard, M.P., Robb, L., Toutain, A., Villard, L., and Moraine, C. (1997). A C2055T transition in exon 8 of the ATP7A gene is associated with exon skipping in an occipital horn syndrome family [letter]. Am. J. Hum. Genet. 61, 233–238.

Rowe, D.W., McGoodwin, E.B., Martin, G.R., Grahn, D. (1977). Decreased lysyl oxidase activity in the aneurisme-prone, mottled mouse 252, 939–942.

Shah, A.B., Chernov, I., Zhang, H.T., Ross, B.M., Das, K., Lutsenko, S., Parano, E., Pavone, L., Evgrafov, O., Ivanova-Smolenskaya, I.A., Anneren, G., Westermark, K., Urrutia, F.H., Penchaszadeh, G.K., Sternlieb, I., Scheinberg, I.H., Gilliam, T.C., and Petrukhin, K. (1997). Identification and analysis of mutations in the Wilson disease gene (ATP7B): population frequencies, genotype-phenotype correlation, and functional analyses. Am. J. Hum. Genet. 61, 317–328.

Shimizu, N., Kawase, C., Nakazono, H., Hemmi, H., Shimatake, H., and Aoki, T. (1995). A novel RNA splicing mutation in Japanese patients with Wilson disease. Biochem Biophys Res Comm 217, 16–20.

Tanzi, R.E., Petrukhin, K., Chernov, I., Pellequer, J.L., Wasco, W., Ross, B., Romano, D.M., Parano, E., Pavone, L., Brzustowicz, L.M., Devoto, M., Peppercorn, J., Bush, A.I., Sternlieb, I., Pirastu, M., Gusella, J.F., Evgrafov, O., Penchaszadeh, G.K., Honig, B., Edelman, I.S., Soares, M.B., Scheinberg, I.H., and Gilliam, T.C. (1993). The Wilson disease gene is a copper transporting ATPase with homology to the Menkes disease gene. Nat Genet 5, 344–357.

Theophilos, M.B., Cox, D.W., and Mercer, J.F.B. (1996). The toxic milk mouse is a murine model of Wilson disease. Human Molecular Genetics 5, 1619–1624.

Thomas, G.R., Forbes, J.R., Roberts, E.A., Walshe, J.M., and Cox, D.W. (1995a). The Wilson disease gene: spectrum of mutations and their consequences. Nat Genet 9, 210–217.

Thomas, G.R., Jensson, O., Gudmundsson, G., Thorsteinsson, L., and Cox, D.W. (1995b). Wilson disease in Iceland: A clinical and genetic study. Am J Hum Genet 56, 1140–1146.

Thomas, G.R., Roberts, E.A., Walshe, J.M., and Cox, D.W. (1995c). Haplotypes and mutations in Wilson disease. Am J Hum Genet 56, 1315–1319.

Tümer, Z., Chelly, J., Tommerup, N., Ishikawa-Brush, Y., Tønnesen, T., Monaco, A.P., and Horn, N. (1992a). Characterization of a 1.0 Mb YAC contig spanning two chromosome breakpoints related to Menkes disease. Hum Molec Genet 1, 483–489.

Tümer, Z., Tommerup, N., Tønnesen, T., Kreuder, J., Craig, I.W., and Horn, N. (1992b). Mapping of the Menkes locus to Xq13.3 distal to the X-inactivation center by an intrachromosomal insertion of the segment Xq13.3-q21.2. Hum Genet 88, 668–672.

Tümer, Z., Tønnesen, T., Böhmann, J., Marg, W., and Horn, N. (1994a). First trimester prenatal diagnosis of Menkes disease by DNA analysis. J Med Genet 31, 615–617.

Tümer, Z., Tønnesen, T., and Horn, N. (1994b). Detection of genetic defects in Menkes disease by direct mutation analysis and its implications in carrier diagnosis. J Inher Metab Dis 17, 267–270.

Tümer, Z., Vural, B., Tønnesen, T., Chelly, J., Monaco, A.P., and Horn, N. (1995). Characterization of the exon structure of the Menkes disease gene using Vectorette PCR. Genomics 26, 437–442.

Tümer, Z., Horn, N., Tønnesen, T., Christodoulou, J., Clarke, J.T.R., and Sarkar, B. (1996). Early copper-histidine treatment for Menkes disease. Nat Genet 12, 11–13.

Tümer, Z., Lund, C., Tolshave, J., Vural, B., Tønnesen, T., and Horn, N. (1997). Identification of point mutations in 41 unrelated patients affected with Menkes disease. Am J Hum Genet 60, 63–71.

Vulpe, C., Levinson, B., Whitney, S., Packman, S., and Gitschier, J. (1993). Isolation of a candidate gene for Menkes disease and evidence that it encodes a copper-transporting ATPase. Nat Genet 3, 7–13.

Waldenström, E., Lagerkvist, A., Dahlman, T., Westermark, K., and Landegren, U. (1996). Efficient detection of mutations in Wilson disease by manifold sequencing. Genomics 37, 303–309.

Wu, J., Forbes, J.R., Chen, H.S., and Cox, D.W. (1994). The LEC rat has a deletion in the copper transporting ATPase gene homologous to the Wilson disease gene. Nat Genet 7, 541–545.

Yamaguchi, Y., Heiny, M.E., and Gitlin, J.D. (1993). Isolation and characterization of a human liver cDNA as a candidate gene for Wilson disease. Biochem Biophys Res Comm 197, 271–277.

8

ANIMAL MODELS OF MENKES DISEASE

Julian F. B. Mercer, Loreta Ambrosini, Sharon Horton, Sophie Gazeas, and Andrew Grimes

Murdoch Institute, Royal Children's Hospital
Flemington Rd.
Parkville, Victoria 3052, Australia

1.0. MENKES DISEASE AND ITS VARIANTS

There are three recognized X-linked copper deficiency disorders in humans: classical Menkes disease (MD), mild Menkes disease and occipital horn syndrome (OHS, also known as X-linked cutis laxa). Since the features of these diseases are so distinct, it was not clear until recently whether the phenotypes were due to mutations in the same gene, or whether OHS in particular is due to mutations in a gene on the X-chromosome, closely linked to the Menkes locus (Danks, 1995). Molecular analysis has now demonstrated that MD, mild Menkes and OHS are indeed due to allelic mutations of the gene affected in Menkes disease (*MNK* or *ATP7A*), however, the basis for the phenotypic differences is still not fully understood. A similar range of phenotypes is also found in the mottled mice mutants, and these are discussed below.

1.1. Clinical Features of Classical Menkes Disease

The clinical features of classical Menkes disease were first described by John Menkes (Menkes et al., 1962). The role of copper in this disorder was not identified until David Danks associated the brittle hair and the tortuous elongated blood vessels found in Menkes disease with similar features in copper deficient animals (Danks et al., 1972). Boys with Menkes disease usually die by the age of three or four years, but occasionally a patient may survive for a longer period (for review see (Bankier, 1995; Danks, 1995; Kaler, 1994; Kodama, 1993). Patients are severely mentally retarded and have marked neurodegeneration. The reason for the neurological abnormalities is not fully understood, but could be a consequence of the low activity of several copper-dependent enzymes required for brain development. Deficiency of the electron transport chain component cytochrome c oxidase, is likely to be a major factor (Kaler, 1994). Menkes patients are hypopigmented due to tyrosinase deficiency and they suffer from hypothermia, presumably due to a reduction in cytochrome c oxidase. Arterial and bony abnormalities are a feature probably due to deficiency of copper-dependent lysyl oxidase required for the normal cross-linking of collagen and elastin.

1.2. Clinical Features of Occipital Horn Syndrome (OHS)

Connective tissue and bone defects are the predominant features of OHS (Byers et al., 1980; Kuinaviemi et al., 1982). Patients have hyper-elastic skin, hernias, bladder diverticulae and multiple skeletal abnormalities. These features can be attributed to deficient activity of lysyl oxidase, resulting from failure of copper to be incorporated into the enzyme during synthesis (Kuinaviemi et al., 1982). Some patients have signs of autonomic dysfunction, such as syncope and diarrhoea, which may be due to dopamine-β-hydroxylase deficiency. They may also be mildly mentally retarded.

1.3. Clinical Features of Mild Menkes Disease

An intermediate phenotype between OHS and MD, termed mild Menkes, has been described in which the neurological abnormalities are far less profound than in classical MD (Procopis et al., 1981). The first patient described had cerebellar ataxia with mild developmental delay. At 10 years of age he had low-normal intelligence but still had problems of ataxia (Danks, 1988). Other cases with similar features have been described (Gerdes et al., 1988; Kaler et al., 1994; Westman et al., 1988).

1.4. Copper Accumulation in Cells from Menkes Patients

Cells from patients with MD and the mottled mouse mutants have been shown to have normal uptake but reduced efflux of copper, leading to an intracellular accumulation (Horn, 1976). The excess copper is bound the to small cysteine-rich proteins, metallothioneins (MTs). Cells from OHS patients accumulate copper to the same extent as those from Menkes patients, which is somewhat surprisingly in view of the milder phenotype (Peltonen et al., 1983).

It may seem paradoxical that copper excess is found in cells from patients suffering from copper deficiency. The explanation may be that the efflux defect manifest in cultured cells is also present in the intestinal enterocytes responsible for uptake of dietary copper. Copper is found to accumulate in these cells in Menkes patients, but cannot be effluxed into the body, thus leading to an overall copper deficiency. This deficiency is exacerbated by the reduced efflux in other cell types, most notably in the cells which constitute the blood-brain barrier. Copper needs to be transported into the brain, and this process is defective in MD patients leading to increased severity of copper deficiency in this organ.

2.0. Cu-ATPases AND GENETIC DISORDERS OF Cu TRANSPORT

The cloning of the Menkes gene provided the first molecular information about the mechanisms of copper transport and this in turn is leading to a better understanding of the features of the disease and its variants (Chelly et al., 1993; Mercer et al., 1993; Vulpe et al., 1993). The Menkes gene encodes a copper pump, termed ATP7A or MNK, which is predicted to have eight transmembrane domains and is a member of the extensive family of cation pumps termed P-type ATPases (Vulpe et al., 1993).

The protein affected in the genetic copper toxicosis disorder, Wilson disease is a very similar molecule, referred to as ATP7B or WND. This Cu-ATPase is thought to function primarily in effluxing copper from the liver into the bile (Cox, 1995). The copper ATPases have been classified into a separate subgroup of P-type ATPases, termed CPX ATPases, and have more in common with bacterial heavy metal transporters than with other known mammalian P-type ATPases (Solioz and Vulpe, 1996). The known function of P-type ATPases is consistent with the copper efflux defects in Menkes cells. Recent work on the cellular localization of MNK is providing some insights into the phenotypic consequences of MNK deficiency (see Section 5)

3.0. THE MOTTLED MICE: ANIMAL MODELS OF MENKES DISEASE AND OCCIPITAL HORN SYNDROME

A large number of mouse mutants at the X-linked mottled locus (Mo) have been described (Doolittle et al., 1996). As with mutations of the Menkes gene in humans, these mutants have a wide range of different phenotypes, but all have copper deficiency in common. Table 1 summarizes the features of the various mutants and the likely human homologues, and Table 2 lists the known mutations in the mottled mice and the associated phenotypes.

3.1. Prenatal Lethal Mutants

There are at least nine mottled mutants which die during gestation, but the phenotypes of these has not been well characterized. The dappled mutant is known to die at the end of gestation, and has skeletal abnormalities (Phillips, 1961). This mutant has been shown to express little if any Mnk mRNA (Levinson et al., 1994) and has a deletion in the promoter region of the gene (Levinson et al., 1997). We have analyzed the Mo^{9H} mutant, which was originally misclassified as dappled, and have shown that, like dappled, the mutant male has no detectable Mnk mRNA (Mercer et al., 1994). The mutants die in just prior to birth and have a range of phenotypes which could be due to connective tissue defects (L. Ambrosini and J. Mercer, unpublished data). Mo^{11H} has been shown to have a missense mutation A1364D, which is likely to disrupt the 7th transmembrane domain by

Table 1. Human and mouse mutants of MNK

Features of mouse mutant	Human equivalent	Mouse mutant
Prenatal lethality.	None described	Mo^{dp}, Mo^{9H}, Mo^{11H}, Mo^{Xm}, MolPub, Mo12DFiOD, Mo3MLP1, Mo7ENURF $Mo^{32DFiOD}$
Death by 3rd postnatal week		
Cu treatment not successful.	Classical Menkes*	Jax Brindled (Mo^{br-J})
treatable with Cu neurological defects.	Classical Menkes*	Brindled(Mo^{br}), macular (Mo^{ml})
not well characterized.	Classical Menkes*	Mo^{13H}, Mo2Acr, Mo8J
Reduced viability, connective tissue defects	Mild Menkes	Viable brindled (Mo^{vbr})
Viable, connective tissue defects	Occipital horn syndrome	Blotchy (Mo^{blo})

*The Menkes homologues are divided into three groups, two based on response to Cu treatment and the other group has not been well characterized.

Table 2. Mutations in the mottled mice

Mutant	Mnk mRNA*	Mutation	Possible consequences
Mo^{dp}	<5%[1]	promoter deletion[2]	
Mo^{9H}	<5%[3]	not characterized	
Mo^{11H}	22%	A1364D[4]	TM7 disrupted
Mo^{Xm}	3.6%[4]	not found	
Mo^{lPub}	44%	splice site[4]	TM5 skipped
$Mo^{12DFiOD}$	?	Not found	
Mo^{3MLP}_l	11%[4]	Not found	
Mo^{13H}	?	Not coding[5]	
Mo^{br}	100%[3]	Deletion of AL[8]	Low activity
Mo^{ml}	100%[1]	P to S	Disrupts TM8
Mo^{vbr}	?	K1036T[4,5]	Reduced phosphorylation
Mo^{blo}	20%[1,3]	Splice site[7]	Reduced protein

*Mnk mRNA refers to the relative amount of mRNA relative to normal. Numbers in superscript are to the following references: 1. Levinson et al., 1994, 2. Levinson et al., 1997, 3. Mercer et al., 1994, 4. Cecchi et al., 1997, 5. Reed and Boyd, 1997, 6. Mori and Nishimura, 1997, 7. Das et al., 1995, 8. Grimes et al., 1997.

insertion of a large negatively charged aspartate residue in place of alanine (Cecchi et al., 1997). Mo^{lPub} has a splice site mutation leading to a reduction in Mnk mRNA and the predicted production of an inactive molecule lacking the 5th transmembrane domain (Cecchi et al., 1997).

The common feature of all the prenatal mutants characterized is that they are predicted to produce little, if any, active MNK. It is of interest that the absence of MNK activity produces prenatal lethality in the mouse. In humans, deletions of MNK cause classical Menkes disease, a postnatal lethal condition (Chelly et al., 1993); there are no reported prenatal deaths associated with Menkes disease. A possible explanation is that mice have a higher prenatal requirement for copper than humans.

3.2. Mutants which Die before the Third Postnatal Week

The next class of mottled mutants are those which usually die between 12 and 17 days after birth. This developmental period in the mouse approximates to the develomental stage at which Menkes children die, and indeed theses mutants are very close homologues of classical Menkes disease. The brindled mouse (Mo^{br}) is typical of this group: the affected male is hypopigmented due to a deficiency of tyrosinase, has severe neurological problems and dies at about 17 days after birth. The affected male can be treated successfully by a single dose of copper, provided the treatment is administered before the tenth postnatal day (Mann et al., 1979); presumably some critical event, probably related to brain development, requires copper in this period. The mutation in the brindled mouse has been identified as a 6 base pair deletion which would lead to the production of a protein lacking alanine and leucine on the C-terminal side of transmembrane 4 (Grimes et al., 1997). The amounts of Mnk mRNA and protein are normal. The macular mouse (Mo^{ml}) has a very similar phenotype to brindled. A point mutation in the coding sequence of Mnk from the macular mouse has been found which predicts a proline to serine substitution in the eight transmembrane domain (Mori and Nishimura, 1997).

Missense mutations of this type are quite interesting since they identify regions of the protein essential for function. The region affected by the brindled mutation is not one of the recognized functional domains. We have proposed, by analogy with mutations introduced into the calcium ATPases, that the brindled mutation will have deleterious effect on the conformational change associated with the hydrolysis of ATP and the translocation of the cation (Grimes et al., 1997). The macular mutation affects the highly conserved eight transmembrane domain and thus might be expected to have a severe effect on activity of the molecule. The effect of these mutations on the copper transport activity of MNK has not been determined, however, it is likely the mutant MNK molecules have some residual activity, since a totally inactive protein would results in death of the mutants prior to birth as occurs with the prenatal lethal group. As discussed below, only one or two percent residual activity may be required to permit survival to the third postnatal week.

3.3. Viable Brindled, a Possible Model of Mild Menkes Disease

The characteristics of the viable brindled mouse suggest that it may be a good model for mild Menkes disease. The mouse has reduced viability and some connective tissue defects. The mutation found in *Mnk* in the viable brindled mouse predicts a lysine to threonine change next to the invariant aspartic acid which is phosphorylated during the reaction cycle of copper transport. Thus it is considered that phosphorylation, and hence copper pumping, will be reduced, but this has not been determined experimentally (Cecchi et al., 1997; Reed and Boyd, 1997). The mutations reported to cause mild Menkes disease in humans have all been splice site mutations (Kaler et al., 1994) which cause a reduction in the amount of normal mRNA. No missense mutations have been described to date, however, we have found a mutation in the 7th transmembrane domain of MNK in a case of mild Menkes disease (L. Ambrosini, unpublished data), which presumably permits some copper transport activity.

3.4. The Blotchy Mouse, a Model for Occipital Horn Syndrome

Another mottled allele, blotchy (Mo^{blo}), has a less severe phenotype characterized by connective tissue abnormalities resembling those found in patients with OHS. The blotchy mouse has a mild splice site mutation which results in the production of two abnormal mRNAs and some normal mRNA. Thus, it is expected that a reduced amount of normal protein will be produced (Das et al., 1995). Similar splice site mutations have been found in patients with OHS (Das et al., 1995). As for mild Menkes disease, the expectation is that a small amount of normal protein will be produced. To explain the pronounced connective tissue defects in this class of proteins, Levinson *et al* have proposed that lysyl oxidase is the most sensitive cupro-enzyme to the reduction in amount or activity of MNK (Das et al., 1995). This sensitivity may relate to the role of MNK in delivering copper to lysyl oxidase in the *trans*Golgi network of the cell (see section 5.0).

3.5. Copper Treatment of the Mottled Mice

As mentioned previously, the brindled and macular mutants will survive if copper salts are administered before the 10th postnatal day. If treatment is delayed beyond this time, the mutant mouse still dies (Mann et al., 1979) The reason for the critical time period for copper treatment is not understood, but there may be a vital period of brain development from 7 to 10 days of age which requires copper. Alternatively, the closure of the blood brain barrier which also occurs around 10 days, may prevent access of copper to the brain, since MNK is needed to pump copper into the brain once the barrier has been established (Fujii et al., 1990; Kodama, 1993). Interestingly, the mutants which respond to therapy, brindled and macular, have mutations of *Mnk* which predict normal amounts of mutant protein and the brindled mutant has been shown to produce normal levels of MNK protein by Western blot analysis. This finding is consistent with Kaler's contention (Kaler, 1996), that Menkes patients who respond to copper therapy are those who retain a small amount of MNK activity sufficient to permit transport some copper to the brain, provided the gut transport block is by-passed by injection The Jax brindled mouse ($Mo^{br J}$) does not respond to copper therapy (Kelley and Palmiter, 1996) but the mutation in this mouse has not been reported. The issue of copper treatment of Menkes disease is controversial and is dealt with in the summary of the Treatment Workshop elsewhere in this book.

4.0. THE TX MOUSE: Cu DEFICIENCY AND TOXICITY

The tx, or toxic milk, mouse has a mutation in the 8th transmembrane domain of the Wilson disease gene homologue, and thus is a proven model of Wilson disease (Theophilos et al., 1996). This mouse accumulates large amounts of copper in the liver, leading to substantial pathological changes (Biempica et al., 1988; Howell and Mercer, 1994; Rauch, 1983). An apparently unique feature of the tx mouse, which led to the name "toxic milk", is that the mutant mother produces copper-deficient milk, and placental transport of copper to the fetus is also reduced. Pups from mutant dams are born copper deficient, and will die in manner similar to the brindled mouse, unless fostered onto a normal mother. In this respect, the tx mouse pups are a model of severe nutritional copper deficiency.

4.1. Double Mutant tx/br Have Increased Severity of Cu Deficiency

To further investigate the physiological roles of WND and MNK, we have produced double mutant mice, which carry both the tx and the brindled mutations. At birth, the double mutant mice are of similar weight and appearance to the single mutants: e.g. mean weight of single mutant male (br/y, tx/+) at birth was 1.55±0.05g (mean±SD) compared with the double mutant male (br/y, tx/tx) 1.72±0.03g; but by day 10 the double mutant average weight was only 3.94±0.27g compared with 5.41±0.33g for the single mutant. The double mutants males had a wasted appearance with dry flaky skin and development of fur was delayed (Fig 1). The double mutant females (br/+, tx/tx) died at about 18 days after birth and double mutant males (br/y, tx/tx) died at about 13 days after birth. The severe effect of the two mutations in combination suggests that the tx mutation affects copper transport in some manner before the second postnatal week, when liver copper begins to accumulate in the tx mutant. Interestingly, copper treatment which saves the brindled male as discussed in Section 3.5, also saves the double mutant female, but not the double mutant male. The success of copper therapy of double mutant female suggests that the br and tx mutations in combination produce a more severe copper deficiency.

Two types of matings were used to obtain the double mutants: cross IIIA in which the female is br/+, tx/+ ; and cross IIIB in which the female is br/+, tx/tx (see Figure 2). In cross IIIA the pups are born copper replete since the female is a tx/+ heterozygote; in cross IIIB, since the female is tx/tx, the pups are copper deficient in utero. Thus, the expectation for cross IIIB was that an intrauterine copper deficiency due to maternal genotype would be superimposed upon the effects of the br and tx mutations, producing a more severe phenotype.

Little difference between the single and double mutant *in utero* survival was expected in cross IIIA, since no prenatal role for WND had been reported, and the copper

Animal Models of Menkes Disease

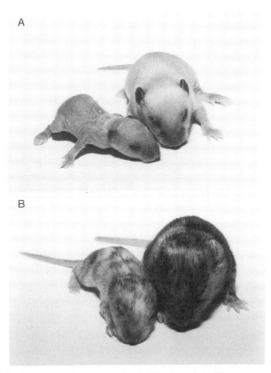

Figure 1. Appearance of single and double mutant mice A. Single and double mutant males at 9 days of age, the br/y, tx/tx double mutant is on the left and the br/y, tx/+ single mutant is on the right. B. Single and double mutant females at 14 days of age, the br/+, tx/tx double mutant is on the left and the br/+, tx/+ single mutant is on the right.

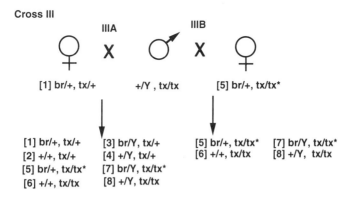

Figure 2. Crosses used to produce double mutant mice. Two types of crosses were used, cross IIIa where the mother was a single mutant and cross IIIB where the mother was a copper treated double mutant. The genotypes of the parents are indicated, as are the expected genotypes of the pups. The genotypes are numbered in square brackets and the asterisks indicate the double mutants.

Table 3. Number and survival of single and double mutant pups from cross IIIA and IIIB

Genotype	Cross	Pups obtained	Pups expected	Age at death
br/Y, tx/+	IIIA	64	73	17.7±2.3 (18)
br/Y, tx/tx	IIIA	23**	73	13.1±2.3 (21)
br/+, tx/+	IIIA	117	73	>200
br/+, tx/tx	IIIA	23**	73	17.6±2.9 (11)
br/Y, tx/tx	IIIB	0**	12.75	<0
br/+, tx/tx	IIIB	1**	12.75	<0

** P<0.001 by χ2 test. Only the single and double mutant genotypes are listed. Cross IIIA and cross IIIB have the same +/Y, tx/tx sire, cross IIIA dam is br/+,tx/+ and cross IIIB dam is br/+, tx/tx. The age of death is shown ±SD with the number of animals listed in brackets.

status of fetuses born to tx heterozygotes is normal (Howell and Mercer, 1994). Contrary to this expectation, the number of both double mutant males and females born was significantly reduced (23 compared to expected 73, p<0.001 by χ2 test, see Table 3) suggesting that the tx gene does in fact have role in fetal copper transport, but the *in utero* effect of the mutation in this gene is only apparent when coupled with a mutation at the mottled locus. The outcome of cross IIIB was striking. No double mutant males were born, and only one double mutant female compared to the expected number of 13 (Table 3). Thus the survival of the double mutants is compromised in the copper deficient environment provided by the tx dam. Recently Kuo et al. presented the developmental profile of both MNK and WND mRNA in mice and showed evidence that the WND gene is expressed in some unexpected tissues such as intestine, lung heart and thymus (Kuo et al., 1997). The authors propose that WND may have a role in providing copper to various cupro-proteins in addition to ceruloplasmin. These finding are consistent with the double mutant phenotype, suggesting that WND does have an important physiological role in the developing mouse, but perhaps MNK can in part substitute for the lack of WND in the single mutant tx.

5.0. POSSIBLE MOLECULAR BASIS FOR THE PHENOTYPIC DIVERSITY IN THE MOTTLED MICE AND MENKES DISEASE

To understand the reasons for the phenotypic diversity of the genetic copper deficiency disorders requires an analysis of the genotype/phenotype correlations, together with patterns of expression of the MNK gene, especially during development. Recent data on the intracellular location of MNK is also important.

5.1. Intracellular Localization of MNK

Recent data on the intracellular localization of MNK is relevant to understanding the phenotypic diversity of the genetic copper deficiency disorders. Petris et al (Petris et al., 1996) showed that in cultured copper resistant CHO cells, MNK is localized primarily at the *trans*Golgi Network (TGN), unless the cell is exposed to high copper, which causes a relocalization of the protein to the plasma membrane of the cell. The distinct locations of MNK, TGN and plasma membrane, are consistent with the two cellular roles of the protein: supply of copper to secreted cupro-enzymes and efflux of excess copper from the cell. Since the TGN is part of the normal secretory pathway for proteins, it is likely that copper is added to the apoenzyme in this compartment. The localization of MNK to the TGN suggests that this copper ATPase is responsible for pumping copper from the cytoplasm into the lumen of the TGN. Copper efflux from the cell is presumably the mechanism by which copper is transferred across the various epithelial barriers in the body, such as absorption across the intestine, resorption of copper in the proximal tubules, and transfer of copper across the blood brain barrier. To achieve this transfer of copper, MNK is presumably located on the plasma membrane, and the amount of this protein on the plasma membrane versus the TGN is regulated by the intracellular concentration of copper. The location of the MNK in the polarized cells that form the various epithelial barriers to copper has not been reported.

Connective tissue defects are a major clinical feature of the milder forms of Menkes disease and the mottled mice, presumably because lysyl oxidase is the most sensitive of the cupro-enzymes to marginal levels of MNK. As noted above, MNK-dependent delivery copper to lysyl oxidase in the TGN is possibly the reason for this sensitivity. In an animal which is already copper deficient, because of the reduction in intestinal uptake of copper, the need for a second MNK-mediated transport step to deliver copper to lysyl oxidase, results in more severe copper deficiency in the TGN lumen than in the cytoplasm.

Since a second transport step across the blood brain barrier is also needed for delivery of copper to the brain, this organ is most severely affected in Menkes disease. It is unclear whether by-passing the gut blockage by copper injection can successfully overcome the deficiency of MNK. The success in copper therapy will presumably depend on how much copper can be delivered to the brain cupro-enzymes at critical developmental stages.

5.2. Model for Phenotypic Diversity of the Mottled Mutants

With the above considerations in mind we propose a model to explain the phenotypic diversity of the mottled mutants. Figure 3 outlines the model which is based on the postulate that different organs or tissues have certain threshold requirements for copper at various stages of development. If the requirement for copper is not met at the relevant stage, failure of that organ or tissue produces a characteristic phenotype. Fig 3 indicates the tissues that would be affected by reduction in MNK activity below a critical percentage of the wild type. The actual percentage residual MNK activity has not been determined, the values indicated are based on the estimated residual mRNA levels from various publications.

For example, if the residual MNK activity is below 2%, the amount of copper reaching the fetus is so low that the connective tissue cross-linking is greatly reduced and vascular rupture occurs in the fetus. Connective tissue failure is consistent with the types of fetal abnormalities observed in Mo^{9H} fetuses (L. Ambrosini and J. Mercer unpublished results). If the amount of MNK is above this 2% threshold level, sufficient copper reaches the fetus to allow survival during gestation. If the activity of MNK is less than 3%, the next critical event is death in the first few weeks of postnatal life, perhaps due to cytochrome c oxidase deficiency causing severe neurological defects, as seen in the brindled mouse. Above 3% MNK, the animal can survive this postnatal critical point, but will develop further possible neurological and connective tissue complications as seen in the viable brindled mouse. Between 10% and 15% MNK activity results in connective tissue failure later in life due to long term effects of poor connective tissue cross-linking (blotchy mouse). Above about 15% MNK activity the animal may be classed as normal.

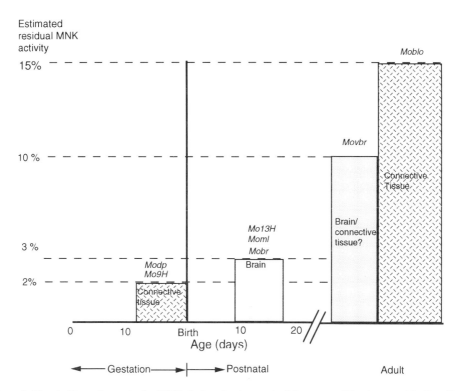

Figure 3. Threshold requirements for MNK during development of the mouse. The tissues critically affected when MNK activity levels fall below the indicated levels at different stages of development, presumably reflecting the copper requirements at that stage. The percentage residual activity is estimated from quoted residual mRNA levels.

The same general explanation can be proposed to explain the different human disease presentations, with the exception of fetal lethality. Since about 15% of Menkes patients have partial or complete deletions of MNK (Chelly et al., 1993), human fetuses, in contrast to mice, can apparently survive development without MNK. Thus, it is unlikely that human prenatal deaths are caused by mutations in *MNK*, but this possibility cannot be ruled out without further investigation.

6.0. CONCLUSIONS

In recent years, knowledge of the genetic copper deficiency disorders has increased markedly. Study of the mottled mice is still providing important information about the clinical course of Menkes disease and variants in humans. Future studies of the mutations responsible for the various mouse mutants, together with assays of the activity of the mutant molecules and the intracellular location of normal and mutant MNK, may provide further understanding of the remarkable phenotypic variation in the diseases caused by mutations in the Menkes gene. This knowledge should aid in the treatment of patients with these diseases.

REFERENCES

Bankier, A. (1995). Menkes disease. J. Med. Genet. *32*, 213–215.

Biempica, L., Rauch, H., Quintana, N., and Sternlieb, I. (1988). Morphological and chemical studies on a murine mutation (toxic milk) resulting in hepatic copper toxicosis. Lab. Invest. *59*, 500–508.

Byers, P. H., Siegel, R. C., Holbrook, K. A., Narayanan, A. S., Bornstein, P., and Hall, J. G. (1980). X-linked cutis laxa: defective cross-link formation in collagen due to decreased lysyl oxidase activity. New Eng. J. Med. *303*, 61–65.

Cecchi, C., Biasotto, M., Tosi, M., and Avner, P. (1997). The *mottled* mouse as a model for human Menkes disease: identification of mutations in the *Atp7a* gene. Hum. Molec. Genet. *6*, 425–433.

Chelly, J., Tumer, Z., Tonnerson, T., Petterson, A., Ishikawa-Brush, Y., Tommerup, N., Horn, N., and Monaco, A. P. (1993). Isolation of a candidate gene for Menkes disease that encodes a potential heavy metal binding protein. Nature Genet. *3*, 14–19.

Cox, D. W. (1995). Genes of the copper pathway. Am. J. Hum. Genet. *56*, 828–834.

Danks, D. M. (1995). Disorders of copper transport. In The metabolic and molecular basis of inherited disease, C. R. Scriver, A. L. Beaudet, W. M. Sly and D. Valle, eds. (New York: McGraw-Hill), pp. 2211–2235.

Danks, D. M. (1988). The mild form of Menkes disease: progress report on the original case. J. Med. Genet. *30*, 859–864.

Danks, D. M., Campbell, P. E., Stevens, B. J., Mayne, V., and Cartwright, E. (1972). Menkes's kinky hair syndrome: an inherited defect in copper absorption with wide-spread effects. Pediatrics *50*, 188–201.

Das, S., Levinson, B., Vulpe, C., Whitney, S., Gitschier, J., and Packman, S. (1995). Similar splicing mutations of the Menkes/mottled copper-tranporting ATPase gene in occipital horn syndrome and the blotchy mouse. Am. J. Hum. Genet. *56*, 570–576.

Doolittle, D. P., Davisson, M. T., Guidi, J. N., and Green, M. C. (1996). *Mo* locus, Chr X. In Genetic variants and strains of the laboratory mouse, M. F. Lyon, S. Rastan and S. D. M. Brown, eds. (Oxford: Oxford University Press), pp. 513–516.

Fujii, T., Ito, M., Tsuda, H., and Mikawa, H. (1990). Biochemical study on the critical period for treatment of the mottled brindled mouse. J. Neurol. *55*, 885–889.

Gerdes, A.-M., Tonnesen, T., Pergament, E., Sander, C., Baerlocher, K. E., Wartha, R., Guttler, F., and Horn, N. (1988). Variability in clinical expression of Menkes syndrome. Eur. J. Pediatr. *148*, 132–135.

Grimes, A., Hearn, C., Lockhart, P., Newgreen, D., and Mercer, J. F. B. (1997). Molecular basis of the brindled mouse mutant (Mo^{br}): a murine model of Menkes disease. Hum. Mol. Genet. *6*, 1032–1042.

Horn, N. (1976). Copper incorporation studies on cultured cells for prenatal diagnosis of Menkes disease. Lancet *1*, 1156–1158.

Howell, J. M., and Mercer, J. F. B. (1994). The pathology and trace element status of the toxic milk mutant mouse. J. Comp. Path. *110*, 37–47.

Kaler, S. G. (1994). Menkes Disease. Advances in Pediatr. *41*, 262–303.

Kaler, S. G. (1996). Menkes disease mutations and response to early copper histidine treatment. Nature Genet. *13*, 21–22.

Kaler, S. G., Gallo, L. K., Proud, V. K., Percy, A. K., Mark, Y., Segal, N. A., Goldstein, D. S., Holmes, C. S., and Gahl, W. A. (1994). Occipital horn syndrome and a mild Menkes phenotype associated with splice site mutations at the MNK locus. Nature Genet. *8*, 195–202.

Kelley, E. J., and Palmiter, R. D. (1996). A murine model of Menkes disease reveals a physiological function of metallothionein. Nature Genet. *13*, 219–222.

Kodama, H. (1993). Recent developments in Menkes disease. J. Inher. Metab. Disease *16*, 791–799.

Kuinaviemi, H., Peltonen, L., Palotie, A., Kaitila, I., and Kivirikko, K. I. (1982). Abnormal copper metabolism and deficient lysyl oxidase activity in a heritable connective tissue disorder. J. Clin. Invest. *69*, 730–733.

Kuo, Y.-M., Gitschier, J., and Packman, S. (1997). Developmental expression of the mouse mottled and toxic milk genes suggests distinct functions for the Menkes and Wilson disease copper transporters. Hum. Mol. Genet. *6*, 1043–1049.

Levinson, B., Packman, S., and Gitschier, J. (1997). Deletion of the promoter region in the Atp7a gene of the *mottled dappled* mouse. Nature Genet. *16*, 224–225.

Levinson, B., Vulpe, C., Elder, B., Martin, C., Verley, F., Packman, S., and Gitschier, J. (1994). The mottled gene is the mouse homologue of the Menkes disease gene. Nature Genet. *6*, 369–373.

Mann, J. R., Camakaris, J., Danks, D. M., and Walliczek, F. G. (1979). Copper metabolism in mottled mouse mutants 2. Copper therapy of brindled (Mobr) mice. Biochem. J. *180*, 605–612.

Menkes, J. H., Alter, M., Stegleder, G. K., Weakley, D. R., and Sung, J. H. (1962). A sex-linked recessive disorder with retardation of growth, peculiar hair, and focal cerebral and cerebellar degeneration. Pediatrics 29, 764–779.

Mercer, J. F. B., Grimes, A., Ambrosini, L., Lockhart, P., Paynter, J. A., Dierick, H., and Glover, T. W. (1994). Mutations in the murine homologue of the Menkes disease gene in dappled and blotchy mice. Nature Genet. 6, 374–378.

Mercer, J. F. B., Livingston, J., Hall, B. K., Paynter, J. A., Begy, C., Chandrasekharappa, S., Lockhart, P., Grimes, A., Bhave, M., Siemenack, D., and Glover, T. W. (1993). Isolation of a partial candidate gene for Menkes disease by positional cloning. Nature Genet. 3, 20–25.

Mori, M., and Nishimura, M. (1997). A serine to proline mutation in the copper-transporting ATPase gene of the macular mouse. Mamm. Genome 8, 407–410.

Peltonen, L., Kuivaniemi, H., Palotie, A., Horn, N., Kaitila, I., and Kivirikko, K. I. (1983). Alterations in copper and collagen metabolism in the Menkes syndrome and a new type of the Ehrlers-Danlos syndrome. Biochemistry 22, 6156–6163.

Petris, M. J., Mercer, J. F. B., Culvenor, J. G., Lockhart, P., Gleeson, P. A., and Camakaris, J. (1996). Ligand-regulated transport of the Menkes copper P-type ATPase efflux pump from the Golgi apparatus to the plasma membrane: a novel mechanism of regulated trafficking. EMBO J. 15, 6084–6095.

Phillips, R. J. S. (1961). 'Dappled', a new allele at the mottled locus in the house mouse. Genet. Res. 2, 209–295.

Procopis, P., Camakaris, J., and Danks, D. M. (1981). A mild form of Menkes' syndrome. J. Pediatr. 98, 97–100.

Rauch, H. (1983). Toxic milk, a new mutation affecting copper metabolism in the mouse. J. Hered. 74, 141–144.

Reed, V., and Boyd, Y. (1997). Mutation analysis provides additional proof that mottled is the mouse homologue of Menkes' disease. Human Mol. Genet. 6, 417–423.

Solioz, M., and Vulpe, C. (1996). CPX-type ATPases: a class of P-type ATPases that pump heavy metals. Trends in Biochem. Sci. 21, 237–241.

Theophilos, M., Cox, D. W., and Mercer, J. F. B. (1996). The toxic milk mouse is a murine model of Wilson disease. Hum. Mol. Genet. 5, 1619–1624.

Vulpe, C., Levinson, B., Whitney, S., Packman, S., and Gitschier, J. (1993). Isolation of a candidate gene for Menkes disease and evidence that it encodes a copper-transporting ATPase. Nature Genet. 3, 7–13.

Westman, J. A., Richardson, D. C., Rennert, O. M., and Morrow, G. (1988). Atypical Menkes steely hair disease. J. Med. Genet. 30, 853–858.

9

DEVELOPMENTAL EXPRESSION OF THE MOUSE MOTTLED AND TOXIC MILK GENES

Yien-Ming Kuo,[1,2] Jane Gitschier,[1,2,3] and Seymour Packman[2]

[1]Department of Medicine
[2]Division of Medical Genetics, Department of Pediatrics
[3]Howard Hughes Medical Institute
University of California–San Francisco
Third and Parnassus Avenues, San Francisco, California 94143-0748

1. INTRODUCTION

In all living organisms, intracellular copper content is regulated and maintained by specialized cellular transport systems mediating copper uptake, intracellular compartmentalization and utilization, and exit . In the face of copper deficiency, impaired catalysis by one or more of approximately thirty cuproenzymes leads to failure of growth, development, or survival of organisms or cells. On the other hand, copper excess may lead to pathology mediated by oxygen free radicals, with resultant damage to cellular components (Harris, 1991; Bull and Cox, 1994; Vulpe and Packman, 1995; Danks, 1995).

Two human genetic disorders, X-linked Menkes disease and autosomal recessive Wilson disease, highlight the importance of intact cellular copper transport mechanisms (Menkes, 1993; Danks, 1995). In Menkes disease, copper significantly accumulates in some tissues (intestinal mucosa and renal tubular epithelium), leading to failure of copper delivery to other tissues, and to systemic copper insufficiency. Clinical manifestations include progressive neurologic degeneration, seizures, growth failure, hypopigmentation, pili torti, arterial aneurysms and skeletal defects. Onset is often in infancy, with death by early childhood. Deficiencies in tissue cuproenzyme activities—e.g., lysyl oxidase (cross-linking of elastin and collagen), cytochrome c oxidase (mitochondrial electron transport), superoxide dismutase (free radical detoxification), tyrosinase (pigmentation), and dopamine ß-hydroxylase (catecholamine production)—are possibly responsible for many of the clinical features (Menkes, 1993; Danks, 1995).

Wilson disease is a later onset disorder, in which children or adults present, variably, with progressive and distinct neurologic findings, chronic liver disease with cirrhosis, renal tubular failure, and pigmented corneal rings (Wilson, 1912; Bull and Cox, 1994;

Copper Transport and Its Disorders, edited by Leone and Mercer.
Kluwer Academic / Plenum Publishers, New York, 1999.

Menkes, 1993; Danks, 1995). The copper content of the liver, brain, kidney and cornea is increased, and the organ failures of Wilson disease reflect the toxicity of copper excess (Sternlieb, 1990).

Menkes disease (Vulpe et al, 1993; Chelly et al, 1993; Mercer et al, 1993) and Wilson disease (Bull et al, 1993; Yamaguchi et al, 1993; and Tanzi et al, 1993) are caused by mutations in related but distinct genes encoding copper-transporting P-type ATPases. The protein encoded by the Wilson disease gene (*WND, ATP7B*) has 56% overall identity to that of the Menkes disease gene (*MNK, ATP7A*). It has been proposed that both gene products mediate cellular export of copper through vesicles derived from the *trans*-Golgi network (Dierick et al, 1997; Petris et al, 1996; Yamaguchi et al, 1996), and mediate the transfer of copper to specific cuproenzymes within that cell compartment (Yuan et al, 1995; Yamaguchi et al, 1996). However, the patterns of *MNK* and *WND* gene expression in the adult are markedly different: the Menkes gene is ubiquitousely expressed in adult tissues, with little or no expression in liver (Vulpe et al, 1993); in contrast, *WND* is expressed in a few cell types, including hepatocytes and, possibly, renal and certain neuronal cells (Bull et al, 1993; Yamaguchi et al, 1993; and Tanzi et al, 1993).

The patterns of expression can, to some degree, be correlated with the distinct clinical manifestations of these disorders. However, such correlations represent both an incomplete understanding of the disorders themselves, and of the cellular and physiologic role(s) of the transport ATPases. Menkes disease has a prenatal onset (Danks, 1995; Ferenci et al, 1996), and the central nervous system manifestations include not only difuse and unspecific neurodegenerative changes, but also selective defects which may be of developmental origin (Menkes et al, 1993). In this context, the effects of mutations of *MNK* on organogenesis, and the prenatal developmental origin of somatic and CNS manifestations, are not known. In the case of Wilson disease, specific hepatic processes—holoceruloplasmin synthesis and biliary copper excretion—are disturbed, and are likely to be mediated by the *WND* gene product. However, the functions of the WND ATPase in extrahepatic cell types have not been elucidated, and the full spectrum and timing of WND gene expression are not known (Sternlieb, 1990; Petrukhin et al, 1994).

With such considerations in mind, and in order to begin to develop a more complete picture of the pathophysiology of human genetic disorders of copper transport, we turned to the mouse models for these disorders, and compared the developmental expression of the mottled and toxic milk genes, the mouse homologues, respectively, of the *MNK* (Levinson et al, 1994; Mercer et all,1994; Cecchi et al, 1997; Reed and Boyd, 1997), and *WND* (Theophilus et al, 1996) genes. We showed that mottled is expressed ubiquitously throughout the embryo during gestation, and that toxic milk is expressed in a limited set of tissues. Our findings (Kuo et al, 1997) are broadly consistent with the notion that the mottled gene product functions in the individual homeostatic maintenance of cellular copper levels in all cell types. In contrast, the toxic milk gene product is more likely involved in a physiologic or biochemical role specific to, and possibly quite different for, each of the several distinct cell types in which it is expressed, and in a way which was previously unsuspected.

2. MOTTLED GENE EXPRESSION

RNA *in situ* hybridizations were performed using specific sense and antisense 33-P–labeled riboprobes (Kuo et al, 1997). In sections of embryos from randomly bred Swiss Webster mice, we observed that the mottled gene is ubiquitously and diffusely expressed in all embryonic tissues, including the liver, from stage E9.5 through E18.5 (Kuo et al, 1997). Expression in the liver at E17.5 had been previously reported (Levinson et al, 1994; Mercer et al, 1994; Paynter et al, 1994). We noted that expression of mottled in embryonic liver was in contrast to the absent or low level of expression in adult liver (Levinson et al, 1994; Mercer et al, 1994).

3. TOXIC MILK EXPRESSION

The pattern of expression of the toxic milk gene was quite different, and limited to a few organs, including the liver, heart ventricle, intestine, thymus, and—by stage E18.5—the lining of the respiratory tract, including nasopharynx, trachea, and bronchi (Kuo et al, 1997). In the lung, toxic milk was expressed most highly in the bronchial epithelium, in contrast to the more diffuse expression of the mottled gene. In both the liver and intestine, the toxic milk and mottled genes were expressed in the same cell types: hepatic parenchyma (and not blood cells) in the liver, and the villous epithelium in the intestine.

4. COMMENTS AND HYPOTHESES ON THE FUNCTIONS OF THE MENKES AND WILSON DISEASE COPPER TRANSPORTERS

The striking mottled expression throughout the embryo and throughout gestation is consistent with a housekeeping role for the mottled gene product, in the maintenance of cellular copper homeostasis and the extracellular microenvironment of multiple cell types during develpment. The difference in the expression of mottled between embryonic and adult liver suggests that the fetal liver cell—with minimal capacity for the release of copper through biliary secretion or ceruloplasmin synthesis—requires the same housekeeping copper export system as any other cell type, and that such copper transport function in fetal liver is served by the mottled/Menkes protein.

The toxic milk gene does not appear to be expressed in the brain or the choroid plexus, either in the embryo or in the adult mouse. In contrast, we observed expression of the mottled gene throughout the brain, with intense expression in the choroid plexuses. This pattern of mottled expression was also seen in adult brain (Kuo, unpublished observations; Murata et al, 1997). We therefore speculate that copper transport regulated by the mottled gene product in the choroid plexuses may be quite important in mediating copper transport into or out of the brain.. In such a construction, copper homeostasis in the central nervous system would be governed by redundant transport systems, mediated by glia and/or capillary epithelium, choroid plexus, and possibly by neuronal cell—autonomous transporters (Kodama, 1993; Kodama et al, 1991; Kodama et al, 1993; Qian et al, 1995; Kaler et al, 1995; Murata et al, 1997).

We were intrigued to find that toxic milk was expressed prenatally in cell types, such as thymus and respiratory epithelium, which appear to be unrelated to clinical manifestations of Wilson disease. In this connection, we note that the ceruloplasmin gene is known to be expressed in a number of extrahepatic tissues (Fleming and Gitlin, 1990; Saenko et al, 1994). For example, fetal rat lung is a major site of extrahepatic ceruloplasmin mRNA during development (Fleming and Gitlin, 1990). Ceruloplasmin is also expressed in lymphocytes and other immune cells (Jaeger et al, 1991), and elevations in ceruloplasmin mRNA are observed in response to inflammation (Klomp and Gitlin, 1996).

Therefore, we suggest that the toxic milk gene product in the fetal lung and other respiratory tissues serves to transport copper into an intracellular compartment for incorporation into ceruloplasmin; and that the toxic milk transporter plays a similar role in the formation of holoceruloplasmin in cells of the thymus. In tissues such as fetal heart and intestine, which do not appear to express ceruloplasmin (Fleming and Gitlin, 1990), we can only speculate that the toxic milk gene might specifically mediate the synthesis of a holocuproenzyme other than ceruloplasmin.

In sum, the comparative prenatal expression patterns of mottled and toxic milk lead to the inference that copper transport mediated by the mottled gene product is required for cellular homeostasis and integrity of copper translocation in a multiplicity of tissues. In contrast, the toxic milk gene appears to function in a distinct subset of cell types, leading to the speculation that the toxic milk gene functions in the biosynthesis of distinct cuproenzymes, such as ceruloplasmin. In a few tissues, notably the fetal liver parenchyma and intestinal epithelium, both genes are expressed in cell types which exhibit polarity in transport functions. It is therefore possible that the two different copper transporters might be performing different functions and be situated in different cellular structures, membranes, or membrane regions in such cells.

In a broader view, it is clear that the capacity to exploit mutations of such copper transport genes in inbred mouse strains will permit the testing of the validity of the physiologic mechanisms that we herein propose. We hope that the data of the present work will also contribute to the design of future tests of rational prenatal therapies that will successfully prevent the devastating multisystem involvement of Menkes disease, and to a more complete understanding of the phenotype and treatment of Wilson disease.

ACKNOWLEDGMENTS

This work was supported by grants from the March of Dimes Birth Defects Foundation and the National Institute of Health. JG is an Associate Investigator of the Howard Hughes Medical Institute.

REFERENCES

Bull PC, Cox DW, 1994. Wilson disease and Menkes disease: new handles on heavy-metal transport. Trends Genet 10:246–252.

Bull PC, Thomas GR, Rommens JM, Forbes JR, Cox DW, 1993. The Wilson disease gene is a putative copper transporting P-type ATPase similar to the Menkes gene. Nature Genet 5:327–337.

Cecchi C, Biasotto M, Tosi M and Avner P, 1997. The mottled mous as a mode for human Menkes disease: identificatin of mutations in the *Atp7a* gene. Hum Molec Genet 6: 425–433.

Chelly J, Tumer Z, Tonnesen T, Petterson A, Ishikawa-Brush Y, Tommerup N, Horn N, Monaco AP, 1993. Isolation of a candidate gene for Menkes disease that encodes a potential heavy metal binding protein. Nature Genet 3:14–19.

Danks D, 1995. Disorders copper transport. In the Metabolic and Molecular Basis of Inherited Diseaxe, eds. Scriver C, Beaudet A, Sly W and Valle D. pp 2211–2235. New York: McGraw Hill.

Dierick HA, Adam AN, Escara-Wilke JF, Glover TW, 1997. Immunocytochemical localization of the Menkes copper transport protein (ATP7A) to the trans-Golgi network. Hum Molec Genet 6:409–416.

Ferenci P, Gilliam TC, Gitlin JD, Packman S, Schilsky ML, Sokol RJ, Sternlieb I, 1996. An international symposium on Wilson's and Menkes' disease. Hepatology 4:952–958.

Fleming R, Gitlin J, 1990. Primary structure of rat ceruloplasmin and analysis of tissue-specific gene expression during development. J Biol Chem 265:7701–7707.

Harris ED, 1991. Copper transport: an overview. Proc Soc Expr Biol Med 196:130–140.

Jaeger, JL, Shimizu N, Gitlin JD, 1991. Tissue-specific ceruloplasmin gene expression in the mammary gland. Biochem J 280:671–677.
Kaler S, Buist N, Holmes C, Goldstein D, Miller R, Gahl W, 1995. Early copper therapy in classic Menkes disease patients with a novel splicing mutation. Ann Neurol 38:921–928.
Klomp LWJ, Gitlin JD, 1996. Expression of the ceruloplasmin gene in the human retina and brain-implications for a pathogenic model in aceruloplasminemia. Hum Mol Genet 12:1989–1996.
Kodama H 1993. Recent development in Menkes disease. J Inherited Metab Dis 16:791–799.
Kodama H, Abe T, Takama M, Takahashi I, Kodama M, Nishimura M 1993. Histochemical localization of copper in the intestine and kidney of macular mice: light and electron microscopic study. J Histochem Cytochem 41:1529–1535.
Kodama H, Meguro Y, Abe T, Rayner MH, Suzuki KT, Kobayashi S, Nishimura M 1991. Genetic expression of Menkes disease in cultured astrocytes of the mascular mouse. J Inherited Metab Dis 14:896–901.
Kuo Y-M, Gitschier J, Packman S, 1997. Developmental expression of the mouse mottled and toxic milk genes suggests distinct functions for the Menkes and Wilson disease copper transporters. Hum Molec Genet 6:1043–1049.
Levinson B, Vulpe C, Elder B, Martin C, Verley F, Packman S, Gitschier J, 1994. The mottled gene is the mouse homologue of the Menkes disease gene. Nature Genet 6:369–373.
Menkes J, 1993. Disorders of copper metabolism. In Rosenberg R, Pruisner S, DiMauro S, Barchi R, Kunkel L (eds). The Molecular and Genetic Basis of Neurological Disease. Butterworth-Heinemann, Boston, pp 325–341.
Mercer JF, Grimes A, Ambrosini L, Lockhart P, Paynter J, Dierick H, Glover T, 1994. Mutations in the murine homologue of the Menkes gene in dappled and blotchy mice. Nature Genet 6:374–378.
Mercer JF, Livingston J, Hall B, Paynter JA, Begy C, Chandrasekharappa S, Lockhart P, Grimes A, Bhave M, Siemieniak D, Glover TW, 1993. Isolation of apartial candidate gene for Menkes disease by positional clonign. Nature Genet 3:20–25.
Murata Y, Kodama H, Abe T, Ishida N, Nishimura M, Levinson B, Gitschier J, Packman S, 1997. Mutation analysis and expression of the mottled gene in the macular mouse model of Menkes disase. Pediatr Res 42:436–442.
Paynter JA, Grimes A, Lockhart P, Mercer JF, 1994. Expression of the Menkes gene homologue in mouse tissues lack of effect of copper on the mRNA levels. Febs Letts 351:186–190.
Petris MJ, Mercer JRB, Culvenor JG, Lockhart P, Gleeson PA, Camakaris J, 1996. Ligand-regulated transport of the Menkes' copper P-type ATPase efflux pump from the Golgi apparatus to the plasma membrane: a novel mechanism of regulated tranffficking. EMBO J 15:6084–6095.
Petrukhin K, Lusenko S, Chernov I, Ross B, Kaplan J, and Gilliam T, 1994. Characterizationof the Wilson disease gene encoding a P-type coper transporting ATPase: genomic organization, alternative splicing, and structure/function predictions. Hum Molec Genet 3:1647–1656.
Qian Y, Tiffany-Castiglioni E, Harris ED 1995. Copper transport and kinetics in cultured C6 rat glioma cells. Am J Physiol 269:C892-C898.
Reed V and Boyd Y, 1997. Mutation analysis provides additional proof that mottled is the mouse homologue of Menkes' disease. Hum Molec Genet 6:417–423, 1997.
Saenko EL, Yaropolov AI, Harris ED, 1994. The biological functions of ceruloplasmin expressed through copper-binding sites and a cellular receptor. J Trace Elem Exp Med 7:69–88.
Sternlieb I, 1990. Perspectives on Wilson's disease. Hepatology 12:1234–1239.
Tanzi RE, Petrukhin K, Chernov I, Pellequer JL, Wasco W, Ross B, Romano DM, Parano E, Pavone L, Brzustowicz LM, Devoto M, Peppercorm J, Bush AI, Sternlieb I, Pirastu M, Gusella JF, Evgrafov O, Penchaszadeh GK, Honig B, Edelman IS, Soares MB, Scheinberg IH, Gilliam TC, 1993. The Wilson disease gene is a copper transporting ATPase with homology to the Menkes diseae gene. Nature Genet 5:344–350.
Theophilos MB, Cox DW, Mercer JF, 1996. The toxic milk mouse is a murine model of Wilson disease. Hum Mol Genet 5:1619–1624.
Vulpe C, Levinson B, Whitney S, Packman S, Gitschier J, 1993. Isolation of a candidate gene for Menkes diseaes and evidence that it encodes copper transporting ATPase. Nature Genet 3:7–13.
Vulpe C, Packman S, 1995. Cellular copper transport. Ann Rev Nut 15:293–322.
Wilson SAK, 1912. Progressive lenticular degeneration: a familial nervous disease associated with cirrhosis of the liver. Brain 34:295–509.
Yamaguchi Y, Heiny ME, Gitlin JD, 1993. Isolation and characterization of a human liver cDNA as a candidate gene for Wilson disease. Biochem Biophys Res Commun 197:271–277.
Yamaguchi Y, Heiny ME, Suzuki M, Gitlin JD, 1996. Biochemical characterization and intracellular localization of the Menkes disease protein. Proc Natl Acad Sci USA 93:14030–14035.

Yamaguchi Y, Heiny ME, Suzuki M, Gitlin JD, 1996. Biochemical characterization and intracellular localization of the Menkes disease protein. Proc Natl Acad Sci USA 93:14030–14035.

Yuan DS, Stearman R, Dancis A, Dunn T, Beller T, Klausner RD, 1995. The Menkes/Wilson disease gene homologue in yeast provides copper to a ceruloplasmin-like oxidae required for iron uptake. Proc of the Natl Acad Sci USA 92:2632–2636.

10

THE TREATMENT OF WILSON'S DISEASE

George J. Brewer

University of Michigan Medical School
1301 Catherine Street
4708 Med. Sci. Bldg. #2
Ann Arbor, Michigan 48109-0618

1. INTRODUCTION

Progress has been made over the last decade or so in the treatment and management of Wilson's disease. This has come about in part through the increasing use of new drugs. Thus, zinc is now accepted as a standard therapy, and tetrathiomolybdate has been introduced, at least in an experimental setting, for the initial treatment of neurologic Wilson's disease. But improved management has also come about through the increasing recognition of the risks of using penicillamine in the patient presenting with neurologic symptoms, and in the increasing awareness of the problems created by poor compliance.

In this paper I will first discuss the four anticopper drugs, then move on to discuss the use of these drugs in various clinical stages of Wilson's disease, and end with a section on follow-up, compliance, and prognosis.

2. CLINICALLY USEFUL ANTICOPPER DRUGS

2.1. Zinc

I have listed the drugs in this section in the rank order of general usefulness in Wilson's Disease, according to my opinion. Zinc, which was approved by the U.S. Food and Drug Administration (FDA) in January of 1997, is the most useful drug in Wilson's disease at the present time. That is because it is effective in all of the various clinical stages of Wilson's disease and has an extremely low toxicity. Schouwink (1961) was the first to use zinc in Wilson's disease, reporting its use in two patients in his thesis. Our work in developing zinc as a practical therapy which led to FDA approval is summarized, and original references cited, in two reviews (Brewer and Yuzbasiyan-Gurkan, 1992; Brewer et al,

1998). The work of Hoogenraad et al, in developing zinc therapy is summarized in his recent monograph (Hoogenraad, 1996).

Zinc acts by inducing metallothionine (MT) in the intestinal cell and blocking the absorption of copper (Yuzbasiyan-Gurkan et al 1992). The intestinal MT, once induced by zinc, has a very high affinity for copper and complexes the copper not only from food, but endogenous secretions, and prevents the transfer of copper into the blood. The complexed copper is sloughed into the stool as the intestinal cells turnover, with about a six day survival time. Thus, the mechanism by which zinc increases the excretion of copper is through the stool, as has been proven in numerous human copper balance studies (Brewer et al 1983; Hill et al 1987; Brewer and Yuzbasiyan-Gurkan, 1992; Brewer et al 1998). Because zinc prevents the reabsorption of relatively large amounts of endogenously secreted copper, it provides for a slow steady reduction in the body burden of copper.

The dose of zinc required to produce a consistent negative copper balance in adult patients is 75mg, divided into at least two daily doses (Brewer et al 1990; Brewer et at 1993a). Thus, 25x3 or 37.5x2 are both effective. (Drug doses will be expressed in shorthand, eg. 25mg three times per day is 25x3). However, these doses are minimally effective and if the patient's compliance is at all incomplete, they will prove to be inadequate. Therefore we recommend as the standard dose, 50x3. It is also clear that the zinc needs to be separated from food and beverages other than water by about one hour because various substances mitigate the ability of zinc to get into the intestinal cell and induce MT.

Zinc has almost no side effects. Wilson's disease patients in our study have received therapeutic doses of zinc for up to fifteen years without difficulty. Occasionally, a patient will exhibit gastric intolerance of zinc. When this occurs it is usually early in therapy and is usually the first morning dose. The simple maneuver of having the patient take the first dose midway between breakfast and lunch usually works quite well. In stubborn cases, we have reduced the first morning dose to 25mg, with the subsequent doses being 50mg. It is also permissible, particularly for short periods, when the patient is suffering from gastric distress from zinc, to take some of the zinc doses with a piece of meat or cheese. Protein seems to be the least damaging food in terms of reducing zinc efficacy. In our work we have used the acetate salt of zinc, which in our experience is much better tolerated than the more acidic zinc sulfate which has been used in the Netherlands (Hoogenraad, 1996). Zinc works well with the other anticopper agents (Brewer et al 1993b), although it is important to keep the doses separated from one another by at least one hour.

Zinc therapy can be monitored very nicely by following the 24 hour urine copper and zinc (Brewer et al 1987a; Brewer and Yuzbasiyan-Gurkan, 1992; Brewer et al 1998). In the absence of a chelating agent, the urinary excretion of copper is a measure of the body loading with copper. As the negative copper balance caused by zinc unloads copper from the body, urine copper excretion goes down. Conversely, if a patient's compliance deteriorates, urine copper will begin going up. Thus, it is not unusual in untreated Wilson's disease to have patients excreting 200 or 300 µg of copper /24 hour (normal is 50 µg or below) in the beginning and gradually decrease. By experience we have learned that values under 125 µg/24 hours indicates good control. Another useful way to monitor the compliance with zinc is to follow the 24 hour urine zinc as well as the copper. In zinc treated patients, 24 hour urine zinc should be at least 2mg. Values below this indicates that the patient is not taking their zinc or is not taking it properly, and is an early warning sign that the urinary copper will likely increase later. All of these changes in urine zinc and copper precede by a considerable margin any risk of symptomatic worsening. Thus, they provide a monitoring tool to alert the medical team that the patient needs counseling about compliance. We recommend monitoring the 24 hour urine copper and zinc every six months during the early periods of treatment (2–3 years) and when patients prove to be reliable, at least annually thereafter. We use a mail in system for this purpose.

Zinc therapy can also be monitored by following the non-ceruloplasmin plasma copper, the potentially toxic copper of the body. This is done by measuring plasma copper and ceruloplasmin on the same plasma sample, and subtracting 3μg of copper for every mg of ceruloplasmin from the plasma copper (Scheinberg and Sternlieb, 1984). (Example: If plasma copper is 40μg/dl, and ceruloplasmin is 10μg/dl, 10x3=30. 40–30=10μg/dl of non-ceruloplasmin plasma copper). Our experience is that this value should be below 25 in well-treated patients. This is a less convenient method than urine copper and zinc because it requires a blood sample, while aliquots from urine collections can be mailed in from home.

2.2. Trientine

The second most useful anticopper drug, in our opinion, is trientine, originally introduced as an alternative for penicillamine intolerant patients (Walshe, 1982). Trientine acts by chelation of copper and increasing its urinary excretion. The standard dose of trientine is 250mgx4 or 500mgx2. Trientine should be separated from food, taking it at least one hour before or at least two hours after meals.

One of the risks from trientine is bone marrow depression or other idiosyncriatic reaction on initial therapy. Initially, patients should be monitored at least weekly by blood counts, a biochemistry panel and a urinalysis. Proteinuria is a fairly common side effect of trientine and should be looked for. After the first month of therapy, if no problems have been seen, these studies can be reduced to every two weeks and then to increasingly less frequent intervals as the patient shows good tolerance of the drug.

One of the special problems of initial therapy of neurologic patients with chelators is that the patient may become neurologically worse. There is not very much reported experience with trientine in the initial therapy of neurologic patients. We are aware of one case in which it did induce neurologic worsening. However, without information on how many patients have been treated, it is not possible to evaluate the risk

The efficacy of trientine can be monitored by 24 hour urine copper measurements. Upon initial therapy in previously untreated patients, daily urine copper excretion may reach several milligrams. Over time, the urine copper gradually decreases, down to 500–1000μg/24 hours. In a well-complying, well-treated patient, this would mean that most of the mobilizeable copper has been eliminated and one is at a steady state where copper balance is being achieved by an increased urinary excretion of copper of this magnitude. However, one of the problems of monitoring chelation therapy with urine copper is that the effect of the drug is to increase the urinary excretion of copper, while the body loading of copper is also reflected in the urinary excretion of copper. Thus, there is a built in ambiguity. For example, a value of 500μg/24 hours could be the steady state level of a well-treated patient. But it could also be the value in a poorly complying patient, in which some of the urine copper is due to inadequate doses of trientine, and the rest is due to increasing body loading with copper.

Using non-ceruloplasmin plasma copper for monitoring is better for trientine treated patients, but of course requires a blood sample. The methodology for this was given previously in the zinc section.

Zinc and trientine work well together in previously untreated patients, possibly because there is an excess of copper available. However, we have shown that the two drugs do not produce an additive effect during maintenance therapy, where there is less copper, perhaps because trientine becomes chelated with zinc (Brewer et al 1993b).

2.3. Penicillamine

Penicillamine was the first practical and effective oral therapy for Wilson's disease, being introduced in 1956 by Walshe. In my opinion, penicillamine is no longer a very useful drug for the treatment of Wilson's disease. Of course, the drug and it's founder, should be given credit for the important historical role of penicillamine. A generation of Wilson's disease patients owe their lives to the therapeutic effect of this drug. However, it is no more effective and much more toxic than the two drugs we have already discussed. Thus, I never prescribe penicillamine anymore and I recommend that it not be used for Wilson's disease if zinc or trientine are available.

Penicillamine, like trientine, acts by copper chelation and increased urinary excretion of copper. Much information on the use of penicillamine in Wilson's disease is given in the monograph of Scheinberg and Sternlieb (1984). The dose of penicillamine is similar to trientine at 250mgx4 or 500mgx2. As with trientine, it should be separated from food, spacing the drug at least one hour before or two hours after meals. The therapeutic effect of penicillamine is evident by a dramatic increase in urinary excretion of copper, which can be as high as 10mg in a previously untreated patient.

The side effects of penicillamine are quite numerous (Scheinberg and Sternlieb, 1984; Brewer and Yuzbasiyan-Gurkan, 1992). There is an initial risk of about 30% of a hypersensitivity reaction consisting of fever, hives, or rash. This reaction can often be overcome by the use of steroids and/or a reduced dose of penicillamine, or by temporarily stopping the drug and restarting it at a lower dose. Other acute side effects are similar to those with trientine but with a higher incidence. Thus, bone marrow depression or other idiosyncratic reaction can occur and it is important to monitor the patient at least once per week with blood counts, a biochemistry panel, and urinalysis. As with trientine, proteinuria is a frequent side effect of penicillamine. Chronic toxicities with penicillamine include induction of autoimmune disorders such as lupus or Goodpasture's syndrome. Connective tissue side effects of penicillamine include skin wrinkling and numerous other dermatologic conditions. In experimental animals, effects on the connective tissues of blood vessels have been seen in numerous studies, raising concerns about possible aneurysm or vessel rupture in chronically treated patients. There is also evidence that the immune system is affected by penicillamine therapy and that resistance to infection may not be normal.

A serious risk of penicillamine is neurological worsening in neurologically presenting patients (Brewer et al 1987b). This risk appears to be about 50%, and 50% of patients who undergo worsening do not recover.

The efficacy of penicillamine can be monitored, as with trientine, by 24 hour urine copper measurements. In untreated patients, penicillamine causes an explosive excretion of urine copper, as high as 10mg. This aggressive mobilization of copper may be related to the neurologic worsening. Again, as the patient's copper load is reduced, urinary copper excretion decreases to the 500–1000µg/24 hours range. However, the same ambiguities exist with penicillamine as was described with trientine, in terms of interpretation, because both the drug's effect, and the body loading of copper, are reflected in these values. As with trientine, non-ceruloplasmin plasma copper is a useful method for monitoring, but requires a blood sample.

Zinc works well with penicillamine during initial therapy if the doses are separated. However, as with trientine, no additive effect of the two drugs is seen during maintenance therapy (Brewer et al 1993b).

2.4. Tetrathiomolybdate

Tetrathiomolybdate (TM) is an experimental drug. It was used in a few patients for maintenance therapy by Harper and Walshe (1986) and they reported, anecdotally, that the drug can produce a reversible anemia. Our use of TM has been quite different than that of Walshe. We have used it for an initial eight week therapy of the neurologically-presenting Wilson's disease patient (Brewer et al 1991; Brewer et al 1994; Brewer et al 1996).

The mechanism of action of TM is to form a tripartite complex between itself, copper and protein (Bremner et al 1982; Gooneratne et al 1981). Such complexes are undigestible and insoluble, and can not be absorbed by the intestine or enter cells. Taking advantage of these properties, we use TM to achieve two mechanisms of action. We give three doses with meals, and the TM complexes food protein with food copper as well as endogenously secreted copper, and prevents the absorption of this copper. This effect puts the patient into an immediate negative copper balance. The second mechanism involves giving the drug between meals, where it is absorbed into the blood and forms a complex with potentially toxic copper in the blood and albumin. This renders the blood copper unavailable for cellular uptake and therefore makes it non-toxic. TM is an extremely potent anticopper agent and is the fastest acting anticopper drug available. Currently, we are using 20mg of TM in each of the six doses.

The efficacy of TM can be followed by measuring the molybdenum and copper levels in the blood. There is a one to one stochiometric relationship between copper and molybdenum in the complex. Thus, one calculates the non-ceruloplasmin plasma copper and determines the molybdenum level in the blood. When these become equal, all the potentially toxic copper in the blood has become complexed. Since this copper is in equilibrium with organ copper, the copper from various organs will flow downhill into the bloodstream, where these copper molecules will also be complexed. Generally, it takes between three to fifteen days to reach a stochiometric equilibrium in the blood. At that point, theoretically at least, further copper toxicity has been halted.

Regarding the potential toxicity of TM, the literature is quite deceptive. TM given to experimental animals at fairly low doses can be quite harmful and produce lethality relatively quickly (Mills et al 1981; Bremner et al 1982). However, all of its side effects are due to copper deficiency. Copper loading the animal prevents all of these toxicities (Mills et al 1981b). Since the Wilson's disease patient is loaded with copper, the patient is not at risk for the long list of toxicities that appear in the literature.

However, we have confirmed that TM is capable of producing a reversible anemia in Wilson's disease patients (Brewer et al 1996). This interesting manifestation is not a side effect, but rather a manifestation of overtreatment. TM is a very potent anticopper agent. If one uses a large enough dose of TM, in order to gain control of copper toxicity quickly, it is possible to deplete the bone of marrow copper, and reduce cell production, since copper is required for hemoglobin synthesis and cellular development. The anemia quickly goes away when the drug is stopped. We saw this result only with higher doses of TM, when we were trying to hasten reaching stochiometric equilibrium in the blood. Using our present dose of TM (20x6), we have avoided this overtreatment effect.

3. DIFFERENT CLINICAL STAGES OF WILSON'S DISEASE AND RECOMMENDATIONS FOR THERAPY

3.1. Initial Neurologic Disease

Perhaps 50% of patients present clinically with manifestations involving the brain. These can be neurologic symptoms of the movement disorder type or psychiatric symptoms.

Psychiatric symptoms can precede the onset of neurologic symptoms by several years. Typically these occur in young teenagers or patients in their early twenties and run the gamut from depression to psychotic behavior (Table 1). Loss of emotional control or temper tantrums are frequent. Patients often find it very difficult to focus on tasks and thus their school or work performance falls off dramatically. Because these behavioral abberations are occurring in a previously normal person, they are often incorrectly attributed to drug abuse.

The neurologic manifestations are classified as a movement disorder (Table 1). Speech is affected in a high proportion of the patients and usually fairly early. This is due to the inability to coordinate speech muscles properly. Tremor is also very common and can be of several different types. One symptom, such as dysarthria or tremor, may predominate for a very long time, perhaps months or years, before other symptoms begin to occur.

Patients presenting with initial neurologic disease are at great risk for neurologic worsening if they are treated with penicillamine, as previously discussed. There is not much published experience with trientine in this setting, and it is therefore difficult to know the extent of the risks of neurologic worsening with this drug. We are aware of one report of trientine induced neurological worsening.

Zinc has been used extensively for initial neurologic treatment by Hoogenrad (1996) in the Netherlands and that group reports good results, although they have reported no quantitative assessments of neurologic function. One disadvantage of using zinc in this setting is that it is relatively slow acting. We estimate that it may take as long as four months for zinc to gain control of neurologic copper toxicity. Thus, the patient is at some risk of the disease progressing as a result of it's own natural history before zinc controls the copper toxicity. We have treated three patients with zinc from the beginning and two did well, while one developed additional tremor during the early periods of zinc therapy. It should be noted that the occasional case report which suggests that zinc makes the disease

Table 1. Some of the behavioral and neurologic manifestations in Wilson's disease

Behavioral	Neurologic
Depression	Dysarthria
Hyperkinetic behavior	Tremor
Irritability or anger	Dystonia
Emotionality	Dysdiadochokinesia
Psychosis	Poor hand writing
Sexual exhibitionism	Incoordination
Personality changes	Rigidity
Falloff in school/work performance	Abnormal eye movement
	Parkinsoniasm features

worse, or fails to control the disease over a longer period of time, is invariably explained by poor compliance with therapy (Shimon et al 1993; Walshe and Munro, 1995).

The therapy we have pioneered for this particular clinical situation is tetrathiomolybdate (Brewer et al 1991; Brewer et al 1994; Brewer et al 1996). As we described in discussing the drug above, TM is extremely fast acting. It has the potential of quickly gaining control over copper toxicity and does not have the risk of inducing additional neurological worsening. So far, we have treated 51 neurologically-presenting Wilson's disease patients with TM and followed neurologic function by quantitative neurologic and quantitative speech examinations. 2 of the 51 patients showed worsening in scores that reached our criteria. These worsenings were transitory and may very well have been part of the natural history of the disease, over which no drug could have had an influence. The main point is that in 96% of the cases the neurologic situation was completely stabilized by TM, and there was no neurologic worsening during the 8 weeks of initial therapy. These patients are transitioned to zinc therapy no later than two weeks before they are taken off TM, and are then sent home on zinc therapy for maintenance purposes.

3.2. Initial Hepatic Disease

About 50% of the patients present with hepatic disease. Again, the age of presentation is in the late teenage years or early twenties, although the age of presentation can be quite broad. A small proportion of the patients, generally the younger ones, present with acute fulminant hepatic failure. These patients tend to go down hill rapidly and the only therapy that will save their lives is hepatic transplantation. The prognostic index of Nazer et al (1986) is useful in determining the likely requirement for transplantation in a given case.

More common than fulminant failure is a history of episodic chronic hepatitis. There may be episodes of jaundice, and apparent recovery. Other patients present with mild compensated hepatic failure. These patients may have a reduced albumin, fluid accumulation in the form of ascites or peripheral edema, and a mildly elevated bilirubin and abnormalities of other liver function tests. Both the hepatitis-like and the mild liver failure patients can be quite effectively treated with medical therapy.

It should be pointed out that among the four anticopper agents that are currently available, there has not been a rigorous comparative study to see which agent, or combination, is optimal for this kind of patient. Our current choice is to use a combination of trientine and zinc. We use the trientine to gain a rather quick negative copper balance and we choose trientine over penicillamine because it is much less toxic. We include zinc in the early phases because of it's effect on inducing hepatic MT. This MT will sequester copper and theoretically help reduce further hepatic damage through removing toxic copper from the liver and storing it in this manner. We use the combination of trientine and zinc for four to six months, and then stop the trientine and have the patient continue on zinc for maintenance therapy.

It is worthy of note that TM given intravenously has been quite effective in saving copper-poisoned sheep (Gooneratne et al, 1981). This suggests that TM could have has a very important role to play in helping the acutely ill human patient who presents with liver disease. However, we are not aware that TM has yet been used in this manner.

3.3. Maintenance Therapy

With its approval by the U.S. FDA, zinc is now by far and away the first choice for maintenance therapy of Wilson's disease, in our opinion. This is because it is 100% effective, and in comparison with the other anticopper drugs, has extremely low toxicity. In adult patients we use a dose of 50x3, although we have shown that 25x3 is also effective. Our reason for using the higher dose is for safety margin reasons. Thus, if compliance reaches the level of only taking two doses per day, the therapy will still be effective. The maintenance of the induction of intestinal MT is quite stable, so that if a patient forgets their zinc on a weekend trip, it is not a problem. Similarly if the patient has gastrointestinal illness or surgery and can not take anything by mouth for a few days, the efficacy of the effect on the intestinal tract of the previous zinc will remain. Obviously, this kind of hiatus should not go on for more than a few days.

We are currently following 175 patients on zinc therapy. Our longest treated zinc patient is out about fifteen years, and we have treated and followed 38 patients for ten years and 92 patients for five years. This is the largest study of any therapy for Wilson's disease with long term follow-up ever reported. From all this it can be concluded that zinc is an effective and very safe therapy for the maintenance treatment of Wilson's disease.

Patients in the maintenance phase can also be treated with either trientine or penicillamine. The dose, monitoring, side effects, and other aspects are similar to those discussed under initial therapy.

3.4. Presymptomatic Patients

Since Wilson's disease is inherited as an autosomal recessive, it is important to work up other members of the family to see if they are affected with Wilson's disease. The full siblings are at a 25% risk, and are of the highest priority to work up. Other relatives such as children, nieces, nephews, and cousins are at an increased risk compared to the general population, but not nearly as high as the siblings. When a sibling is diagnosed with Wilson's disease, they are generally in what we call the presymptomatic phase. That is, they have not yet presented symptomatically with either neurologic or hepatic disease. However, the disease is believed to be almost 100% penetrant. Thus, those patients who are homozygous affected will almost always present symptomatically at some point in their life. It is important that these patients go on anticopper therapy to prevent becoming ill.

Our first choice for therapy is zinc. We view these patients as comparable to the maintenance phase of a previously symptomatic patient and begin them on zinc from the beginning. The dose, the monitoring and other aspects of zinc therapy are identical to those previously discussed under maintenance therapy.

Presymptomatic patients can also be treated with either trientine or penicillamine as maintenance therapies. The dose, monitoring, side effects and other aspects are identical to those previously discussed.

3.5. Pregnant Patients

When a Wilson's disease patient becomes pregnant it is critically important that she continue on anticopper therapy. In the past, because of the known teratogenic effects of penicillamine, patients have been known to stop their therapy to protect their fetus. This has led to a series of case reports of pregnant women becoming critically ill from worsening of their Wilson's disease during pregnancy, due to their being off therapy (reviewed in Brewer and Yuzbasiyan-Gurkan, 1992).

The therapy we choose for this type of patient is zinc. Both penicillamine and trientine have been shown to be teratogenic in animals (Keen et al 1982) and penicillamine has produced a teratogenic effect in humans (Mjolnerod et al 1971; Solomon et al 1977). Zinc, on the other hand has been specifically studied, and has been found to be free of teratogenic effects

Table 2. Recommended zinc doses for pediatric patients

Age	Zinc dose
Ages 1-5 years	25x2
Ages 6-16 years, or until a body weight of 125 pounds is reached	25x3
Ages 16 and up, or when 125 pounds body weight is reached	50x3

(Food and Drug Administration, 1974). We have used both 25x3 and 50x3 doses during pregnancy. Because some of the 25x3 patients have had an increase in urine copper during pregnancy (perhaps due to inadequate compliance), we currently recommend 50x3.

3.6. Pediatric Patients

The near 100% penetrance of this disease, and the finding of liver disease in very young children, leads one to recommend therapy at any time the diagnosis is made after age one. (We don't recommend treating during the first year of life). Our recommended therapy in pediatric patients is zinc. The doses we use are shown in Table 2.

Our experience with patients under age five years is limited to one patient. She seems to be doing well on the 25x2 dose. We have treated a total of 30 patients in the pediatric age group and so far these dose range recommendation seem to work reasonably well.

Pediatric patients can also be treated with trientine or penicillamine if one elects to use these drugs. The recommended initial dose of trientine is 500–750mg/day, given in divided doses (Physicians Desk Reference). I have not seen official dose recommendations for the use of penicillamine in children, but one would assume that lower than adult doses would be used on some type of weight or age formula.

A summary of therapy recommendations for the various clinical stages is given in Table 3.

Table 3. Recommended therapies for various clinical stages of Wilson's disease

Clinical stage	Recommended therapy in order of choice
Initial neurologic	1. Tetrathiomolybdate, if available 2. Zinc 3. Trientine
Initial hepatic	1. Zinc and trientine 2. Trientine alone 3. Zinc alone
Maintenance	1. Zinc 2. Trientine
Presymptomatic	1. Zinc 2. Trientine
Pregnant	1. Zinc
Pediatric	1. Zinc 2. Trientine

4. COMPLIANCE, FOLLOW-UP AND PROGNOSIS

4.1. Compliance

It can't be emphasized too much that in a situation of therapy of this type there is a major compliance problem. There are many reasons for this. One of them is that therapy is lifelong. The longer the therapy has to be given, the greater the risk that at some point, perhaps related to changes in life style or family tensions, the patient will not comply adequately with the regimen. Second, and tied into the first, is the fact that many patients have little or no symptoms during their maintenance phase and nothing to continually remind them that they need to take medicine. The rate of serious, chronic, problems with compliance we have observed in the 175 patients we follow runs about 10%. There is an additional 25% of the patients who have had temporary lapses in compliance. These rates are in a group of patients whose drug and medical care for Wilson's disease have been provided free, and who are asked to provide a urine sample every six months so that we can check for compliance. Thus, these levels of compliance will be minimal compared to that which will be encountered in a general clinical population followed by physicians offering service only when the patient comes to see them. These problems with compliance mean that life long follow-up is a critically important component of the management of Wilson's disease patients.

4.2. Follow-Up

As suggested above, due to the problems with compliance, patients with Wilson's disease should have a consistent follow-up program for their entire life. The system that we recommend is that in the early phases of therapy, patients be followed every three to six months, monitoring whatever it is that requires monitoring in that particular patient. At a minimum, this would be 24 hour urine copper and if the patient is on zinc, 24 hour urine zinc. Additionally, during these early phases it is useful to at least every six months look at liver function, at hematologic values, and non-ceruloplasmin plasma copper in order to get a good baseline on the patient. The pattern of monitoring every six months should probably occur for the first two or three years in all patients. After that, in well complying patients, the frequency of monitoring can be reduced to annually. However, in our opinion it should not be reduced less than this, because as suggested above in the section on compliance, patients may eventually lose the discipline that has maintained their therapy in the past.

4.3. Prognosis

During initial therapy, symptomatic patients can expect improvement to begin approximately five to six months after the initiation of therapy and to continue for a period of an additional eighteen months. Thus, in neurologically presenting patients, quantitative speech exams and quantitative neurologic exams both improve, on the average, between the periods of six to twenty-four months after the initiation of therapy. The same is true of hepatic abnormalities as measured by liver function tests. After that, in our experience, most of the abnormalities that are present at the two year point are permanent. They are due to irreversible damage.

The degree of improvement during the first two years can be very dramatic. Generally the psychiatric manifestations disappear and many if not all of the neurologic manifestations either improve significantly or go away completely. Thus, it is important to maintain a positive attitude toward the patient during that two year period, because the attitude of the patient is important in terms of maintaining as much physical ability as possible, so that when recovery does begin it will be within the milieu of as well conditioned a body as possible.

With respect to liver disease, it should first be pointed out that all patients with Wilson's disease have a certain amount of underlying liver disease. This is true of whether the patient presented with hepatic manifestations, neurologic manifestations or was a presymptomatic patient. This early copper damage usually results in some degree of cirrhosis. The patient and the physician both need to be aware that many of these patients, while their liver function tests may be normal, have underlying cirrhosis and may exhibit some of the complications of portal hypertension. For example, a very high proportion of patients have thrombocytopenia and leukopenia as a result of the hypersplenism that is part of the portal hypertension picture. This does not progress and need not be of much concern, except that if the patient and physician are unaware of it's cause, it may lead to unnecessary workup. The patient is also at risk for varicele bleeding. This risk is greatest in the patients that present with liver disease, but is also present in those with underlying cirrhosis and portal hypertension who present with neurologic disease or who are presymptomatic. Patients should be counseled that if they have any manifestations of GI bleeding, such as black tarry stools or vomiting of blood, they should go immediately to an emergency room.

REFERENCES

Bremner, I., Mills, C.F., Yong, B.W. (1982). Copper metabolism in rats given di-or trithiomolybdates. Journal of Inorganic Biochemistry, 16:109–19.

Brewer, G.J., et al. (1998). The Treatment of Wilson's Disease with zinc XV. Long-term follow-up studies. Journal Laboratory Clinical Medicine. In Press.

Brewer, G.J., Johnson, V., Dick R.D., Kluin, K.J., Fink, J.K., Brunberg, J.A. (1996). Treatment of Wilson's disease with ammonium tetrathiomolybdate: II. Initial therapy in 33 neurologically affected patients and follow-up on zinc therapy. Archives of Neurology, 53 (10): 1017–1025.

Brewer, G.J., Dick, R.D., Johnson, V., Wang, Y., Yuzbasiyan-Gurkan. V., Kluin, K., Fink, J.K., Aisen, A. (1994). Treatment of Wilson's disease with tetrathiomolybdate I. Initial therapy in 17 neurologically affected patients. Archives of Neurology, 51: 545–554.

Brewer, G.J., Yuzbasiyan-Gurkan, V., Johnson, V., Dick, R.D., Wang, Y. (1993a). Treatment of Wilson's Disease with zinc: XII. Dose regimen requirements. American Journal of Medical Sciences, 305 (4),199–202.

Brewer, G.J., Yuzbasiyan-Gurkan, V., Johnson, V., Dick, R.D., Wang, Y. (1993b). Treatment of Wilson's disease with zinc: XI. Interaction with other anticopper agents. Journal of the American College of Nutrition, 12(1), 26–30.

Brewer, G.J., Yuzbasiyan-Gurkan, V. (1992). Wilson's disease. Medicine, 71:139–164.

Brewer, G.J., Dick, R.D., Yuzbasiyan-Gurkan, V., Tankanow, R., Young, A.B., Kluin, K.J. (1991). Initial therapy of Wilson's disease patients with tetrathiomolybdate. Archives of Neurology, 48: 42–47.

Brewer, G.J., Yuzbasiyan-Gurkan, V., Dick, R.D. (1990). Zinc therapy of Wilson's disease: VIII. Dose response studies. Trace Elements in Experimental Medicine, 3:227–234.

Brewer, G.J, Hill, G.M., Prasad, A.S., and Dick, R.D. (1987a). Treatment of Wilson's disease with zinc: IV. Efficacy monitoring using urine and plasma copper. Proceedings of the Society of Experimental Biology in Medicine, 7:446–455.

Brewer, G.J., Terry, C.A., Aisen, A.M., Hill, G.M. (1987b). Worsening of neurologic syndrome in patients with Wilson's disease with initial penicillamine therapy. Archives of Neurology, 44:490–93.

Brewer, G.J., Hill, G.M., Prasad, A.S., Cossack, Z.T., Rabbani, P. (1983). Oral zinc therapy for Wilson's disease. Annals of Internal Medicine ,99:314–320.

Food and Drug Administration. (1973 and 1974). Teratologic evaluation of FDA 71–49 (zinc sulfate). Food and Drug Research Laboratories, Inc. Prepared for Food and Drug Administration, United States Department of Commerce Publications PD-221 805, and PB 267.

Gooneratne, S.R., Howell, J.M., Gawthorne, J. (1981). An investigation of the effect of intravenous administration of thiomolybdate on copper metabolism in chronic CU-poisoned sheep. British Journal of Nutrition, 46:469–80.

Harper, P.L., Walshe, J.M. (1986). Reversible pancytopenia secondary to treatment with tetrahiomolybdate. British Journal of Haematology, 64:851–3.

Hill, G.M., Brewer, G.J., Prasad, A.S., Hydrick, C.R., Hartmann, D.E. (1987). Treatment of Wilson's disease with zinc: I. Oral zinc therapy regimens. Hepatology, 7:522–528

Hoogenraad, T. (1996). Wilson's disease. In Major Problems in Neurology, van Gijn, J., ed. (Philadelphia, Pennsylvania: W.B. Saunders Company).

Keen, C.L., Lonnerdal, B., Hurley, L.S. (1982). Teratogenic effect of copper deficiency and excess. In Inflammatory Diseases and Copper, Sorenson, ed., (Humana Press).

Mills, C.F., El-Gallad, T.T., Bremner, I. (1981a). Effects of molybdate, sulfide, and tetrathiomolybdate on copper metabolism in rats. Journal of Inorganic Biochemistry, 14: 189–207.

Mills, C.F., El-Gallad, T.T., Bremner, I., Weham, G. (1981b). Copper and molybdenum absorption by rats given ammonium tetrathiomolybdate. Journal of Inorganic Biochemistry, 14:163–75.

Mjolnerod, O.K., Dommerud, S.A., Rasmussen, K., Gjeruldsen, S.T. (1971). Congenital connective tissue defect probably due to D-penicillamine treatment in pregnancy. Lancet I, 673–675.

Nazer, H., Ede, R.J., Mowat, A.P., Willaims, R. (1986), Wilson's disease: clinical presentation and use of prognostic index. Gut, 27:1377–81.

Scheinberg, I.H. and Sternlieb, I. (1984). Wilson's disease. In Major Problems in Internal Medicine, Smith, M.D., L.E., ed. (Philadelphia, Pennsylvania: W.B. Saunders Company).

Schouwink, C. (1961). De Hepato-Cerebrale Degeneratie. Biemond, A., ed. (Arnhem, Amsterdam: G.W. Van Der Wiel and Company).

Shimon, I., Sela, B.A., Moses, B., Doley, E. (1993). Hemolytic episode in a patient with Wilson's disease treated with zinc. Israel Journal of Medical Sciences, 29: 646–7, 1993.

Solomon, L., Abrams, G., Dinner, M., Berman, L. (1977). Neonatal abnormalities associated with D-penicillamine treatment during pregnancy. New England Journal of Medicine, 296: 54–55.

Walshe, J.M., Munro, N.A.R. (1995). Zinc-induced deterioration in Wilson's disease aborted by treatment with penicillamine, dimercaprol, and a novel zero copper diet. Archives of Neurology, 52: 10–11.

Walshe, J.M. (1982). Treatment of Wilson's disease with trientine (triethylene tetramine) dihydrochloride. Lancet, 2:1401–2.

Walshe, J.M. (1956). Penicillamine. A new oral therapy for Wilson's disease. American Journal of Medicine, 21: 487–95.

Yuzbasiyan-Gurkan, V., Grider, A., Nostran,t T., Cousins, R.J., Brewer, G. (1992). The treatment of Wilson's disease with zinc: X. Intestinal metallothionein induction. Journal of Laboratory Clinical Medicine, 120:380–386.

INDIAN CHILDHOOD CIRRHOSIS AND TYROLEAN CHILDHOOD CIRRHOSIS

Disorders of a Copper Transport Gene?

M. S. Tanner[*]

University of Sheffield, Children's Hospital
Western Bank
Sheffield S10 2TH, UK
m.s.tanner@shef.ac.uk

1. INTRODUCTION

Indian Childhood Cirrhosis (ICC) (Tanner 1997) and Tyrolean Childhood Cirrhosis (TCC) (Müller et al 1996) have both now virtually disappeared. In ICC, and putatively in TCC, greatly increased levels of ingested dietary copper in infancy were associated with greatly increased hepatic copper concentrations and, hence, cirrhosis. Cessation of this feeding habit has been associated with disappearance of the disease in each case.

The imperative to debate whether they were due in part to a genetic abnormality is the continued sporadic occurrence of "ICC-like" cases in other parts of the world. For some of these, there is no clear evidence of excess copper ingestion. The hypothesis to be examined therefore is that in both conditions there was a genetic susceptibility to the effects of excess copper ingestion.

It is tempting to draw a comparison with the autosomal recessive disorder, genetic haemochromatosis, which (a) is common, (b) clinically manifests at different ages depend-

---

[*] Tel 44 114 271 7228 Fax 44 114 274 5364

ing upon iron balance, (c) is treatable and presumably preventable by accelerating iron loss, and (d) does not produce the disease in every genetically affected individual.

2. DEFINITIONS

The literature about ICC and the ICC-like disorders has become confusing as case reports multiply and terminology proliferates. The following conditions need to be defined and distinguished:

1. Indian Childhood Cirrhosis
2. Tyrolean Childhood Cirrhosis
3. Other cirrhoses in Indian children
4. The ICC-like disorders
5. Wilson's disease

2.1. Indian Childhood Cirrhosis

This progressive liver disorder of infancy is defined on the histological criteria

1. necrosis of hepatocytes with ballooning and Mallory's hyaline
2. pericellular intralobular fibrosis
3. inflammatory infiltrate
4. poor hepatocyte regenerative activity
5. absence of fatty change
6. granular orcein staining

(Nayak and Ramalingswami, 1975; Portman et al, 1978) and the clinical criteria

1. occurrence in Indian subcontinent
2. children 6 months to 4 years
3. presenting either with gross abdominal distension or with jaundice
4. found on examination to have gross hepatomegaly with an extraordinarily hard liver texture and sharp liver edge
5. without treatment a rapidly downhill and ultimately fatal clinical course

2.2. Tyrolean Childhood Cirrhosis

This label is attached to a group of 138 infants who died in an area of the Tyrol between 1900 and 1974 (Muller et al,1995). In their clinical course and epidemiology they bore a striking resemblance to ICC. They were also defined pathologically, the histological features being described in detail in a monograph (Gogl, 1947). Features 1 - 5 as for ICC above were well described. Unfortunately, no copper stains were performed on these samples and no retrospective copper staining can be done because paraffin blocks were not kept. Copper accumulation in TCC is, therefore, inferential.

The clinical and epidemiological features of ICC and TCC are compared in Tables 1 and 2.

Table 1. Similarities between Indian cHildhood Cirrhosis and Tyrolean Childhood Cirrhosis

Feature	Description
Epidemiology	Endemic in a defined geographical area. Affected rural farming families rather than urban communities. Has (almost) disappeared
Clinical features	Infancy, approximately 6 months to 4 years Failure to thrive, followed by abdominal distension, hepatomegaly, and usually later jauundice, ascites. Rapid progression, fatal.
Pathological features	See text for description of ICC TCC had similar H&E appearances, but copper stains not done
Feeding history	Breast feeding brief or not at all. Did not use a proprietary milk formula but own animals' milk Animal milk was copper contaminated No evidence implicating copper in drinking water
Genetics	Affected siblings High rate of parental consanguinity Parents overtly healthy Some children with the same feeding pattern remained healthy

2.3. Other Cirrhoses in Indian Children

Amongst a large series of children with progressive liver disease seen in India are those with a disorder clearly distinguishable from ICC (eg. infantile cholestasis, autoimmune hepatitis, chronic hepatitis B infection, fulminant hepatic failure, storage disorders) and those with cryptogenic cirrhosis. There is a suggested relationship between cryptogenic micronodular cirrhosis and ICC. The occurrence of ICC and cryptogenic cirrhosis among sibships, and the evolution of ICC to an inactive cirrhosis during penicillamine treatment, suggests that some inactive cirrhoses represent "burnt out" ICC (Bavdekar et al, 1996; Pradhan et al, 1995). In epidemiological or genetic studies of, however, it is important to define the ICC study group tightly as above.

2.4. The ICC-Like Disorders

There are an increasing number of case reports of liver disease clinically and histologically resembling ICC in young children outwith India.

Table 3 lists the reported instances of ICC-like liver disease in which there has been a clear history of excessive copper ingestion in infancy. Unlike ICC, the copper has in each case been derived from the water used to reconstitute formula feeds. In each case, the family were rural, drew water from a private well, and fed the infant on artificial formulae made up with water which had been contaminated with copper from plumbing. In some

Table 2. Differences between Indian Childhood Cirrhosis and Tyrolean Childhood Cirrhosis

	Indian Childhood Cirrhosis	Tyrolean Childhood Cirrhosis
Epidemiology	Occurred all over India	Probably localised
Clinical features	Male predominance Penicillamine treatment	Sexes equal Effect of penicillamine not known
Pathology	Liver copper high Evolution to inactive cirrhosis with treatment	Liver copper not known Evolution not known
Feeding history	Cow or buffalo alone	Cows' milk mixed with flour, sugar and water
Milk Vessels	Brass, poorly tinned	Copper, never tinned, occasionally brass
Genetics	Male predominance contradicts simple autosomal recessive, ? cultural factors	Probaby autosomal recessive

Table 3. Cases of infantile cirrhosis in which water used to make up infant feed had high copper concentration

Country	Age	Sex	Outcome	Liver Cu microg/g dry wt	Water Cu mg/l.	Reference
Australia	1	M	d in 6 w	3360	6.75	Walker-Smith et al 1973
Germany	7m	F	d in 3m	684	0.430	Muller-Hocker et al 1987
Germany	9m	F	d in 4m	2154	2.2–3.4	Muller-Hocker et al 1988
	5m	M	survived	2094		
Germany	5–9m	F	died	1240	0.43–3.4	Schramel et al 1988
		F	died	1870		
		M	survived	1450		
Eire	10m	M	d age 11m	1245	8.0	Baker et al 1995
	14m	M	d age 15m	1320	6.7	
Germany	13m	M		high	12–28.6	Bent et al 1995
Australia	20m	M	liver transplantation	10-fold increased	7.8	Price et al 1996

cases the source of the copper was quite obvious. In one of the cases reported by Baker et al (1995) the water blackened the inside of the kettle, and in at least 2 cases the infant feed was made up with water which had been heated in a copper cistern.

There is a single case report of an adult who ingested 30 mg copper for 2 years, then 60 mg/day for 1 year before presenting with fulminant liver failure; his explanted liver at transplantation had a copper concentration of 3250 microgram/g dry weight (O'Donohue et al 1993). The lack of other such adult cases suggests that the infantile liver may be particularly susceptible to copper loading.

Table 4 lists those published cases of infants and children with a liver disease histologically resembling ICC in whom there has been no history of excessive dietary copper. In several, there has been a family history suggestive of a genetic disorder. Note that the mean age at presentation in this group (4.45 years, SD 2.46y, range 2.5–10y) is greater than in the cases listed in Table 3 (10 months, SD 4.46 m, range 5–20m)

Table 5 lists reported instances of children with liver disease associated with high hepatic copper concentration in whom the histology differed significantly from ICC.

There are other reported cases described as having Indian childhood cirrhosis in whom histology was atypical and liver copper only moderately elevated (Klass et al 1980,

Table 4. Cases of infantile cirrhosis in which liver histology resembled ICC but no dietary source of excess copper was identifiable

Country	Age	Sex	Outcome	Liver Cu microg/g dry wt	Family history	Reference
Singapore	4	F	died	?	Siblings	Lim et al. 1979
	4	F	died	?		
	6	F	died	?		
	5	M	survived	1200		
USA	6	F	d in 12 d	1031	Siblings. Aunt died age 10 y with cirrhosis	Lefkowitch et al. 1982
	5	M	d in 6 m	2083		
	4	M	d in 6m	708		
	5	F	d in 6m	992		
Italy	10	M	treated	1970		Maggiore et al. 1987
USA	2	M	d 2.7	1500	Consanguinity	Adamson et al. 1992
UK	29m	F	transplanted	1100		Baker et al. 1995

Table 5. Cases of childhood liver disease with very high liver copper, not attributable to Wilson's disease or chronic cholestasis, with histology different from Indian childhood cirrhosis

Country	Age	Sex	Outcome	Liver Cu microg/g dry wt	Family history	Reference
UK	7	M	treated with penicillamine	2319	Brother had similar illness	Horslen et al 1994

Aljajeh et al 1994), or in whom subsequent analysis proved an alternative diagnosis (e.g. Bartok et al 1971).

2.5. Wilson's Disease Is Described in Detail Elsewhere in This Volume

3. ENVIRONMENTAL CAUSE OF ICC AND TCC

The evidence that ICC was associated with the early introduction of animal feeds contaminated with copper by contact with brass utensils had been reviewed recently (Tanner 1997). Müller et al (1995) produced convincing evidence that in Tyrolean Childhood Cirrhosis the infants were fed with milk which had been heated in a copper container. The most persuasive evidence that these feeding practices were aetiologically causative is the disappearance of both disorders as infant feeding practices changed (Bhave et al 1992). The "well-water" cases in Table 3 further support this contention.

3.1. Evidence that Genetic Factors May Have Been Operative in ICC and TCC

3.1.1. Susceptibility. In both geographical , families reported that other infants who had exactly the same feeds remained well. Although there are multiple possible confounding factors which might explain this, this apparent variability in susceptibility suggests a genetic variation. In some of the "well-water" cases in Table 3, the exposure to copper applied only to the affected child, but in others other children may have had the same copper intake.

3.1.2. Animal Models. There is no animal model in which chronic ingestion of copper in any form has produced cirrhosis of the liver. There are multiple examples of acute liver injury of varying severity and time course in different species. A reported cirrhosis induced by cupric nitriloacetate in rats (Toyokuni et al 1989) could not be replicated.

In those animal models where excessive liver copper storage occurs the aetiology appears to be genetic rather than environmental. In the white perch, for example, (Morone Americane) liver copper concentrations are enormously high but cause no apparent hepatic inflammation, fibrosis or necrosis (Bunton et al 1987). In the Bedlington terrier, dogs affected by an inherited copper toxicosis develop progressive or fulminant liver disease. It is important to note that Bedlington terrier disease differs from ICC on histological parameters and from Wilson's disease biochemically.

3.1.3. Müller et al. (1995) Produced Convincing Data Supporting an Autosomal Mode of Inheritance in TCC. Parents were clinically normal with no history of liver disease. some but not all siblings were affected. there was a high degree of parental consanguinity.

3.1.4. Family Data in ICC Is Conflicting. Family data from Pune showed that of 136 children with ICC 36 (26%) were from consanguinous parents, compared with a consanguinity rate of 13% in children with other liver disorders. The segregation ratio for the families with ICC was 0.155, SD 0.03.

In ICC, though not TCC, there is a marked male predominance. Again, it is easy to suggest possible cultural reasons for this. Perhaps the favoured male infants were more likely to receive the family's precious cow or buffalo milk. This is stoutly denied by affected families. If this is a genuinely innate difference between the Indian and Tyrolean cases it suggests that more than one gene may be implicated.

3.1.5. Amongst the sporadic ICC-like disorders in whom there was no history of excessive copper ingestion, the occurrence within sibships and the presence of parental consanguinity suggests an inherited disorder.

4. MODELS

From the above data, a number of different models may be advanced:

Toxicological: That all copper related cirrhosis in infancy is environmental in origin. Clustering within families simply results from common environmental or cultural practices

Ecogenetic: an inherited enzymatic defect becomes manifest only when the organism is stressed by a xenobiotic, in this case copper. Phenylketonuria or glucose 6 phosphate dehydrogenase deficiency are examples.

Multifactorial: Several different mechanisms may lead to excess copper storage and, hence, liver disease.

Epiphenomenal: hepatic copper accumulation is simply an epiphenomenon of an unrelated or toxicological hepatopathy - an exaggeration possibly of the copper accumulation seen in chronic active hepatitis or the iron accumulation seen in alcoholic cirrhosis. Certainly copper accumulates in the liver during chronic cholestasis, but hepatic copper concentrations in Indian cases of extrahepatic biliary atresia not reach levels found in ICC.

It is logical to seek for a genetic component to models 2 or 3.

5. CANDIDATE GENES

5.1. The Wilson's Gene

Wilson's disease and ICC are quite dissimilar (Table 6). Nevertheless, it is necessary to exclude the possibility that a mutation at or near the Wilson's disease locus is responsi-

Table 6. Difference between Indian Childhood Cirrhosis and Wilson's disease

	Indian Childhood Cirrhosis	Wilson's disease
Age	6 m - 4 y	after 5 years
Geography	India; sporadic ICC-like cases elsewhere	Worldwide
Clinical features		
Hepatic	Failure to thrive, followed by abdominal distension, hepatomegaly, and usually later jauundice, ascites. Rapid progression, fatal.	Variable presentation fulminant hepatic failure chronic hepatitis aymptomatic hepatomegaly cirrhosis
Neurology	Hepatic encephalopathy	Neuropsychiatric abnormalities after age 10 years
Eyes	Ocular features not described	Kayser-Fleischer rings, cataracts
Other features	respiratory distress, diarrhoea, failure to thrive	haemolysis skeletal abnormalities
Biochemistry		
Plasma copper	Normal or high	Low, normal, or high
Plasma non-caeruloplasmin Cu	High	High
Plasma caeruloplasmin	Normal or high	low
Urine copper	High	High
Hepatic copper	High	High
Histopathology	Gross hepatocyte damage, Fat absent	Cu stains strongly positive
Treatment	Penicillamine effective in preterminal cases	Cu chelators, zinc

ble for ICC. Since there are now more than 100 described Wilson's disease mutations this cannot be done by direct mutational analysis. Haplotype analysis, however, has been used in ICC families where the parents were consanguineous. In the child whose parents were first cousins the probability of homozygosity at any gene locus is 1/16, the probability depending upon the degree of consanguinity. If ICC is due to a mutation at the Wilson's locus one would expect homozygosity by haplotype at that locus in the child of consanguineous parents.

Polymorphic microsatellite markers adjacent to the WD gene were studied (Figure 1). D13S314 and D13S301 are respectively approximately 900kb and 300kb proximal to ATP7B. D13S296 is approximately 400 kb distal to ATP7B. Of five families with varying degrees of consanguinity in one was there homozygosity by haplotype. This is convincing evidence that the Wilson's disease locus cannot be implicated.

In the paediatric liver clinic at KEM Hospital, Pune, 70 families with Wilson's disease and >400 cases of ICC have been seen during the period 1980 - 1997. No family has been encountered in which both diseases occurred. This makes it vanishingly unlikely that ICC results from the heterozygous state for a Wilson's disease mutation.

5.2. Metallothionein

In one case of an ICC-like disorder occurring in an American child, the defect in metallothionein synthesis was demonstrated in cultured skin fibroblasts (Hahn et al 1994). However, this defect was not found in three cases of Indian Childhood Cirrhosis from Pune, nor in an Irish child with well-water associated infantile cirrhosis (case 1 from Baker et al 1995), nor in an older child with a non-Wilsonian copper associated cirrhosis (Hahn et al 1995).

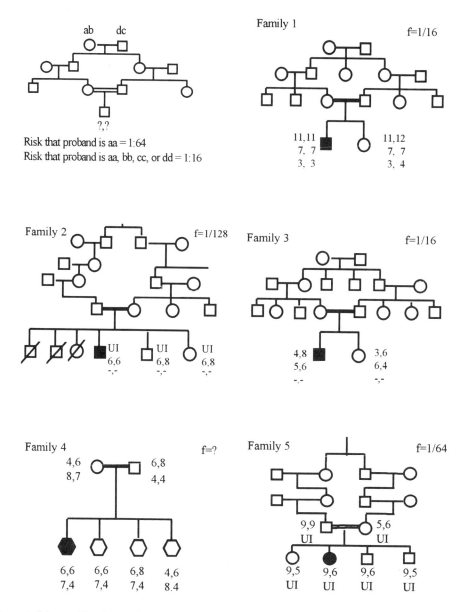

Figure 1. Polymorphic microsatelite markers were studied in 5 families in whom the parents of a child with histologically proven ICC were consanguinous. Although the proband was homozygous by haplotype at the Wilson's disease locus in family 1, this was not the case in families 3–5. Family 2 was uninformative. The ATP7B locus is therefore highly unlikely to be implicated in the causation of ICC, and, by implication, TCC.

Metallothionein is readily identifiable in ICC.

In a study of German children with infantile cirrhosis with copper storage attributed to excessive copper ingestion from contaminated water, Summer et al (1994) showed a relative impairment of metallothionein production when compared to data from children of a similar age and foetuses at mid-term. This has not been studied in ICC or TCC.

Metallothionein knockout mice do not develop an ICC-like disorder.

5.3. Caeruloplasmin

Since plasma levels of caeruloplasmin are normal or elevated in ICC it seems unlikely that this protein is aetiologically implicable. In the condition acaeruloplasminaemia, excessive iron accumulation occurs but this is not seen in ICC (Harris et al, 1995; Takahashi et al 1996). Indeed, some of the "well-water" cases (Table 3) had an apparently reduced hepatic iron concentration.

5.4. Copper Chaperones

Defects in the recently identified intracellular pathways of delivery of copper to specific proteins are possible causes of disease. These pathways of copper trafficking include a soluble copper carrier, yeast ATX1 and its human homologue HAH1, that specifically function in the delivery of copper to transport ATPases in the secretory pathway (Klomp et al 1997, Lin et al 1997). Glerum et al (1996a, b) have identified a pathway involving a soluble yeast factor, COX17, that specifically delivers copper to cytochrome oxidase in the mitochondria. Culotta et al (1997) have characterised a "copper chaperone" for superoxide dismutase.

5.6. "Cuprochromatosis?"

Genetic haemochromatosis results from a failure of regulation of iron absorption. Might the primary defect in ICC/TCC lie within the enterocyte? ie. might unregulated iron uptake cause copper overload? This seems intrinsically unlikely since the primary regulator of copper balance is biliary excretion.

6. CONCLUSION

In the absence of data it remains possible to speculate upon a possible genetic cause of infantile copper cirrhosis. Excretion from the hepatocyte rather than absorption by the enterocyte is the likely site of a possible inborn error of copper metabolism. ATP7B, caeruloplasmin, and metallothionein are unlikely to be implicated, but one of the newly described copper chaperones may be a candidate. A definitive answer awaits the study by homozygosity mapping of a number of families in which there is consanguinity.

REFERENCES

Adamson M, Reiner B, Olson JL, Goodman Z, Plotnick L, Bernardini I, Gahl WA. (1992) Indian childhood cirrhosis in an American child. Gastroenterology 102: 1771–1777

Aljajeh IA, Mughal S, Al-Tahou B, Ajrawl T, Ismail EA, Nayak NC (1994) Indian childhood cirrhosis - like liver disease in an Arab child. A brief report. Virchows Archiv 424: 225–227

Baker A, Gormally S, Saxena R, Baldwin D, Drumm B, Bonham J, Portmann B, Mowat AP (1995). Copper-associated liver disease in childhood. J Hepatol 23: 538–543

Bartok I, Szabo L, Horvath E, Ormos J (1971) Juvenile Cirrhose mit hochgradiger kupferspeicherung in der leber. Acta Hepato-Splenologica 18: 119–128

Bavdekar AR, Bhave SA, Pradhan AM, Pandit AN, Tanner MS. (1996) Long term survival in Indian childhood cirrhosis treated with D-penicillamine. Archives of Disease in Childhood 74: 32–35

Bent S, Bohm K (1995) Copper-induced liver cirrhosis in a 13-month old boy Gesundheitswesen 57:667–669

. Bhave SA, Pandit AN, Singh S, Walia BNS, Tanner MS. (1992) The prevention of Indian Childhood Cirrhosis. Annals of Tropical Paediatrics 12: 23–30

Bunton TE, Baksi SM, George SG, Frazier JM. (1987)Abnormal hepatic copper storage in a teleost fish (*Morone americana*) Vet Pathol; 24: 515–524

Culotta VC, Klomp LWJ, Strain J, Ruby Leah B. Casareno, Krems B, Gitlin JD (1997). The Copper Chaperone for Superoxide Dismutase. J Biol Chem; 272: 23469–23472

Glerum, D. M., Shtanko, A., and Tzagoloff, A. (1996a) J. Biol. Chem. 34, 20531–20535

Glerum, D. M., Shtanko, A., and Tzagoloff, A. (1996b) J. Biol. Chem. 271, 14504–14509

Gogl H, (1947) Pathologisch-anatomische Untersuchungen uber Leberzirrhose bei Sauglingen und Kleinkindern (infantile Leberzirrhose) mit endemischer Haufung. In Priesel R, ed: Wiener Beitrage zur Kinderheilkunde, vol 1. Wien: Wilhelm Maudrich.

Hahn SH, Brantly ML, Oliver C, Adamson M, Kaler SG, Gahl WA. (1994) Metallothionein synthesis and degradation in Indian childhood cirrhosis fibroblasts. Pediatric Research,; 35: 197–204

Hahn SH, Tanner MS, Danks DM, Gahl WA. (1995) Normal metallothionein synthesis in fibroblasts obtained from children with Indian childhood cirrhosis or copper-associated childhood cirrhosis. Biochemical and Molecular Medicine 54: 142–145

Harris ZL, Takahashi Y, Miyajima H, Serizawa M, MacGillivray RT, Gitlin JD (1995) Aceruloplasminemia: molecular characterization of this disorder of iron metabolism. Proc Natl Acad Sci U S A 92:2539–2543

Horslen SP, Tanner MS, Lyon TDB, Fell GS, Lowry MF (1994) Copper associated childhood cirrhosis. Gut 1994; 35 1497–1500.

Klass HJ, Kelly JK, Warnes TW (1980) Indian childhood cirrhosis in the United Kingdom. Gut 21: 344–50

Klomp, L. W. J., Lin, S. J., Yuan, D., Klausner, R. D., Culotta, V. C., and Gitlin, J. D. (1997) Identification and functional expression of HAH1, a novel human gene involved in copper homeostasis J. Biol. Chem. 272, 9221–9226.

Lefkowitch JH, Honig CL, King ME, Hagstrom JWC (1982) Hepatic coper overload and features of Indian childhood cirrhosis in an American sibship. New Engl J Med 307:271–7

Lim CT, Choo KE. (1979) Wilson's disease - in a 2 year old. J Singapore Paediatric Society. 21: 99–102

Lin, S. J., Pufahl, R., Dancis, A., O'Halloran, T. V., and Culotta, V. C. (1997) J. Biol. Chem. 272, 9215–9220

Maggiore G, de Giacomo C, Sessa F, Burgio GR. (1987) Idiopathic hepatic toxicosis in a child. J Ped Gastroenterol Nutr 6:980–983

Muller-Hocker J, Weiss M, Meyer J et al(1987). Fatal copper storage disease of the liver in a German infant resembling Indian childhood cirrhosis. Virchows Arch A; 411: 379–385.

Muller-Hocker J, Meyer U, Wiebecke B, et al. (1988) Copper storage disease of the liver and chronic dietary copper intoxication in two further German infants mimicking Indian Childhood Cirrhosis. Path Res Pract; 183: 39–45.

Muller T, Feichtinger H, Berger H, Muller W. (1996) Endemic Tyrolean infantile cirrhosis: An ecogenetic disorder. Lancet; 347: 877–880

Nayak NC, Ramalingaswami V. (1975) Indian childhood cirrhosis. Clin Gastroenterol;4: 333–349.

O'Donohue JW, Reid MA, Varghese A, Portmann B, Williams R. (1993) Micronodular cirrhosis and acute liver failure due to chronic copper self-intoxication. Eur J Gastroenterol Hepatol; 5: 561–562

Portmann B, Tanner MS, Mowat AP, Williams R. (1978) Orcein positive liver deposits in Indian childhood cirrhosis. Lancet; i: 1338–40

Pradhan AM, Bhave SA, Joshi VV, Bavdekar AR, Pandit AN, Tanner MS (1995). Reversal of Indian Childhood Cirrhosis by D-penicillamine therapy.. J Ped Gastroenterol Nutr;20: 28–35

Price LA, Walker NI, Clague AE, Pullen ID, Smits SJ, Ong TH, Patrick M (1996) Chronic copper toxicosis presenting as liver failure in an Australian child Pathology 28:316–320

Schramel P, Muller-Hocker J, Meyer U, Weiss M, Eife R. (1988) Nutritional copper intoxication in three German infants with severe liver cell damage (features of Indian Childhood Cirrhosis). J Trace Elem Electrolytes Health Dis; 2: 85–89.

Summer K (1994) Metallothionein und die chronische Toxizitat von kupfer. (Metallothionein and copper toxicity). In: Institut fur wasser-, boden-, und lufthygiene. Kupfer und fruhkindliche Leberzirrhose. (Copper and Childhood Liver Cirrhosis). Dieter, H.H. and Seffner, W. editors. Umwelt Bundes Amt. Berlin.; 38 p38–42

Tanner (1997) The role of copper in Indian childhood cirrhosis Amer J Clin Nutr 1997 in press

Takahashi Y, Miyajima H, Shirabe S, Nagataki S, Suenaga A, Gitlin JD (1996) Characterization of a nonsense mutation in the ceruloplasmin gene resulting in diabetes and neurodegenerative disease. Hum Mol Genet 5:81–84

Toyokuni S, Okada S, Hamazaki S, Fujioka M, Li JL, Midorikawa O. (1989) Cirrhosis of the liver induced by cupric nitrilotriacetate in Wistar rats. An experimental model of copper toxicosis. Amer J Path; 134: 1263–1274

Walker-Smith J, Blomfield J. (1973) Wilson's disease or chronic copper poisoning? Arch Dis Child; 48: 476–479

Weiss M, Muller-Hocker J, Wiebecke B, Belohradsky BH. (1989) First description of "Indian Childhood Cirrhosis" in a non-Indian infant in Europe. Acta Paediatr Scand; 78: 152–156

12

ANIMAL MODELS OF WILSON'S DISEASE

J. McC. Howell

Division of Veterinary and Biomedical Sciences,
Murdoch University
Western Australia 6150

1. WILSON'S DISEASE

Wilson's disease is an autosomal recessive, copper overload disorder, characterised by the co-existence of progressive neurological findings, chronic liver disease with cirrhosis, renal tubular dysfunction and pigmented corneal rings. The concentration of copper in the liver may be of the order of 200 - 3000 µg/g dry weight whereas the concentration in the liver of normal adults varies from 20 - 50 µg/g dry weight. (Danks 1995; Seymour 1987). The gene for Wilson's disease (ATP7B) maps to the long arm of chromosome 13 (Bowcock et al, 1987; Yuzbasiyan-Gurkan et al., 1987) and encodes a putative P type copper transporting ATPase (Bull et al., 1993; Tanzi et al., 1993). In humans the gene is expressed in liver, kidney, placenta and brain (Bull et al., 1993; Tanzi et al., 1993), and the mutations causing Wilson's disease have been found to be single base changes or small deletions causing frame shifts and protein truncations (Thomas et al., 1995). The disease was first described by Wilson (1912) as a familial neurological disorder, which he called lenticular degeneration. He was far sighted enough to note that "the morbid agent is probably a toxin associated with hepatic cirrhosis". It took thirty years for the association with copper overload to be established and it took eighty years before the gene was determined. It has been found useful to break Wilson's disease down into 4 stages which are : Presymptomatic, in which copper accumulates and there are no clinical signs; Copper Toxic, in which signs are present and, if untreated, the patient will die; Symptomatic but treated, in which the patient is recovering and Maintenance therapy, in which there is clinical stability but anticopper medication must continue (Brewer and Yuzbasian-Gurkan, 1989).

The two major modes of presentation are hepatic and neurological. About 40 to 45% of patients present with symptoms and signs of hepatic disease and only 35% present with neurological symptoms. A mixed picture of neurologic and hepatic involvement is more common between the ages of 12 and 25 years. The signs of hepatic disease may be insidious or acute and may be associated with haemolysis. The neurologic signs may be those of motor defects or psychiatric abnormalities, and the corneal Keyser-Fleischer rings are always present when there is neurologic damage (Walshe 1984; Seymore 1987). Copper overload syndromes are best regarded as multisystem diseases in which the liver has a very important role to play (Howell and Gooneratne, 1987). Walshe in 1982 stated that "Wilson's disease is neither a hepatic nor a nervous system affliction, but a disturbance of intermediary metabolism induced by an error of copper transport and storage; it is in truth a multisystem disease".

The basis of Wilson's disease is likely to be impaired intracellular transport of copper in the hepatocyte with associated reduction in the biliary excretion of copper. In almost all patients, caeruloplasmin concentrations and serum copper concentrations are below the normal range and urinary copper excretion is almost always increased. In the early stages of Wilson's disease, copper accumulates diffusely in the cytoplasm of hepatocytes, probably bound to metallothioneins, and as copper loading increases, the copper is sequested into lysosomes as copper associated protein. There is an increased concentration of copper in liver, kidney and brain and the Kayser-Fleischer pigmented corneal rings are due to the deposition of copper in the region of Descemet's membrane (Danks 1995; Seymour 1987; Sternlieb 1989; Walshe 1984).

The common treatments are all based on the use of compounds which remove copper, such as d-penicillamine, triethylene tetramine, zinc sulphate and ammonium tetrathiomolybdate. The de-coppering treatments are coupled with the symptomatic treatment of any existing liver disease, and in some cases liver transplantation has been used. Early diagnosis and the institution of de-coppering therapy is a most effective way of ensuring an excellent prognosis, but a poor prognosis is likely in patients with acute neurological disease (Brewer and Yuzbasiyan-Gurkan 1989; Danks 1995).

2. MODELS FOR WILSON'S DISEASE

Some inherited human diseases have precise animal models. These animal diseases have defects in the same gene, similar clinical signs and similar morphological and biochemical changes as the disease in humans. They have proved to be most useful for studies of pathogenesis and treatment (Howell 1992). There are a number of conditions in humans and animals in which excessive deposition of copper occurs, and which show hepatic signs sometimes accompanied by neurologic signs and haemolysis. These models are useful for studies of the pathogenesis and treatment of copper overload conditions but it should be pointed out that none are exact replicas of Wilson's disease.

2.1. Other Copper Overload Syndromes in Man

2.1.1. Indian Childhood Cirrhosis (ICC) and ICC- Like Disorders. These syndromes have a nutritional component but, there maybe two types, inherited and non inherited. The presenting signs are those of liver disease, but haemolysis and neurological involvement have not been reported. The onset occurs in children between 6 and 18 months of age, and death, which is usually secondary to hepatic failure, occurs within a year of diagnosis. The concentration of copper in the liver may be as high as 6,654 mg/g dry weight. It is a much more rapidly progressing disease than Wilson's disease and the changes in the liver are reported to have some unique features. These conditions have recently been reviewed by Tanner 1998a and 1998b.

2.1.2. Biliary Atresia. This condition is due to a congenital inability to excrete bile because of a malformed biliary tree. It is an obstructive jaundice syndrome with copper deposition within hepatocytes and Kuppfer cells in the later stages of the disease. Liver copper concentrations are increased, but not in the same order of magnitude as in Wilson's disease. The common presenting sign is hepatic. Haemolysis and neurological involvement are not associated with the syndrome (see Danks 1995; Seymour 1987).

2.1.3. Alpha Antitrypsin Deficiency. Alpha antitrypsin is synthesised in the liver and is responsible for the inhibition of proteolytic enzymes such as trypsin, plasmin, elastase and some of the leucocyte proteases. Alpha antitrypsin deficiency is an inherited condition which is associated with damage to the liver and lung which culminates in cirrhosis and emphysema. It presents as longstanding neonatal jaundice. Copper accumulates in the liver because of the acquired abnormalities in biliary excretion. There is neither haemolysis nor neurologic involvement (see Seymour 1987).

2.1.4. Primary Biliary Cirrhosis. This liver disease is not inherited and arises in adults due to the progressive destruction of hepatic bile ducts. It is a slowly progressive disease, which is associated with jaundice, increased circulating cholesterol levels, skin lesions on the face and hands and occasionally a peripheral neuropathy. The concentration of copper in the liver increases and is sometimes as high as that found in Wilson's disease. Treatment with penicillamine has sometimes been associated with improvements in the condition. The etiology is unclear, but there seems to be an association with autoimmune disease. The role of copper in the production of tissue damage is likely to be secondary to the immunological induced changes (see Danks 1995; Seymour 1987).

2.2. Copper Overload Syndromes in Other Animals

2.2.1. Dogs. Copper overload syndromes have been reported in Bedlington Terriers, West Highland White Terriers, Skye Terriers and Doberman Pinschers.

2.2.2. Bedlington Terriers and West Highland White Terriers. Copper toxicosis of Bedlington terriers (Owen and Ludwig 1982; Howell and Gooneratne 1987) and West Highland terriers (Thornburg et al., 1986) are similar syndromes, both are inherited as autosomal recessive conditions but in neither has the gene been characterised. Haemolysis has been reported in the Bedlington terriers but not in the West Highland Whites. In neither has neurological involvement been reported.

Most work has been done on the condition in Bedlington terriers. The accumulation of copper in the liver appears to be caused by impaired biliary copper excretion (Su et al., 1982). Studies by Haywood et al. (1996) using electron microscopy and X-ray electron probe microanalysis indicated that copper was taken up into lysosomes but following saturation of that compartment it accumulated in the nucleus. The condition can be classified into three forms, Asymptomatic, in which copper accumulates and liver damage occurs but in which clinical signs are absent; the Fulminating Form, which is usually seen in young adult dogs which die after 2 to 3 days of illness and which may be initiated by stress; and the Chronic Form, in which there is a long period of liver disease progressing to liver failure and death.

This disease and Wilson's disease are inherited autosomal recessive, progressive copper overload syndromes with marked changes in the liver. The usefulness of this model may well depend upon the nature of the responsible gene defect.

2.2.3. Skye Terriers and Doberman Pinschers. The etiology of the Skye terrier (Heywood et al., 1988) and Doberman Pinscher (Thornburg et al., 1984) copper toxicosis syndromes is unclear. They are probably not inherited but they show hepatic damage and interference with bile flow, which is probably why copper accumulates in the liver. Neither haemolysis nor neurological involvement have been reported in affected dogs.

2.2.4. Ruminants. Small ruminants, such as sheep and goats, seem particularly susceptible to copper overload. In these animals copper overload is a nutritional disease, but there is a significant genetic component, some breeds being much more susceptible than others (Weiner 1987). The sheep homologue of the Wilson's gene has recently been isolated (P. Lockhart and J. Mercer personal communication). Haemolysis is the common presenting sign and neurological involvement may be seen as a terminal phenomenon. The disease can be classified into three phases, Prehaemolytic, in which the animals accumulate copper, liver damage occurs but the animals remain symptom free; Haemolytic in which haemolysis suddenly occurs and the animal may die or recover; if the animal survives it goes into the Posthaemolytic phase during which further haemolytic episodes may occur. The animals die of either liver failure or renal failure. It has been shown that individual hepatocytes die during the copper accumulation phase, but that a short time before haemolysis there is a significant increase in the number of cells which become necrotic. Copper is liberated from the dead cells and is transported in the plasma, where it is taken up by erythrocytes which then haemolyse. If there is a significant haemolytic anaemia there is marked necrosis of cells around the central veins of the liver, which is probably due to a combination of copper overload and decreased oxygen tension in this area (see Howell and Gooneratne 1987). The changes in the kidney induced by the toxicity are largely confined to the cells of the proximal convoluted tubules where copper is stored in lysosomes. At haemolysis these cells accumulate more copper and take in haemoglobin breakdown products. The onset of haemolysis is closely followed by degeneration and necrosis of some of the cells (Gooneratne et al., 1986; Howell and Gooneratne 1987).

In small ruminants the liver changes are much more acute than those in Wilson's Disease and fibrosis is not a common feature, but in sheep which are copper loaded for long periods of time fibroblast proliferation does occur (Ishmael et al., 1971). The neurological changes seen in sheep are those of hepatic encephalopathy (Howell et al., 1974), a condition which may accompany liver failure due to any cause in all animals.

2.2.5. Rodents. Copper overload syndromes are seen as inherited conditions in toxic milk mice, the Long Evans Cinnamon (LEC) rat, and as nutritionally induced copper toxicosis following the use of copper salts given by injection or added to the feed or drinking water.

2.2.6. Inherited Conditions. The toxic milk mouse (Rauch 1983; Howell and Mercer 1994) and the LEC rat (Li et al., 1991; Sugawara et al., 1993; Takeichi et al., 1988) are autosomal recessive copper overload conditions. Both of these models have mutations which affect the Wilson's gene homologue. The LEC rat has a deletion which removes the 3' end of the gene (Wu et al., 1994), and the toxic milk mouse has a point mutation leading to an amino acid substitution in the 8th transmembrane domain (Theophilus et al., 1996). In both conditions the common presenting signs are those associated with liver damage and haemolysis occurs in both (Howell and Mercer 1994; Yoshida et al., 1987). Neurological signs have not been reported in the toxic milk mice but convulsions have been reported in LEC rats (Li et al., 1991). In both models the pattern of copper metabolism is similar and both have been promoted as animal models of Wilson's disease (Theophilus et al., 1996; Suzuki et al., 1995) The histological appearance of the hepatocytes is similar in both conditions with many enlarged and abnormally shaped cells and nuclei (Howell and Mercer 1994; Takeichi et al., 1988; Yoshida et al., 1987). In addition in the LEC rat an abnormally high number of multinucleate cells and polyploidy have been reported (Fujimoto et al., 1989).

2.2.7. Nutritionally Induced Injury. Hepatic damage has been induced in rats by feeding them a diet containing large quantities of copper. The presenting signs related to liver damage, but neither haemolysis nor neurological signs were found. After a period of time the rats appeared to became resistant to a high concentration of copper in the diet, the concentration of copper in the liver fell and regeneration and tissue repair occurred. Necrosis of hepatocytes occurred without enlargement of nuclei (Haywood 1980). Copper nitrilotriacetate was reported to have produced necrosis of hepatocytes and subsequently cirrhosis of the liver when given by intraperitoneal injection to rats. Haemolysis occurred but there were no neurological signs (Toyokuni et al., 1989). Diets containing a high concentration of copper were fed to C57 mice and produced degeneration and necrosis of hepatocytes with no evidence of haemolysis and no neurologic signs. Copper has also been fed to toxic milk mice and this has induced significant liver damage and death in these animals (D. Hughes, P. Dorling and J. Howell unpublished observations). In neither the rats nor the mice was enlargement of hepatocyte nuclei reported.

3. GENERAL CONSIDERATIONS

Some of these animal models of Wilson's disease are due to genetic abnormalities, others are due to the administration of copper salts. They are copper overload syndromes but none of them are the exact equivalent of Wilson's disease in people. Changes in the brain are not common in these models, but changes in the liver are. In the chronic progressive disorders, such as that seen in Bedlington terriers the gross and microscopic lesions in the liver may be similar to those of Wilson's disease. However, the histological appearance of the liver in toxic milk mice and LEC rats is markedly different from that seen in both Wilson's disease and nutritionally induced copper overload in other animals including rodents. The similarities and the differences have been, and will continue to be, used to provide information which will help develop a better understanding of copper metabolism, the pathogenesis of copper overload injury and the development of treatments for copper overload syndromes.

Animal models have already contributed significantly to the development of treatments for Wilson's disease. Tetrathiomolybdate was developed as an agent which would prevent and treat chronic copper poisoning in sheep (Gooneratne et al., 1981a; 1981b; Humphries et al., 1988), has been used in LEC rats (Sugawara et al., 1995) and in toxic milk mice to reduce the concentration of copper in the liver (J. Howell and J. Mercer unpublished observations) and is now used for the treatment of Wilson's disease in people (Walshe 1986; Brewer and Yuzbasiyan-Gurkan 1989).

3.1. Changes in the Liver

3.1.1. The Nature of the Lesions. The gross appearance of the liver varies from model to model but in general terms if the liver damage is of some standing the liver may be firmer than normal, nodular and may be reduced in size. These changes are an indication that there is longstanding damage and loss of hepatocytes with resulting fibrosis and the development of nodules of hyperplastic, regenerating hepatocytes. This is the appearance that may be seen in longstanding cases of Wilson's disease, in the toxic milk mouse and in Bedlington terriers. It is the result of a combination of damage, repair and regeneration. It is not specific for a particular disease. If animals die as the result of acutely induced hepatocyte damage, as they often do in chronic copper poisoning in sheep, there is insufficient time for fibrosis to occur.

There are similarities and differences in the histological appearance of the liver in the copper overload syndromes. The basic changes in the liver in all copper overload syndromes are degeneration and necrosis of hepatocytes. Fibrosis and the development of nodules of hyperplastic hepatocytes will only occur if the disease process is ongoing and long standing. Such changes would be expected to occur in the genetically induced diseases in which the disease processes start before birth and progress throughout life, for in these diseases there is chronic, progressive liver injury. The changes would only be seen in the nutritional copper overload syndromes if they were allowed to progress. It would be interesting to perform a detailed, time related investigation of the progress of the changes in morphology and distribution of copper in the liver of Wilson's disease and several animal models. This would avoid the confusion that sometimes occurs in the literature when lesions from differing stages of the disease are compared.

3.1.2. Changes in Hepatocytes. 3.1.2.1. Differences in Copper Loading. In the copper overload syndromes the major copper storage cell is the hepatocyte but not all hepatocytes are involved to the same extent. In a copper loaded liver adjacent hepatocytes may be either substantially loaded with or contain very little copper This has been shown in the sheep Fig. 1 (Kumaratilake and Howell 1989) (Fig. 2), in the Bedlington Terrier Fig. 3, and in the LEC rat (Fujii et al., 1993). The reason for this is not known but combined his-

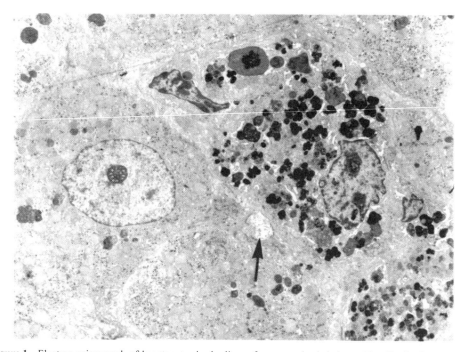

Figure 1. Electron micrograph of hepatocytes in the liver of a copper loaded sheep. The distribution between hepatocytes of copper loaded electron-dense lysosomes between hepatocytes is unequal and the hepatocytes that are packed with them are degenerating. The arrow indicates a bile canaliculus. x 2953.

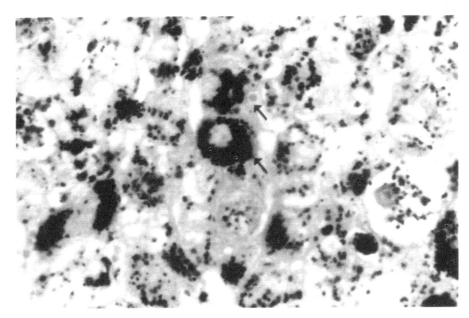

Figure 2. Section of liver of a copper loaded sheep stained for copper. There is an unequal distribution of positively stained granules between hepatocytes. The arrows indicate two of the hepatocytes that are packed with copper containing, positively stained granules. Ferricyanide method. x 732 (From Kumaratilake, J. and Howell, J. McC., 1987, Res. Vet. Sci., 42, 73–81).

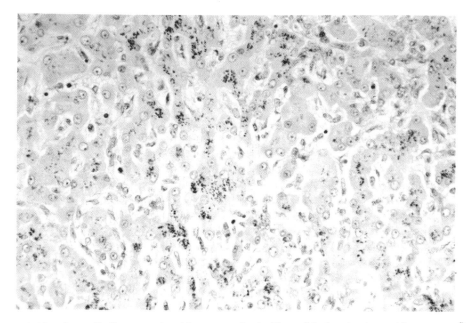

Figure 3. Liver from a Bedlington terrier with copper toxicosis. Many of the hepatocytes contain numerous fine, copper positive, dark granules. These cells are heavily copper loaded. Some adjacent hepatocytes contain very few granules. P-dimethylaminobenz-ylidene rhodamine stain. x 260.

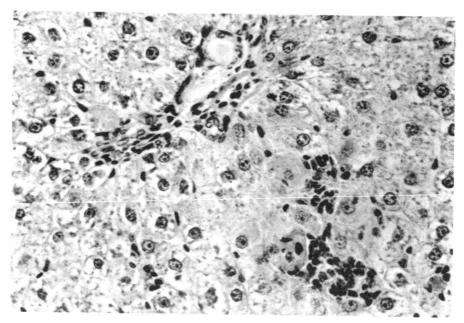

Figure 4. Liver from a male rat which received 2,000 mg per kg copper in the diet. Some necrotic hepatocytes and inflammatory cells are present. Markedly enlarged nuclei are absent. H + E. x 505 (From Haywood, S. 1979, J. Comp. Path., 89, 481–487).

tochemical, immunohistological, *in situ* hybridisation and other studies utilising the recent molecular biological findings may help to develop an understanding of why some hepatocytes become overloaded with copper and die, whereas those adjacent to them do not.

3.1.2.2. Differences in the Morphology of Hepatocytes in Inherited and Nutritionally Induced Copper Loading. The histological appearance of hepatocytes in the toxic milk mouse and the LEC rat is markedly different from that seen in the hepatocytes of other copper overload syndromes. When rats, other than the LECs, are nutritionally copper overloaded individual hepatocytes die and foci of necrosis occur. There is little increase in size of hepatocyte nuclei (Fig. 4) (Haywood 1980). Similar changes have been seen in copper overloaded C57 mice (D. Hughes, P. Dorling and J. Howell unpublished observations). However in the toxic milk mouse and the LEC rat the changes are those of increase in size of hepatocytes, with some developing a foamy cytoplasm, and increase in size of hepatocyte nuclei, some of the nuclei showing marked enlargement, bizarre shapes with inclusion bodies and infoldings of the nuclear membrane Figs. 5, 6 and 7 (Howell and Mercer 1994; Fujii et al., 1993). These changes are different from the megalocytosis of hepatocytes which is induced by pyrrolizidene alkaloids and is seen when copper and pyrrolizidene alkaloids act together (Howell et al., 1991) They are also different to the vacuolation of hepatocytes which is sometimes seen in copper poisoning in sheep (Ishmael et al., 1971), in copper toxicosis in Bedlington terriers (Eriksson and Peura 1989) and in Wilson's disease (Walshe 1984 . This change is probably an involution of the nuclear membrane. These differences may relate to the way in which the hepatocytes handle the copper. It may be that the nucleus has a greater degree of involvement in copper storage in the toxic milk mouse and the LEC rat than it does in the nutritionally overloaded rodent.

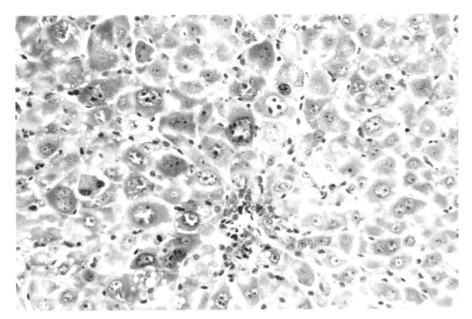

Figure 5. Liver from a 87 day old female LEC rat, Uneven nuclear size is evident. Some hepatocytes have greatly enlarged nuclei. H + E. x 260. (The tissue was supplied by Dr K H Summer, GSP, National Research Centre for Environmental Health, Neuherberg, Germany).

Studies of copper storage in the various compartments of the cell, including the nucleus, of the movement of copper between the compartments, and of the interactions of the newly discovered genes and their proteins may help in our understanding of the differences in behaviour between hepatocytes in the models of copper loading outlined above and will provide new information on transport, storage, mobilisation and excretion of copper. An extension of the recent work of Kuo et al. 1997 for example may be of value. These workers used RNA *in situ* hybridisation to determine the distribution of mottled and toxic milk transcripts during mouse embryonic development. Their studies indicated that the toxic milk gene product may be specifically involved in the biosynthesis of distinct cuproteins in different tissues.

4. CHANGES IN THE BRAIN

Wilson's disease was first described as lenticular degeneration, a neurological disorder and Wilson (1912) drew attention to the cystic changes within the anterior brain stem nuclei, particularly the lenticular nucleus. These changes together with the formation of the so called Opalsky cells, which are PAS positive altered glial cells, are characteristic of the brain changes in Wilson's disease (Duchen and Jacobs 1984). The changes in the brain are clearly associated with increased concentrations of copper in the brain (see Table 1) (Cummings 1948). Kodoma et al. 1988 reported the concentration of copper found in the cerebrospinal fluid (CSF) of four patients with Wilson's disease. In 2 of the patients without neurological signs the concentration of copper in the CSF was within normal limits, but the concentration in the 2 patients with neurological signs was significantly elevated. Following penicillamine therapy the neurological signs diminished and the concentration

Table 1. Copper concentration in liver and anterior brain stem nuclei (mg Copper/100 g tissue)

	Normal	Jaundice	Cirrhosis	Wilson's Disease		
				Normal	Jaundice	Cirrhosis
Liver	3.7 - 17.2	7.3	20.3	156.4	55.0	39.4
Caudate	3.4 - 9.4	4.2	6.1	10.1	31.8	13.8
Putamen	6.1 - 12.0	9.8*	8.8*	-	60.5	69.3
Globus pallidus	-	-	-	8.4	23.0	39.9

*Putamen and Globus pallidus were measured together.
Adapted from Cummings, 1948

of copper in the CSF was reduced. The reason for the elevations of copper concentration in the brain and cerebrospinal fluid of patients with Wilson's disease is not known. It has been suggested that the elevations occur when copper overflows from a saturated, damaged liver, but it has been reported that "occasionally, even in the presence of an advanced neurological lesion, the liver histology may show only minor abnormalities" (Walshe 1984). One would not expect much copper outflow from liver to blood in these cases. It has also been reported that "occasionally the illness (Wilson's disease) is chronic, but it is more common for it to run a fluctuating course over a few months, at most two to three years. In the latter case, neurologic signs are likely to appear before the terminal episode, which may be liver failure, haemorrhage, or general inanition from widespread destruction of motor centres of the brain" (Walshe 1982).

The clinical fluctuations are likely to be associated with periods of an increased rate of death of copper loaded hepatocytes in a heavily copper loaded liver. This may lead to elevations of readily exchangeable copper in the blood and presumably increased deposition in brain, but it would not account for the neurological cases where there is little liver damage. There may be an alternative explanation to that of overflow of copper from a damaged liver which could account for the elevated concentration of copper in the brain of Wilson's disease patients. In this disease the concentration of copper in plasma is usually reduced, caeruloplasmin is greatly reduced and there is a much greater proportion of plasma copper present as non-caeruloplasmin direct reacting copper (Danks 1995; Cartright et al., 1960 . In this Wilson's disease appears to be unique among the copper overload syndromes. It could be that the high concentrations of copper in brain and CSF are brought about by the changes in the blood, which exist over a number of years, coupled with a copper receptor system present in brain but not in other long surviving cells, such as muscle. Perhaps an isoform of the gene which expresses in brain is responsible for either a failure to keep copper out or to remove it once it gets in.

In copper overload syndromes other than Wilson's disease, elevations of brain copper are not commonly recorded and when they are, the elevations do not often approach those seen in Wilson's disease. However Danks, 1995 records the occurrence of neurological symptoms due to lenticular copper deposition in a 5 year old girl with Aagenes syndrome and in a 19 year old woman with an unclassified cholestasis syndrome. The concentration of copper in the brain and spinal cord of copper loaded sheep was found to be significantly elevated but not to the same degree as in the brain of Wilson's disease (Howell and Gooneratne 1987) neither was the copper concentration in the cerebrospinal fluid of copper loaded sheep elevated during or subsequent to the haemolytic crisis (Gooneratne and Howell 1979). Sugawara et al., 1992 found that the concentration of copper in various parts of the brain of LEC rats and Howell and Mercer 1994 found that the concentration of copper of the brain of toxic milk mice was significantly greater than the concentration in the brain of control animals but it was

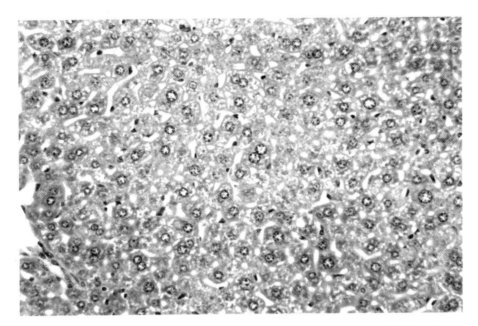

Figure 6. Liver from a control mouse. There is slight variation in the size of hepatocyte nuclei. Some hepatocytes contain two nuclei. H + E. x 260.

not elevated to the same extent as in Wilson's disease. In most affected Bedlington terriers the concentration of copper in the brain is not much higher than normal (26.4 ± 11.8 standard deviation µg copper/g dry brain with a range of 11 - 50 µg copper/g). However Twedt et al. 1979 reported that three older affected Bedlington Terrier dogs with well developed liver lesions had concentrations of 237, 331 and 2,621 µg copper/g dry brain tissue. Neurological signs and lesions were not reported.

Neurological symptoms are rarely reported in copper overload syndromes in animals, but they do occur in the terminal stages of copper overload in sheep (Howell et al., 1974). In all animals suffering from liver failure, as opposed to liver disease, neurological signs and changes in the central nervous system may be seen. The syndrome is called hepatic encephalopathy (Victor et al., 1965) and is due to ammonia induced changes in glial cells. When the liver fails it can no longer efficiently metabolise ammonia into urea for excretion by the kidney. The concentration of ammonia in the blood and in the brain increases. This is followed by the development of changes in astrocytes, neurons and oligodendrocytes and therefore myelin. The neuropathological changes associated with liver disease, including Wilson's disease, have been summarised by Duchen and Jacobs (1984).

The neurological lesions seen when the liver fails in terminal chronic copper overload in sheep are the lesions of hepatic encephalopathy and do not have the distribution and some of the characteristics of Wilson's disease (Howell et al., 1974). The neurological signs, and therefore the brain lesions, are present when some patients first present with Wilson's disease. These patients do not yet have liver failure and these brain changes are unique to Wilson's disease. However hepatic encephalopathy must have some influence on the brain lesions and symptoms of Wilsons disease patients if they develop liver failure.

A satisfactory explanation for the excessive accumulation of copper and the distribution and type of changes in the brain in Wilson's disease patients has not been found. This

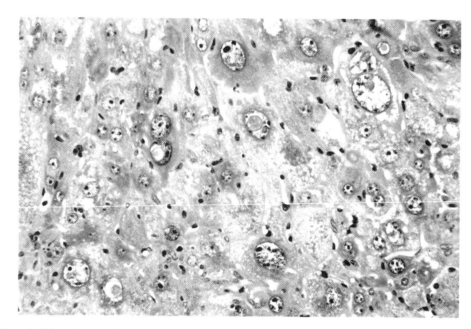

Figure 7. Liver from a male toxic milk mouse. Some hepatocyte nuclei are of normal size others are greatly enlarged and some of them contain inclusion bodies. H + E. x 260.

is an area that needs to be further explored in animal models. The models that would be most suitable for such a study are the Bedlington terrier, the toxic milk mouse and the LEC rat and of these the latter two are likely to be more accessible. It may be that the application of molecular biological investigations including *in situ* hybridisation would help by showing which cells are expressing the Wilson's homologue and whether the areas dependent upon the activity of these cells are more heavily Cu loaded and damaged. It would also be of interest to compare the lesions in the brains of animal models in copper induced liver failure with those of unfortunate cases of Wilson's disease that die due to liver failure being too acute or too advanced for successful treatment.

ACKNOWLEDGMENTS

I wish to thank Drs J F B Mercer and P R Dorling for useful discussions during the preparation of this paper.

REFERENCES

Bowcock, A.M., Farrer, L.A., Cavalli-Sforza, L.L., Herbert, J.M., Kidd, K.K., Frydman, M. and Bonne-Tamir, B. (1987). Mapping the Wilson's disease locus to a cluster of linked polymorphic markers on chromosome 13. Am J Hum Genet, 41, 27–35

Brewer, G.J., Yuzbasiyan-Gurkan, V. (1989). Wilson's Disease : An update, with emphasis on new approaches to treatment. Dig Dis 7, 178–193

Bull, P.C., Thomas, G.R., Rommens, J.M., Forbes, J.R. and Cox, D.C. (1993). The Wilson disease gene is a putative copper transporting P-type ATPase similar to the Memkes gene. Nature Genet, 5, 327–337

Cartwright, G.E., Markowitz, H., Shields, G.S. and Wintrobe, M.M. (1960). Studies on copper metabolism xxix. A critical analysis of serum, copper and cellular plasma concentrations in normal subjects, patients with Wilson's disease and relatives of patients with Wilson's disease. Am J Med 28, 555-?

Cumings, J.N. (1948). The copper and iron content of brain and liver in the normal and in hepato-lenticular degeneration. Brain 71, 410–415.
Danks, D.M. (1995). Disorders of copper transport In: the Metabolic Basis of Inherited Disease. C.R. Scriver, A.L. Beaudet, W.M. Sly, and D.E. Valle, eds (McGraw-Hill, New York), pp 2211–2235.
Duchen, L.W. and Jacobs, J.M. (1984). Nutritional deficiencies and metabolic disorders In Greenfield's Neuropathology, 4th edition, J. Hume Adams, J.A.N. Corsellis and L.W. Duchen, eds (Edward Arnold, London), pp. 595–602.
Eriksson, J. and Peura, R. (1989). Verifying copper toxicosis in Bedlington Terriers by analytical electron microscopy of nedle biopsies of liver. J Comp Path 100, 443–448.
Fujii, Y., Shimizu, K., Satoh, M., Fujita, M., Fujioka, Y., Li, Y., Togashi, Y., Takeichi, N. ad Nagashima, K. (1993). Histochemical demonstration of copper in LEC rat liver. Histochemistry 100, 249–256.
Fujimoto, Y., Takahashi, H., Dempo, K., Mori, M., Wirth, P.J., Nagao, M. and Sugimura, T. (1989). Hereditary hepatitis in LEC rats : accumulation of abnormally high ploid nuclei. Cancer Detect and Prevent 14, 235–237.
Gooneratne, S.R. and Howell, J. McC. (1979). Copper, zinc and iron levels in the cerebrospinal fluid of copper poisoned sheep. Res Vet Sci, 27, 384–385.
Gooneratne, S.R., Howell, J. McC. and Aughey, E. (1986). An ultrastructural study of the kidney of normal, copper poisoned and thiomolybdate treated sheep. J Comp Path, 96, 593–612.
Gooneratne, S.R., Howell, J,McC. and Gawthorne, J.M. (1981a). Intravenous administration of thiomolybdate for treatment and prevention of chronic copper poisoning in sheep. Brit J Nut, 46, 457–467
Gooneratne, S.R., Howell, J. McC. and Gawthorne, J.M. (1981b). An investigation of the effects of intravenous administration of thiomolybdate on chronic metabolism in chronic copper poisoned sheep. Brit J Nut, 46, 469–480
Haywood, S. (1980). The effect of excess dietary copper on the liver and kidney of the male rat. J Com Path 90, 217–232.
Haywood, S., Fuentealba, I.C., Foster, J. and Ross, G. (1996). Pathobiology of copper-induced injury in Bedlington terriers : ultrastructural and microanalytical studies. Analyt Cell Path. 10, 229–241.
Haywood, S., Rutgers, H.C. and Christian, M.K. (1988). Hepatitis and copper accumulation in Sky Terriers. Vet Pathol, 25, 408–414.
Howell, J. McC. (1992). Relevance of animal models for human disease In: Models for Duchenne Muscular Dystrophy. B.A. Kakulas, J.McC Howell and A.D. Roses eds, (Raven Press, New York), pp 89–94.
Howell, J. McC., Deol, H.S., Dorling, P.R. and Thomas, J.B. (1991). Experimental copper and heliotrope intoxication in sheep : Morphological changes. J Comp Path, 105, 49–74.
Howell, J. McC. and Gooneratne, S.R. (1987). The pathology of copper toxicity In Copper in Animals and Man. J. McC. Howell and J.M. Gawthorne eds (CRC Press, Boca Raton, Florida), pp 53–78.
Howell, J. McC. and Mercer, J.F.B. (1994). The pathology and trace element status of the toxic milk mutant mouse. J Comp Path, 110, 37–47.
Howell, J. McC., Blakemore. W.F., Gopinath, C., Hall, G.A. and Parker, J.H. (1974). Chronic copper poisoning and changes in the central nervous system of sheep. Acta Neuropath, 29, 9–24.
Humphries, W.R., Morrice, P.C. and Bremner, I. (1988). A convenient method for the treatment of chronic copper poisoning in sheep using subcutaneous ammonium tetrathiomolybdate. Vet Rec, 123, 51–53.
Ishmael, J., Gopinath, C. and Howell, J.McC. (1971). Experimental chronic copper toxicity in sheep. Res Vet Sci, 12, 358–366.
Kodama, H., Okabei, I., Yanagisawa, M., Nomiyama, H., Nomiyama, K., Nose, O and Kamoshita, S. (1988). Does CSF copper level in Wilson's disease reflect copper accumulation in the brain? Pediat Neurol, 4, 35–37.
Kumaratilake, J.S. and Howell J.McC. (1989). Lysosomes in the pathogenesis of liver injury in chronic copper poisoned sheep : An ultrastructural and morphometric study. J Comp Path, 100, 381–390.
Kuo, Y-M., Gitschier, J. and Packman, S. (1997). Developmental expression of the mouse mottled and toxic milk genes suggests distinct functions for the Menke's and Wilson disease copper transporters. Hum Mol Genet, 6, 1043–1049.
Li, Y., Togashi, Y., Sato, S., Emoto, T., Kang, J-H., Takeichi, N., Kobayashi, H., Kojima, Y., Une, Y. and Uchino, J. (1991). Spontaneous hepatic copper accumulation in Long-Evans chinnamon rats with hereditary hepatitis. J Clin Invest, 87, 1858–1861.
Owen, C.A. and Ludwig, J. (1982). Inherited copper toxicosis in Bedlington Terriers : Wilson's disease (Hepatolenticular Degeneration). Am J Path, 106, 432–434.
Rauch, H. (1983). Toxic milk, a new mutation affecting copper metabolism in the mouse. J Hered, 74, 141–144.
Seymour, C.A. (1987). Copper toxicity in man. In Copper in Animals and Man Volume II, J.McC. Howell, J.M. Gawthorne eds (CRC Press, Boca Raton, Florida) pp 79–106.

Schaefer, M., Hofmann, W.J., Dijkstra, M., Kuipers, F., Stemmel, W., Gitlin, J.D. and Vonk, J.R. (1988). Bile cannicular location of the Wilson disease protein in normal human liver. In : Copper Transport and Its Disorders : Molecular and Cellular Aspects. A. Leone and J.B.F Mercer eds, Advances in Experimental Medicine and Biology (Plenum Press, New York), pp....

Sternlieb, I. (1989). Hepatic copper toxicosis. J Gast and Hep, 4, 175–181.

Su, L-C., Ravanshad, S., Owen, C.A., McCall, J.T., Zollman, P.E. and Hardy, R.M. (1982). A comparison of copper-loading disease in Bedlington terriers and Wilson's disease in humans. Am J Physiol, 243, G226-G230.

Sugawara, N., Yuasa, M. and Sugawara, C. (1995). Excretion of copper in copper-overloaded Fischer and Long-Evans cinnamon rats treated with tetrathiomolybdate. J Trace Elements in Ex Med, 8, 1–10.

Sugawara, N., Ikeda, T., Sugawara, C., Kohgo, Y., Kato, J. and Takeichi, N. (1992). Regional distribution of copper, zinc and iron in the brain in Long-Evans cinnamon (LEC) rats with a new mutation causing heridatory hepatitis. Brain Res, 588, 287–290.

Suzuki, K.T. (1995). Disordered copper metabolism in LEC rats, an animal model of Wilson disease : Roles of metallothionein. Res Com in Mol Path Pharm, 89, 221–240.

Takeichi, N., Kobayashi, H., Yoshida, M.C., Sasaki, M., Dempo, K. and Mori, M. (1988). Spontaneous hepatitis in Long-Evans rat. A potential animal model for fulminant hepatitis in man. Acta Pathol Jpn, 38, 1369–1375.

Tanner, M.S. (1998a). The role of copper in Indian childhood cirrhosis In : Genetic and Environmental Determinants of Copper Metabolism. Ed B. Lonnerdahl. Am. J. Clin. Nutrition, in press.

Tanner, M.S. (1998b). Indian childhood cirrhosis and Tyrrolean cirrhosis, disorders of copper transport gene? In: Copper Transport and Its Disorders : Molecular and Cellular Aspects. A Leonie and J F B Mercer eds, Advances in Experimental Medicine and Biology (Plenum Press, New York), pp.....

Tanzi, R.E., Petrukhin, K., Chernov, I., Pellequer, J.L., Wasco, W., Ross, B., Romano, D.M., Parano, E., Pavone, L., Brzustowicz, L.M., Devoto, M., Peppercorn, J., Bush A.I., Sternlieb, I., Pirastu, M., Gusella, J.F., Evgrafov, O., Penchaszadeh, G.K., Honig, B., Edelman, I.S., Soares, M.B., Scheinberg, I.H. and Gilliam, T.C. (1993). The Wilson disease gene is a copper transporting ATPase with homology to the Menkes disease gene. Nature Genet, 5, 344–350.

Theophilos, M.B., Cox, D.W. and Mercer, J.F.B. (1996). The toxic milk mouse is a murine model of Wilson disease. Human Mol Gen, 5, 1619–1924.

Thomas, G.R., Forbes, J.R., Roberts, E.A., Walshe, J.M. and Cox, D.W. (1995). The Wilson disease gene : Spectrum of mutations and their consequences. Nature Genet, 9, 210–216.

Thornberg, L.P., Rottinghaus, G., Cogh, J., and Hause, W.R. (1984). High liver copper levels in two Doberman Pinschers with subacute hepatitis. J Am Hosp Assoc 20, 1003.

Thornberg, L.P., Shaw, D., Dollan, M., Raisbeck, M., Crawford, S., Dennis, G.L. and Olwin, D.B. (1986). Hereditary copper toxicosis in West Highland white terriers. Vet Pathol, 23, 148–154.

Toyokuni, S., Okada, S., Hamarzaki, S., Fujioka, M., Li, J-L. and Midorikawa, O. (1989). Cirrhosis of the liver induced by cupric nitrilotriacetate in Wistar rats. Am J Path, 134, 1263–1274.

Tweedt, D.C., Sternlieb, I. and Gilbertson, S.R. (1979). Clinical, morphologic and chemical studies on copper toxicosis of Bedlington terriers. J Am Vet Med Assoc, 175, 269–275

Victor, M., Adams, R.D. and Cole, M. (1965). The acquired (non-Wilsonian) type of chronic hepatocerebral degeneration. Medicine, 44, 345–396.

Walshe, J.M. (1982). The liver in Wilson's disease (hepatolenticular degeneration). In Diseases of the Liver, L Schiff and E R Schiff eds., (J B Lippencott Company, Philadelphia), pp. 1043–1059.

Walshe, J.M. (1984). Copper : Its role in the pathogenesis of liver disease. In Seminars in Liver Disease, 4, 252–263.

Walshe, J.M. (1986). Tetrathiomolybdate (MoS_4) as an "anti-copper" agent in man. In Orphan Diseases and Orphan Drugs. I H Scheinberg and J M Walshe, eds Filbright Papers, No 3, (Manchester University Press) pp. 76–85.

Wiener, G. (1987). The genetics of copper metabolism in animals and man. In Copper in Animals and Man, J McC Howell and J M Gawthorne eds, (CRC Press, Boca Raton, Florida), pp 45–61.

Wilson, S.A.K. (1912). Progressive lenticular degeneration : A familial nervous disease associated with cirrhosis of the liver. Brain 34, 295–509.

Wu., J., Forbes, J.R., Chen, H.S. and Cox, D.W. (1994). The LEC rat has a deletion in the copper transporting ATPase homologous to the Wilson's disease gene. Nature Genet, 7, 541–545.

Yazbasujan-Gurkan, W.A., Brewer, G. and Boerwinkle, E. (1987). Linkage of the Wilson's disease gene and polymorphic loci on chromosome 13 in North American pedigrees. Am J Hum Genet, 41, A193.

Yoshida, M.C., Masuda, R., Sasaki, M., Takeichi, N., Kobayashi, H., Dempo, K. and Mori, M. (1987). New mutation causing hereditary hepatitis in the laboratory rat. J Hered, 78, 361–365.

13

COPPER-BINDING PROPERTIES OF THE N-TERMINUS OF THE MENKES PROTEIN

Paul Cobine, Mark D. Harrison, and Charles T. Dameron[*]

National Research Centre for Environmental Toxicology
The University of Queensland
39 Kessels Rd
Coopers Plains, QLD 4108 Australia

1. INTRODUCTION

Copper is an essential element that functions as a cofactor in a wide variety of enzymes including cytochrome C-oxidase, Cu-Zn superoxide dismutase, lysyl oxidase, amine oxidase and dopamine-β-hydroxylase. Copper deficiencies are characterised by skeletal abnormalities, severe mental retardation, neurologic degeneration and mortality in extreme cases. These symptoms of deficiency are evident in newborns diagnosed with Menkes disease. The symptoms of Menkes disease result from the decreased activity of copper-dependent enzymes (Kodama, 1993). The redox properties of copper that make it so effective as a cofactor in single electron transfer also contribute to its toxicity when it is present in excess. The Menkes protein is one part of the mechanism by which the human body maintains cellular levels of copper between the limits of deficiency and excess.

1.1. The Menkes Protein

The Menkes protein, MNK, (1500 amino acids) has within its ATPase domain (850 amino acids) all the characteristic elements of a P-type ATPase; a phosphorylation site, a phosphatase domain, an ATP binding site and a transmembrane cation channel (Chelly et al., 1993; Mercer et al., 1993; Vulpe et al., 1993). P-type ATPases are a family of membrane proteins that use an aspartyl phosphate intermediate in their reaction cycle and transport cations across both the plasma membrane and other intracellular membranes. At the amino terminus of the ATPase their is a 650 amino acid domain (MNKr) that is postu-

---

[*] Please address all correspondence to: Dr. Charles T. Dameron, National Research Centre for Environmental Toxicology, The University of Queensland, 39 Kessels Rd, Coopers Plains, QLD 4108, Australia.

Copper Transport and Its Disorders, edited by Leone and Mercer.
Kluwer Academic / Plenum Publishers, New York, 1999.

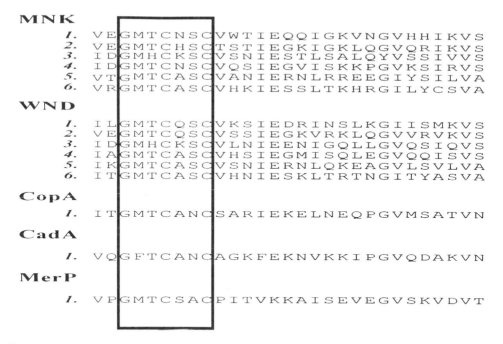

Figure 1. Alignment of the copper-binding motifs in the Menkes protein and the metal-binding motifs in selected metal-ion detoxification proteins. MNK = The Menkes protein, human, copper export. WND = The Wilsons disease protein, human, copper export. CopA = *E. hirae* (Odermatt and Solioz, 1995), copper export. CadA = *S. aureus* (Yoon et al., 1991), MerP = *T. ferrooxidans*, periplasmic mercury scavenger (Eriksson and Sahlman, 1993).

lated to be a regulatory domain. The Menkes protein, especially the ATPase domain, has strong homology with the copper-exporting ATPase deficient in Wilsons disease. It also has significant homology to prokaryotic copper, cadmium, mercury and calcium ATPases (Silver and Ji, 1994). The putative regulatory domain of the Menkes protein contains six homologous 70 amino acid modules (MNKr1–6), each containing a single -Gly-Met-X-Cys-X-X-Cys- sequence (where X is any amino acid). The module structure and metal-binding motif have been conserved in the Wilsons protein, CopA and several prokaryotic metal-ion detoxification proteins, including those involved in mercury and cadmium clearance (Solioz et al., 1994). The modules are found as components of ATPases like the MNK and the mercuric reductases (Schiering et al., 1991) and as independent transport proteins for mercury (merP (Eriksson and Sahlman, 1993)) and copper (CopZ (Odermatt and Solioz, 1995) and Atx1 (Pufahl et al., 1997)). The homology between the Menkes protein amino-terminal repeats and these metal-ion detoxification proteins is illustrated in Figure 1. Structural studies conducted on merP, a periplasmic Hg binding protein (Steele and Opella, 1997), show that the metal binding site is located at one end of the molecule. As might be expected for proteins whose function is to transfer metal ions the proposed metal binding site is exposed. In between the conserved modules of these ATPases there are regions of conserved length with unknown functions that display no predicted repetitive secondary structure (Protein Predict - ANGIS). The conservation of length but not identity suggest they may be involved as spacers in the formation of tertiary structures but not as direct structural elements.

1.2. Metalloregulation of the Menkes Protein

1.2.1. Metalloregulatory Models. The -Cys-X-X-Cys- segment of the putative metal binding site -Gly-Met-X-Cys-X-X-Cys- is common to other well characterised copper(I)-binding proteins such as metallothionein and the copper regulated fungal transcription factors ACE1 and AMT. There is no homology between MNK, the metallothioneins and the transcription factors other than the -Cys-X-X-Cys motif. Metallothioneins (MTs), ACE1 and AMT1 have numerous -Cys-X-X-Cys-, -Cys-X-Cys- and -Cys-Cys- sequences which bind the three coordinate copper(I) ions in polymetallic clusters through the cysteinyl thiolates (Pickering et al., 1993; Winge et al., 1993). The sulphydrals serve as terminal or bridging ligands for the Cu(I) ions. The number of Cu(I)-ions binding in such a polymetallic cluster cannot be determined empirically from the number of cysteine residues. Copper(I) binding promotes changes in the tertiary fold of these proteins. Typically copper is incorporated into a protein for catalytic purposes but in the metallothioneins, which function to sequester or detoxify excess transition metals, the copper ions serve as structural elements. The copper(I)-cluster forms the core of the MT structure and in the absence of metals these small proteins exist as random coils with no discernable secondary structure (Kille et al., 1994). Copper(I) complexation also leads to the formation of distinct conformations in ACE1 (Dameron et al., 1993) and AMT1 (Thorvaldsen et al., 1994) with concomitant increases in the affinity of these proteins for the yeast metallothionein promoter. In effect excess copper in effect induces, through ACE1, the synthesis of MT that subsequently sequesters the excess copper. In this case copper assumes the unusual semistructural role of being a transducer and ACE1 of being a "molecular switch". Whether the regulation takes place at transcriptional, translational or enzymatic levels the induction or stabilisation of distinct tertiary folds upon metal binding is a hallmark of metalloregulatory proteins (O Halloran, 1993).

1.2.2. Regulation of the Menkes Protein. 1.2.2.1. Putative Regulatory Domain. In an analogy to the ACE1 metalloregulatory mechanism it has been postulated that copper binding to the -Gly-Met-X-Cys-X-X-Cys- motifs may be involved in the regulation of the Menkes and Wilson's ATPases. Alternatively, the homology of the 6 modules in the N-terminus of MNK to the transport proteins merP, CopZ and Atx1 suggests the modules may be involved in the transfer of copper to the ATPase domain. Conceivably, both processes could be performed simultaneously. Translocation of copper across the plasma membrane would not seem to require six modules as the highly homologous P-typeATPase CopA (*Enterococcus hirae*) has only one module. This single module-ATPase is very effective at transporting copper (Odermatt et al., 1993). It is clear that the cellular concentration of MNK is not transcriptionally or translationally regulated by copper (Camakaris et al., 1995). An increase in the levels of extracellular copper does, however, cause a dramatic increase in the efflux of copper from cells. The efflux is dependent on MNK and its maximum is directly linked to the cellular concentration of MNK. This increase in efflux upon copper exposure is an indication that the ATPase activity is being modulated by copper. The mechanism by which the activity or apparent activity is increased is not understood. Recently it has been shown that copper appears to modulate the rapid translocation of the protein from the *trans*-Golgi network (TGN) to the plasma membrane (Petris et al., 1996). Homology searches have not revealed any of the known trafficking signals in MNK. Potentially, these metalloregulatory behaviours could be attributed to conformational changes induced in MNK upon copper binding.

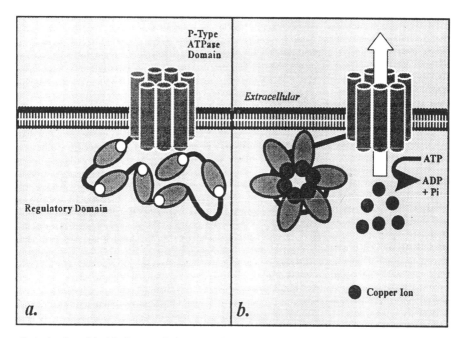

Figure 2. Activation of the Menkes protein by copper. Panel a depicts the Menkes protein in its inactive state. The copper-binding sites in the six amino-terminal sub-units are vacant. Panel b depicts the result of copper(I) binding to the six amino-terminal sub-domains. Copper(I)-binding causes a conformational change in the amino terminal region of the Menkes protein and results in the activation of the ATPase.

1.2.2.2. *Regulation of the Ca^{2+}-ATPase.* The sarcoplasmic reticulum Ca^{2+}-ATPase is a P-type ATPase activated by calcium binding to the carboxy-terminal domain of the protein. The ATPase region of the protein is highly homologous to the Menkes ATPase region. The regulatory domain in the Ca^{2+}-ATPase is found at the carboxy-terminal region in contrast to the amino-terminal regulatory domain present in the Menkes protein. Calcium-binding is facilitated by the protein calmodulin. It has been proposed that in the absence of Ca^{2+}-calmodulin the regulatory domain of the protein interacts with the energy transduction domain (ATPase region) and denies the protein structure the mobility required to bring the catalytic aspartic acid residue and the ATP binding site into close proximity (Falchetto et al., 1992). In the presence of Ca^{2+}-calmodulin the carboxy-terminal regulatory domain dissociates from the energy transduction region of the ATPase leading to phosphorylation of the aspartic acid residue and subsequently transport of Ca^{2+} by the pump.

1.2.2.3. *Regulatory Model.* Activation of the Menkes protein in response to increased intracellular copper levels could be achieved by a mechanism similar to that of the Ca^{2+}-ATPase. Considering the clustering of copper ions in other Cu(I) binding/regulatory proteins it is also plausible that an analogous structure could be formed between the modules. In this case it would be expected that copper binding would induce a significant conformational change in the amino-terminal domain that would in turn lead to the activation of the ATPase. The formation of such a cluster could then fulfil two distinct but related roles as both a "molecular switch" to activate the ATPase in the presence of increased intracellular copper and as a signal for the cell to form vesicles enriched with the protein and transport them to the plasma membrane. The purpose of this investigation is to develop an understanding of the copper binding properties of MNK, its activation and the mechanism of activation. To facilitate the characterisation of the metal binding properties of MNK we have over expressed and purified the N-terminal 648 amino acids (MNKr) and have studied its interactions with metals. Here we provide biochemical evidence that copper MNKr can induce multiple distinct conformations that contain a Cu(I) in solvent shielded environments, probably as copper-thiolate clusters.

2. SPECTROSCOPIC ANALYSIS OF Cu-MNKr

The N-terminal domain of the Menkes protein (1–1944 bp) was amplified using the polymerase chain reaction and cloned into the S-tag expression vector pET29-a (Novagen). This system allowed MNKr to be over expressed as a fusion protein with the S-Tag as the fusion partner. The S-tag is a 15 amino acid fragment of Ribonuclease S. This expression system produced 5–10 mg per litre of MNKr-S-tag fusion protein. The hydrophobic nature of the fusion protein resulted in its precipitation with the cell debris during purification. The purification of the insoluble fusion to homogeneity was readily achieved by dissolving the precipitated fusion protein in guanidine hydrochloride followed by ammonium sulfate fractionation and size exclusion chromatography (Figure 3). After its initial solubilisation in GdHCl the protein is easily maintained in buffers with 0.3 M or more NaCl and a pH of 7.6 or higher. The pI of the fusion protein is approximately 6.7 and should be avoided. The purified MNKr fusion was denatured in 6 M GdHCl and incubated at 40°C with 150 mM dithiothreitol prior to acidification and buffer exchange into 0.025 M hydrochloric or 0.1% trifluoroacetic acid by gel filtration, Sephadex G-25 (Pharmacia). This apo-protein does not contain bound metal ions and the cysteine residues are in their reduced form, as assessed by the diothiopyridine (DTDP) assay for reduced sulphydryl groups (Grassetti and Murray, 1967). Samples with reduction state >95% were used in spectral studies and metal titrations.

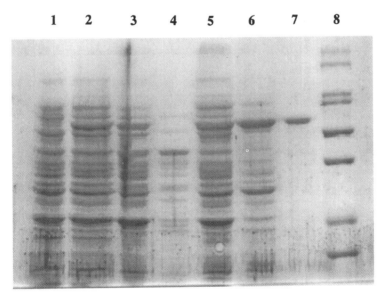

Figure 3. SDS-PAGE purification gel of the MNKr S-tag fusion. Lane 1, Uninduced cell lysate; Lane 2, Induced cell lysate; Lane 3, Size Standards; Lane 4, Insoluble fraction; Lane 5, Soluble fraction; Lane 6, 30%-50% Ammonium Sulfate fraction; Lane 7, Pooled and concentrated Gel Filtration fractions.

Amino acid analysis was used to verify the stoichiometry of the protein samples. Direct amino acid analysis is much preferred over indirect methods such as comparisons of coomasie staining intensities on SDS-Page (Lutsenko et al., 1997). In a similar vein flame atomic absorption spectroscopy was used to determine the copper concentration as opposed to the chemical assay developed by Harris *et al* (Brenner and Harris, 1995) which is not suitable for samples in the nanamole range. These methods of determining copper and protein concentrations are crucial for stoichiometry analysis.

2.1. Copper Titration into MNKr: Constant Protein

Based on the homology between the N-terminal modules of the Menkes fusion and other copper binding metalloproteins and the *in vivo* response of the Menkes protein to copper, metal titrations were performed using copper(I). Copper reconstitutions into MNKr were performed anaerobically with $[CuCl_2]^-$ (stabilised in hydrochloric acid and sodium chloride) as previously described (Byrd et al., 1988). The *in vitro* Cu(I) stoichiometry was determined by monitoring the formation of a Cu(I)-thiol charge-transfer band (254 nm) in the UV-visible spectrum (220 nm–820 nm) at sequential equimolar concentrations of Cu(I). There were no absorption bands in the visible range. The optical transition at 254 nm peaks at 8 molar equivalents of Cu(I) per MNKr (Figure 4).

Cu(I) has the unique spectral property of forming luminescent species at room temperature when the copper is ligated to thiols and shielded from the solvent. This property has been demonstrated and exploited in the study of other Cu(I) binding proteins such as the metallothioneins (Byrd et al., 1988), ACE1 (Dameron et al., 1991; Winge et al., 1993) and AMT (Thorvaldsen et al., 1994). The luminescence data was collected by monitoring emission at 600 nm with an excitation wavelength of 295 nm. The relative luminescence increases with additions of equimolar equivalents of Cu(I) to a stoichiometry of 8 mole equivalents of Cu(I) per mole of MNKr (Figure 5). The emission band was sensitive to oxygen and was quenched by cyanide but not by EDTA (data not shown). The emission peak red shifted 5–10 nm and decreased in intensity at stoichiometries above 8 moles of Cu(I) per mole of MNKr (Figure 5). This decrease is consistent with data obtained on the MTs and ACE1; shielding of the Cu(I) is lost as excess Cu(I) binds within and expands the binding site. The binding of 8 molar equivalents of Cu(I) per MNKr under limiting concentrations of copper was confirmed by both the UV-visible and luminescence spectra. The formation of the copper-specific luminescence shows that the copper(I) ions are pro-

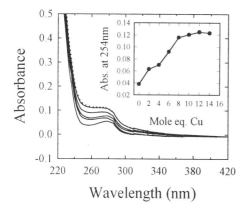

Figure 4. The UV-visible spectrum of Cu(I) titrations into apo-MNKr. The absorption spectra of 2 nmoles apo-MNKr titrated with increasing molar equivalents of Cu(I) was determined from 220 nm - 420 nm. (——) Cu(I) equivalents to 8 Cu(I) per MNKr (·······) Cu(I) equivalents above stoichiometric concentrations. {Inset: Absorption at 254 nm versus molar equivalents of Cu(I)}.

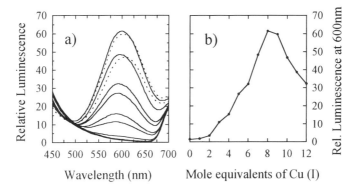

Figure 5. Luminescence of Cu(I)-MNKr. a. Emission Spectra of Cu(I)-MNKr at room temperature, 23 °C (excitation 295 nm). Relative Intensity versus Emission Wavelength. b. Relative Intensity at 600 nm vs Mole equivalents of Cu(I). Performed with 2 nmoles apo-MNKr using 350 nm filter. (----) Concentrations of Cu(I) to stoichiometry. (.....) Concentrations of Cu(I) above a stoichiometry of 8 Cu per mole equivalent.

tected from solvent, potentially the metal could be organised into a copper(I)-cluster within MNKr, analogous to that found in ACE1.

2.3. Copper Titration into MNKr: Constant Copper

The *in vitro* stoichiometry was also determined by titrations performed at a fixed concentration of copper with varying equimolar equivalents of apo-MNKr. Constant copper titrations have been used to investigate the nature of metal binding, random vs cooperative binding, in the metallothioneins (Byrd et al., 1988). In these reconstitutions the stoichiometry of copper binding, as assessed by luminescence at 600 nm, was 4 moles of Cu(I) per mole of MNKr (Figure 6). The luminescence spectra display the same red shift and decrease in intensity at concentrations of copper above the stoichiometry as is observed in the constant protein titrations. The discrepancy in the maximum metal binding

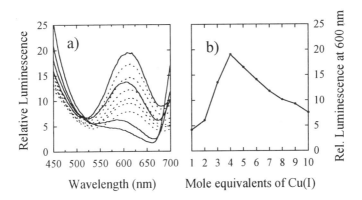

Figure 6. Constabt Copper(I) titrations of MNKr: a. Relative intensity vs. emission wavelength (nm); b. realtive luminescence at 600 nm versus molar equivalents of Cu(I). Titrations performed with a fixed protein concentration of 2 nmoles/1.5ml. Spectra were collected with 350 nm filter. Cu(I) equivalents to stoichiometry (—) Cu(I) equivalents above stoichiometry (······).

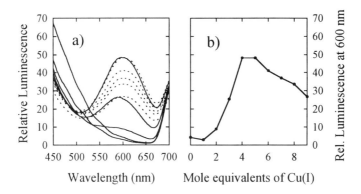

Figure 7. Heated reconstitutions of MNKr with Cu (I). a. Relative Intensity vs. Emission Wavelength (nm); b. Relative luminescence at 600 nm versus molar equivalents of Cu(I). Titrations performed with a fixed protein concentration of 2 nmoles/1.5 ml. Spectra were collected with a 350 nm filter.

stoichiometry is unprecedented as this was not seen in the MT and ACE1 studies. One would expect that the total number of metals bound per protein would be the same whether they bound under copper or protein limiting conditions and/or whether the binding was cooperative or non-cooperative. The differences suggest the reconstitutions of MNKr are more complicated. In an attempt to probe for thermodynamic differences the copper(I) proteins were exposed to short, 30 min, incubations at elevated temperatures. Heat treatment of the mammalian metallothioneins causes rearrangement of metals into their highest affinity binding sites (Green et al., 1994). By heat treating either the apo-MNKr or 8Cu-MNKr in the presence of Chelex 100 (BioRad) the stoichiometry of the copper-binding to MNKr, determined by luminescence and UV-visible spectra, is 4 mole equivalents of Cu(I) per MNKr (Figure 7). Both the 8 Cu-MNKr and the 4 Cu-MNKr configurations were stable to removal of copper ions by Chelex-100.

3. SUMMARY

The *in vivo* stoichiometry of copper binding to MNK has yet to be determined but *in vitro* reconstitutions of the N-terminal domain of the Menkes protein suggest copper(I) can induce the protein to adopt distinct conformations. Stability analyses, in the form of heat treatments and protein limited conditions, suggest that a stoichiometry of 4 moles of Cu (I) per mole of MNKr is the more stable conformer (Figure 8). The binding of copper has been studied under conditions of varied pH and buffering systems and also in the presence of reductants (cysteine and glutathione) and none of these conditions affected the stoichiometry. The stoichiometry of the MNKr protein has been determined using the most precise analytical methods available; quantitative amino acid analysis for determination of protein concentration and flame atomic absorption spectroscopy for the determination of metal concentration. In other metal binding studies, MNKr displays no specific binding of the heavy metals Cd and Zn and no charge transfer band was formed for the Cd-MNKr (Data not shown). Additions of either Cd or Zn led to the rapid precipitation of the protein. The precipitation Cd or Zn could be reversed through the addition of EDTA.

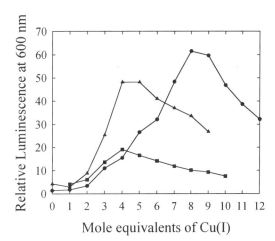

Figure 8. Summary of copper titrations of the N-terminal domain. Relative Luminescence at 600 nm vs Mole Equivalents of Cu(I): ●, constant protein titration; ▲, heat treated constant protein titration; ■, constant copper titration.

4. DISCUSSION

In vitro metal binding studies suggest that MNKr, the putative regulatory portion of MNK, can exist in multiple conformations, a prerequisite for its involvement in metalloregulatory mechanism. Specifically, the titration data suggests the protein can form and convert between 4 and 8 copper(I) structures, demonstrating some flexibility in the tertiary folding of the molecule. This flexibility of the cluster may be important in metal transfer, translocation or regulatory processes. The homology of the modules in MNKr to the intracellular copper transport proteins CopZ, Atx1 and HAH1 implies that the Menkes modules may be involved in the transfer of copper to the ATPase domain of the Menkes protein. On the other hand, cell biology studies clearly suggest the protein (Camakaris et al., 1995) and/or its cellular location (Petris et al., 1996) are metalloregulated. Presumably, the roles of modules in metalloregulation and metal transfer are not exclusive. Metalloregulation would likely be caused by the copper induced conformational changes of the modules or interactions between the modules. Indeed the homologue of the MNK modules, merP, has distinct apo- and Hg-merP conformations as shown by NMR (Steele and Opella, 1997). Mercury stabilises the -M-X-C-X-X-C-loop and promotes shifts of the alpha helices and beta sheets. It is plausible, therefore, that the modules in MNKr will undergo similar molecular re-arrangements and thereby promote the activation or change in cellular position of the ATPase. Simple re-arrangements of the module level are, however, not consistent with the fluorescence and titration data. As expected from the merP model the individual modules are not luminescent (data not shown). To provide the solvent protection necessary to achieve room temperature luminescence it is likely that Cu(I) would need to be shielded between two or more modules. In the other poly-cysteine copper(I)-binding proteins, ACE1 and MT, copper is bound in a three coordinate distorted trigonal planer conformation. If each binding site in the MNKr1–6 modules were able to trigonally coordinate Cu(I) through the -M-X-C-X-X-C- binding site we would expect the protein to bind at least 6 metals and perhaps more if the cysteines could serve as bridging ligands. Methionines are not strong ligands for copper and are readily displaced by cysteines and other strong ligands. Recent EXAFS and mutation studies on Atx1 argue for the involvement of the two cysteines but against the involvement of the methionine in the coordination of the copper(I) in Atx1 (Pufahl et al., 1997). The third ligand in Atx1, an independent molecule/module, is postulated to be from the involvement of a weak distant, undefined, thiolate. In the case of MNKr, however, the modules are not independent. In MNKr the six modules are tethered closely together by the chain of intervening amino acids. Tethering of the modules effectively increases the local concentration of the thiolates and the chances the sites could interact. If only the 12 cysteines were involved in the ligation of the Cu(I) then we expect, as observed in the more stable MNKr conformer, to bind 4 Cu(I) ions. In this case, an interesting counter point to MT and ACE1, the Cu(I) ions would serve as bridges between modules. More Cu(I) could be incorporated if the ligands were also bridging. Synthetic inorganic 8 Cu-12S clusters with bridging thiolates are readily formed and have been well characterised (Dance and Calabrese, 1976). In the case where the cysteines were not bridging it would be expected that 1 Cu(I) ion would be bound per 3 modules in a conformation that might resemble a 3 bladed propeller with a stack of 2 Cu(I)s as the connecting shaft. MNKr would have two such propellers. It would be likely in this case that a polymetallic copper(I)-thiolate cluster analogous to those in MT and ACE1 would form. Large conformational changes would be induced if a cluster of this type were formed. Cluster formation could serve as the conformational switch to activate the protein in a manner analogous to that observed in the Ca^{2+}-regulated ATPases. In this scenario the cluster formation would allow the critical aspartic acid to be phosphorylated and Cu(I), not in the cluster, to be translocated across the membrane. Similarly, the conformational change could be involved in export of the ATPase from the TGN to the plasma membrane. Additionally, the tertiary folds responsible for the shielding of copper may provide protection from copper induced oxidative damage in the cell while activating the protein. The kinetic lability of clusters would be unlikely to impede the translocation of the copper if the copper in the cluster was the transported species.

5. CONCLUSIONS

The *in vitro* studies of the N-terminal domain of the Menkes protein have confirmed that it serves as the metal binding domain of the Menkes protein. Here we present the plausible formation of a polynuclear copper cluster similar to those observed in other Cu(I)-binding regulatory and detoxifying proteins. We propose that the tertiary folding of the N-terminal domain around a metal cluster provides the structural changes necessary to activate and/or alter the cellular location of the Menkes protein.

ACKNOWLEDGMENTS

We would like to acknowledge Julian F.B. Mercer and Paul Lockhart for supplying the Menkes the AT7A clone, Shadi Mogahadas for technical support and the support of the Australian Research Council for funding PC and operation expenses. CTD and MDH acknowledge the support of the National Health and Medical Research Council, Queensland Health, Griffith University and The University of Queensland.

REFERENCES

Brenner, A., and Harris, E. D. (1995). A quantitative test for copper using bichinchioninc acid. Anal. Biochem. 226, 80–84.

Byrd, J., Berger, J. M., McMillin, D. R., Wright, C. F., Hamer, D., and Winge, D. R. (1988). Characterization of the copper-thiolate cluster in yeast metallothionein and two truncated mutants. J. Biol. Chem. *263(14)*, 6688–6694.

Camakaris, J., Petris, M. J., Bailey, L., Shen, P., Lockhart, P., Glover, T. W., Barcroft, C. L., Patton, J., and Mercer, J. F. B. (1995). Gene amplification of the menkes (MNK;ATP7A) P-type ATPase gene of CHO cells is associated with copper resistance and enhanced copper efflux. Hum. Mole. Genetics 4, 2117–2123.

Chelly, J., Tumer, Z., Tonnesen, T., Petterson, A., Ishikawa-Brush, Y., Tommerup, N., Horn, N., and Monaco, A. P. (1993). Isolation of a candidate gene for Menkes disease that encodes a potential heavy metal binding protein see comments. Nat. Genet. 3, 14–19.

Dameron, C. T., Winge, D. R., George, G. N., Sansone, M., Hu, S., and Hamer, D. (1991). A copper-thiolate polynuclear cluster in the ACE1 transcription factor. Proc. Natl. Acad. Sci. USA 88, 6127–6131.

Dameron, C. T., Arnold, P., Santhanagopalan, V., George, G., and Winge, D. R. (1993). Distinct Metal Binding Configurations in ACE1. Biochem. 32, 7294–7301.

Dance, I. G., and Calabrese, J. C. (1976). The Crystal and Molecular Structure of the Hexa-(u2-Benzenethiolato)tetracuprate(I) Dianion. Inorganica Chim. Acta 19, 141-142.

Eriksson, P. O., and Sahlman, L. (1993). 1H NMR studies of the mercuric ion binding protein MerP: sequential assignment, secondary structure and global fold of oxidized MerP. J. Biomol. NMR 3, 613–626.

Falchetto, R., Vorherr, T., and Carafoli, E. (1992). The calmodulin-binding site of the plasma membrane Ca2+ pump interacts with the transduction domain of the enzyme. Prot. Sci. 1, 1613–1621.

Grassetti, D. R., and Murray, J. F., Jr. (1967). . Arch. Biochem. Biophys. 119, 41–49.

Green, A. R., Presta, A., Gasyna, Z., and Stillman, M. J. (1994). Luminescent probe of copper-thiolate cluster formation within mammalian metallothionein. Inorg. Chem. 33, 4159–4168.

Kille, P., Hemmings, A., and Lunney, E. A. (1994). Memories of Metallothionein. Biochim. Biophys. Acta 1205, 151–161.

Kodama, H. (1993). Recent developments in Menkes disease. J. Inherit. Metab. Dis. 16, 791–799.

Lutsenko, S., Petrukin, K., Cooper, M. J., Gilliam, C. T., and Kaplan, J. H. (1997). H-terminal Domains of Human Copper-transporting Adenosin Triphosphatases (the Wilson's and Menkes Disease Proteins) Binding Copper Selectively in Vivo and in Vitro with Stoichiometry of one Copper per Metal-binding Repeat. J. Biol. Chem. 272, 18939–18944.

Mercer, J. F., Livingston, J., Hall, B., Paynter, J. A., Begy, C., Chandrasekharappa, S., Lockhart, P., Grimes, A., Bhave, M., Siemieniak, D., et al. (1993). Isolation of a partial candidate gene for Menkes disease by positional cloning see comments. Nat. Genet. 3, 20–25.

Odermatt, A., and Solioz, M. (1995). Two trans-acting metalloregulatory proteins controlling expression of the copper-ATPases of enterococcus hirae. J. Biol. Chem. 270, 4349–4354.

Odermatt, A., Suter, H., Krapf, R., and Solioz, M. (1993). Primary structure of two P-type ATPases involved in copper homeostasis in Enterococcus hirae. J. Biol. Chem. 268, 12775–12779.

O Halloran, T. V. (1993). Transition metals in control of gene expression. Science 261, 715–725.

Petris, M. J., Mercer, J. F. B., Culvenor, J. G., Lockhart, P., Gleeson, P. A., and Camakaris, J. (1996). Ligand-regulated Transport of the Menkes Copper P-type ATPase Efflux Pump from the Golgi Apparatus to the Plasma Membrane: A Novel Mechanism of Regulated Trafficking. EMBO 15, 6084–6095.

Pickering, I. J., George, G. N., Dameron, C. T., Kurtz, B., Winge, D. R., and Dance, I. G. (1993). X-ray Absorption Spectroscopy of Cuprous-Thiolate Multinuclear Clusters in Proteins and Model Systems. J. Am. Chem. Soc. 115, 9498–9505.

Pufahl, R. A., Singer, C. P., Peariso, K. L., Lin, S.-J., Schmidt, P. J., Fahrni, C. J., Culotta, V. C., Penner-Hahn, J. E., and O'Halloran, T. V. (1997). Metal Ion Chaperone Cunction of the Soluble Cu(I) Receptor Atx1. Science 278, 853–856.

Schiering, N., Kabsch, W., Moore, M. J., Distefano, M. D., Walsh, C. T., and Pai, E. F. (1991). Structure of the detoxification catalyst mercuric ion reductase from Bacillus sp. strain RC607. Nature 352, 168–172.

Silver, S., and Ji, G. (1994). Newer systems for bacterial resistances to toxic heavy metals. Env. Health Perspect. 102 Suppl 3, 107–113.

Solioz, M., Odermatt, A., and Krapf, R. (1994). Copper pumping ATPases: common concepts in bacteria and man. FEBS Lett. 346, 44–47.

Steele, R. A., and Opella, S. J. (1997). Structures of the Reduced and Mercury-Bound Forms of MerP, the Periplasmic Protein from the Bacterial Mercury Detoxification System. Biochem. 36, 6885–6895.

Thorvaldsen, J. K., Sewell, A. K., Tanner, A. M., Peltier, J. M., Pickering, I. J., George, G. N., and Winge, D. R. (1994). Mixed Cu+ and Zn2+ Coordination in the DNA-Binding Domain of the AMT1 Transcription Factor from Candida glabrata. Biochem. 33, 9566–9577.

Vulpe, C., Levinson, B., Whitney, S., Packman, S., and Gitschier, J. (1993). Isolation of a candidate gene for Menkes disease and evidence that it encodes a copper-transporting ATPase published erratum appears in Nat Genet 1993 Mar;3(3):273. see comments. Nat. Genet. 3, 7–13.

Winge, D. R., Dameron, C. T., George, G. N., Pickering, I. J., and Dance, I. G. (1993). Cuprous-Thiolate Polymetallic Clusters in Biology. In Bioinorganic Chemistry of Copper, K. D. Karlin, and Z. Tyeklar, eds. (New York, NY: Chapman & Hall), pp. 110–123.

Yoon, K. P., Misra, T. K., and Silver, S. (1991). Regulation of the cadA cadmium resistance determinant of Staphylococcus aureus plasmid pI258. J. Bacteriol. *173*, 7643–7649.

14

EXPRESSION, PURIFICATION, AND METAL BINDING CHARACTERISTICS OF THE PUTATIVE COPPER BINDING DOMAIN FROM THE WILSON DISEASE COPPER TRANSPORTING ATPASE (ATP7B)

Michael DiDonato,[1,2] Suree Narindrasorasak,[2] and Bibudhendra Sarkar[1,2]

[1]Department of Biochemistry, University of Toronto
Toronto, Ontario, Canada, M5S 1A8
[2]Division of Structural Biology
The Research Institute of the Hospital for Sick Children
Toronto, Ontario, Canada, M5G 1X8

1. INTRODUCTION

Copper in humans is an absolutely essential trace element. It is an integral part of many enzymes, the majority of which are developmentally important (Table 1) (DiDonato and Sarkar, 1997). However, copper which is present in excess of the trace amounts that are needed, is very toxic to cells. In humans, the two major genetic disorders of copper transport are Wilson and Menkes diseases. In many ways these diseases are two sides of the same story. While Wilson disease is characterized by an accumulation of copper in various tissues, global copper deficiency is the hallmark of Menkes disease. Our understanding of the molecular mechanisms underlying the transport and trafficking of copper in the body has increased greatly with the identification of both the Wilson (ATP7B) (Bull et al., 1993; Tanzi et al., 1993), and Menkes (ATP7A) (Chelly et al., 1993; Mercer et al., 1993; Vulpe et al., 1993) disease genes. Sequence analysis of the predicted gene products has indicated that they code for putative copper transporting P-type ATPases, and share a high degree of sequence homology (Bull et al., 1993; Tanzi et al., 1993; Vulpe et al., 1993).

Both proteins have large N-terminal domains which are thought to be responsible for ligating the copper atoms prior to transport across the membrane. The N-terminal domains contain six repeats of a highly conserved motif (GMTCXXC) which is thought to directly ligate the metal. This domain is also found in various bacterial heavy metal transporters

Table 1. Copper-containing enzymes in humans

	Reaction catalyzed
Oxidoreductases	
Cu, Zn Superoxide Dismutase (EC 1.15.1.1)	$O_2^{\cdot-} + O_2^{\cdot-} + 2\,H^+ \rightarrow O_2 + H_2O_2$
Lysyl Oxidase (EC 1.4.3.13)	Peptidyl-L-lysyl-Peptide + H_2O + O_2 $\rightarrow$ Peptidyl-allysyl-Peptide + NH_3 + H_2O_2
Amine Oxidase (EC 1.4.3.6)	$RCH_2NH_2 + H_2O \rightarrow RCHO + NH_3 + H_2O_2$
Cytochrome c Oxidase (EC 1.9.3.1)	4 Ferrocytochrome c + O_2 $\rightarrow$ 4 Ferricytochrome c + 2 H_2O
Ceruloplasmin (EC 1.16.3.1)	4 Fe(II) + 4 H^+ + O_2 $\rightarrow$ 4 Fe(III) + 2 H_2O
Monooxygenases	
Tyrosinase (EC 1.14.18.1)	L-Tyrosine + L-Dopa + O_2 $\rightarrow$ L-Dopa + Dopaquinone + H_2O
Dopamine β-Hydroxylase (EC 1.14.17.1)	Dopamine + Ascorbate + O_2 $\rightarrow$ Noradrenaline + Dehydroascorbate + H_2O
Peptidylglycine monooxygenase (EC 1.14.17.3)	Peptidylglycine + Ascorbate + O_2 $\rightarrow$ Peptidyl(2-hydroxyglycine) + Dehydroascorbate + H_2O

(Odermatt et al., 1994; Phung et al., 1994), and more recently in the yeast cytosolic protein Atx1, which has been shown to deliver copper to the yeast homologue of the Wilson/Menkes disease proteins (Pufahl et al., 1997). This motif does not correspond to any known copper(I) or copper(II) ligating motifs in proteins and EXAFS studies on the Atx1 protein, which contains one copy of the motif, show that both linear and trigonal coordination of copper(I) is possible in this protein (Pufahl et al., 1997).

In an effort to characterize the metal binding domain of the Wilson disease protein (WCBD), we have expressed and purified it as a fusion to glutathione-S-transferase (GST). Further experiments have shown that the domain does bind copper very tightly in addition to binding other transition metals with varying affinity. Stoichiometric analysis of metal binding using Neutron Activation Analysis (NAA) has indicated that each repeat in the motif is able to bind one metal atom giving a ratio of approximately 6:1 moles of copper/mole of protein (DiDonato et al., 1997).

2. EXPERIMENTAL PROCEDURES

2.1. Construction and Expression of GST-WCBD Fusion Protein

The 2-kilobase cDNA fragment encoding the Wilson disease copper binding domain (WCBD) was cloned into the GST fusion expression vector pGEX-4-T-2 to create the expression vector pGEX-WCBD. The sequence of the insert was confirmed by dideoxy sequencing prior to expression. After transforming *E. Coli* with the pGEX-4-T-2, expression of the fusion protein was initiated by treatment of midlog cultures of bacteria with IPTG. Following a three hour induction period the cells were harvested and lysed and the fusion protein was isolated.

2.2. Purification of GST-WCBD

Analysis of bacterial lysates indicated that the majority of the fusion protein (~70%) was partitioning to inclusion badies while the remainder was in the soluble fraction. A refolding method was developed to recover the fusion protein contained in the inclusion bodies (DiDonato et al., 1997). The combined soluble and refolded fractions were then purified by glutathione sepharose-4B affinity chromatography, followed by denaturing anion exchange chromatography. Using this method, it was possible to purify the fusion protein to greater than 90% homogeneity. Typical yields of fusion protein are 8–10 mg of purified fusion protein per litre of bacterial culture. When needed the GST moiety was removed by thrombin cleavage. The identity of the protein was confirmed by N-terminal sequence analysis.

2.3. Characterization of Metal Binding Properties of WCBD

2.3.1. Immobilized Metal Ion Affinity Chromatography (IMAC). Purified samples of GST-WCBD were applied to chelating sepharose columns charged with different transition metals. The bound protein was eluted by lowering the pH or by the addition of chelators. The experiment was performed under denaturing and non-denaturing conditions with very similar results.

2.3.2. $^{65}Zn(II)$ Blotting and Competition $^{65}Zn(II)$ Blotting Analysis. The metal binding properties of the domain were also analyzed using $^{65}Zn(II)$ blotting analysis as well as competition $^{65}Zn(II)$ blotting analysis. The $^{65}Zn(II)$ blotting experiments were performed essentially as described in (Schiff et al., 1988) with some minor modifications (DiDonato et al., 1997). Competition $^{65}Zn(II)$ was performed in the same way except the nitrocellulose strips were probed with $^{65}Zn(II)$ in the presence of a cold competitor metal. The signals were quantified using NIH Image 1.61 and compared to controls. In some cases the strips were stained with amido black to ensure that each strip contained equal amounts of protein.

2.3.3. Neutron Activation Analysis. Urea-denatured samples of the fusion protein were refolded in the presence of copper and then subjected to extensive dialysis against 1% formic acid to remove excess and weakly bound copper atoms. Following dialysis the protein concentration was determined and the sample was lyophilized and submitted for copper determination by neutron activation.

3. RESULTS AND DISCUSSION

3.1. Expression and Purification of the Fusion Protein

Expression of the GST fusion protein in *E. Coli* resulted in the protein partitioning between the soluble fraction and the inclusion body fraction. Approximately 60–70% of the fusion protein is localized to inclusion bodies while the remainder is found in the soluble fraction. If the amount of aeration is increased, either by increased agitation of the culture or through the use of a fermentor, the amount of fusion protein in the soluble fraction is increased (Data not shown). Figure 1 illustrates the results of a typical purification of both the fusion protein and the cleaved domain. Fusion protein which was eluted from the glutathione affinity column using urea in the absence of reducing agents and then treated with bathocuproinedisufonic acid (BCS) gave rise to a reddish-orange color (λ_{max} = 480 nm) indicative of the $Cu(I)(BCS)_2^-$ complex (Chen et al., 1996). This indicates that copper bound to the domain is most likely in the +1 oxidation state and that copper is being incorporated in the domain as it is produced in *E. Coli*.

Furthermore, we have found that refolded fusion protein obtained from inclusion bodies also gives rise to the $Cu(I)(BCS)_2^-$ when treated with BCS regardless of whether

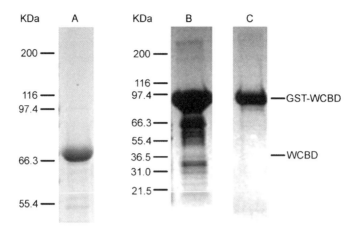

Figure 1. Summary of typical purification of WCBD and GST-WCBD. GST-WCBD was expressed and purified as described under "Experimental Procedures". A, purified WCBD eluted from glutathione-Sepharose 4B affinity resin following cleavage with thrombin on the column. B, GST-WCBD eluted with urea from glutathione affinity resin following solubilization and refolding from inclusion bodies. C, purified GST-WCBD eluted from DEAE-Sephacel anion exchange resin. (DiDonato et al., 1997.)

copper is present or absent from the refolding buffer. This suggests that copper is able to remain bound to the domain under denaturing conditions in the +1 oxidation state. Metal binding under denaturing conditions has also been observed for the estrogen receptor DNA binding domain (2 zinc finger protein having 4 Cys in each finger) where low pH and chelators are required to liberate the metal (Conte et al., 1996). Using the combination of glutathione affinity chromatography and anion exchange chromatography we are able to purify the fusion protein to >90% homogeneity.

3.2. Analysis of Metal Binding Characteristics of WCBD

3.2.1. Immobilized Metal Ion Affinity Chromatography (IMAC) and Neutron Activation Analysis. The metal binding ability of the domain was first analyzed by IMAC. Samples of the fusion protein were applied to chelating sepharose columns charged with various metals and then eluted by lowering the pH and then by the addition of chelating agents (Figure 2). The experiments were run under both denaturing and non-denaturing conditions in the absence of reducing agents with very similar results. Preliminary studies showed that GST had some interaction with the columns under non-denaturing conditions but did not bind at all under denaturing conditions. Results obtained in the IMAC experiments were similar regardless of whether denaturing or non-denaturing conditions were used suggesting that the major binding interactions are from the WCBD and not GST. Specific binding of proteins with internal high affinity metal binding sites to IMAC columns under denaturing conditions has also been demonstrated for troponin T, which contains four repeated metal binding motifs (Jin and Smillie, 1994).

The fusion protein was found to have varying affinities for the transition metals tested. Based on the elution conditions the observed order of affinity was as follows: Cu(II)>>Zn(II)>Ni(II)> Co(II). No binding was observed to columns charged with either Fe(II) or Fe(III). The binding of copper to the domain was by far the strongest. Neither the

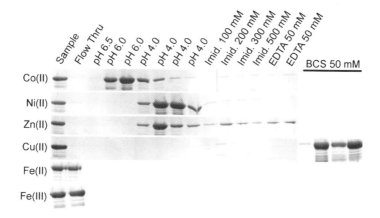

Figure 2. Comparison of GST-WCBD binding to various metal columns in IMAC under non-denaturing conditions. Samples of GST-WCBD were applied to chelating sepharose resin charged with the indicated metal. Various elution conditions appear across the top of the gel. The metals are arranged in order of increasing strength of binding. No binding was observed to columns charged with iron. Note that elution from the Cu(II) column is achieved only through the use of the cuprous chelator BCS. Very similar results were obtained under denaturing conditions. (DiDonato et al., 1997.)

low pH with the addition of chelators (imidazole and EDTA), or excess reducing agents (10 mM DTT, data not shown) were able to release it from the copper column. Interestingly, the application of BCS to the column resulted in the formation of a reddish-orange band accompanied by the elution of the domain. This not only suggests that the bound copper is in the +1 oxidation state but that Cu(II) atoms may undergo reduction to Cu(I) upon binding to the domain. The varying affinities observed for the other metals may be a result of the inability of the metal binding sites in the domain to conform to the preferred ligation geometries of certain metals while they are bound to the column matrix.

Neutron activation analysis was used to assess the stoichiometry of metal binding to the domain. After refolding of the fusion protein in the presence of copper, the refolded protein was extensively dialyzed, lyophilized and submitted for copper determination by neutron activation. NAA revealed a copper to protein ratio of 6.5:1 indicating that each metal binding motif is responsible for the ligation of one copper atom. The ligating atoms may reside in each domain or may be a combination of ligands from the motif and elsewhere in the protein. In addition to the Cys residues in each motif, the domain contains six other Cys residues interspersed between the metal binding regions which may participate in metal ligation or disulfide bridge formation.

3.2.2. ^{65}Zn(II) Blotting and Competition ^{65}Zn(II) Blotting Analysis. To further investigate the metal binding properties of the domain, ^{65}Zn(II) blotting and competition ^{65}Zn(II) blotting assays were employed. These assays have been used to probe the metal binding specificity of zinc binding sites in other proteins (Schiff et al., 1988). Radioactive zinc was chosen over copper because of its longer half-life and is more easily obtainable. Preliminary assays showed that the domain is able to bind Zn(II) in this assay and that pretreatment of the membranes with DTT is needed to observe Zn(II) binding (Figure 3).

The requirement of DTT pretreatment suggests that cysteine residues are directly involved in metal ligation and that a free sulfhydryl is required for metal ligation to occur.

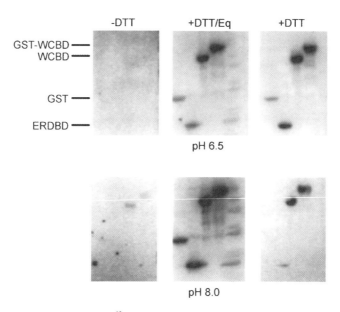

Figure 3. Effect of DTT on the binding of ^{65}Zn(II). Samples of GST-WCBD, WCBD, GST, and the estrogen receptor DNA binding domain (ERDBD) were subjected to ^{65}Zn(II) blotting analysis either in the absence of DTT, pretreatment with DTT during the equilibration step only, or in the presence of DTT throughout the assay. The experiments were performed at pH 6.5 and 8.0. The ERDBD is a two zinc finger protein used as a positive control in the assay. (DiDonato et al., 1997.)

Following SDS-PAGE mixed disulfides may form due to the presence of β-mercaptoethanol in the loading buffer which is known to react with protein sulfhydryls (Scopes, 1994; Tombs, 1985). DTT would then be required to ensure the reduction of these residues. Increased non-specific binding is observed at the higher pH value due to the increased proportion of negative charge on the proteins at this pH. Inclusion of DTT throughout the assay significantly reduced the amount of non-specific binding at either pH value most likely by chelating weekly bound or non-specifically bound Zn(II) atoms. These results support the notion that the GMTCXXC motif, which is strictly conserved in each of the metal binding domains of ATP7B as well as many bacterial heavy metal transporters, is crucial for metal binding (O'Halloran, 1993; Silver et al., 1989).

The effect of pH on the binding of Zn(II) to the domain was investigated and the results presented in figure 4. Both the fusion protein and the free domain are able to bind

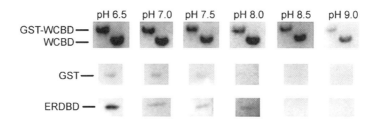

Figure 4. Effect of pH on the binding of ^{65}Zn(II). Samples of GST-WCBD, WCBD, GST and the estrogen receptor DNA binding domain (ERDBD) were subjected to ^{65}Zn(II) blotting analysis at various pH values. The assay was performed in the presence of DTT. (DiDonato et al., 1997.)

Zn(II) over a range of pH values from 6.5 to 9.0. A significant decrease in the amount of bound Zn(II) is only observed at the highest pH value tested. This may be caused by a change in the protonation state of the sulfhydryl residues which may affect their ability to ligate metal atoms. The fusion protein appears to bind slightly less zinc than the free domain because there is less of it blotted on the membrane relative to the free domain. This was confirmed by staining the membranes post-autoradiography with amido black.

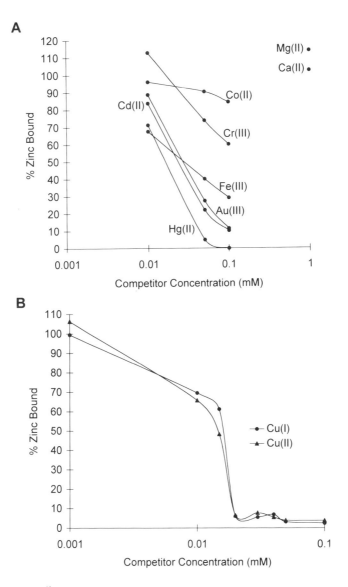

Figure 5. Competition of ^{65}Zn(II) binding to WCBD. Samples of purified WCBD were subjected to competition ^{65}Zn(II) blotting analysis. In each case the final ^{65}Zn(II) concentration was 30 μM. A, competition of ^{65}Zn(II) binding with various transition metals. Each competitor was added as the chloride salt at the indicated concentration. Successful competition resulted in a decreased signal relative to the control. Ni(II) and Mn(II) showed little or no affinity for the domain relative to Zn(II). B, competition of ^{65}Zn(II) binding with Cu(II) (▲) and Cu(I) (●). Cu(II) was presented as CuCl$_2$, while Cu(I) was presented as the tetrakis(acetonitrile)copper(I) hexafluorophosphate complex. Similar results were obtained when Cu(I) was generated *in vitro* using CuCl$_2$ and ascorbate. (DiDonato et al., 1997.)

To investigate the possibility that the domain is able to bind other transition metals, a competition ^{65}Zn(II) blotting assay was employed. The results of this analysis are presented in figure 5A. As seen, several metals are able to compete successfully with Zn(II) for binding to the domain. In particular, Cd(II), Au(III), and Hg(II) seem to have the highest affinities for the domain relative to Zn(II), whereas Mn(II) and Ni(II) had little or no affinity relative to Zn(II). The results are what would be expected since Zn(II), Cd(II) and Hg(II) are in the same group and therefore have the similar ligation geometries with ionic radii being the only difference. Hg(II) is also a soft metal and is therefore expected to have a high affinity for sulhydryl ligands which is what is observed. In contrast to the results found with IMAC, Fe(III) was able to act as a competitor in this experiment. This may result from an inability of the protein to conform to the preferred ligation geometry of Fe(III) while it is bound to the column matrix. The reverse may be true for Ni(II), which was unable to act as a competitor in the blotting experiments but bound to the domain tightly in the IMAC experiments. The binding of metal to the domain is specific since both Mg(II) and Ca(II) did not compete at all for Zn(II) binding.

The results of the competition experiments involving copper is summarized in figure 5B. At low concentrations relative to Zn(II), Cu(I) and Cu(II) are able to decrease Zn(II) binding by about 30%. However, as the concentration is raised, the affinity for additional copper seems to increase rapidly. This pattern was only observed for copper and may suggest that copper ligation by the domain is to some degree cooperative. However, further experiments are needed before any conclusions about cooperativity can be made. The competition pattern is reproducible and is independent of whether copper is presented in the +1 or +2 oxidation state suggesting that the domain has similar affinities for both Cu(I) and Cu(II). If cooperativity of binding is present only for copper and for none of the other metals, it may serve as a selection criteria for binding copper exclusively *in vivo*. Furthermore, conformational changes which may accompany cooperative binding could act as a trigger for other events such as translocation of the protein as has been demonstrated for the Menkes protein (Petris et al., 1996).

4. CONCLUSIONS

We have demonstrated, by two independent methods that the metal binding domain from the Wilson disease copper transporting P-type ATPase (ATP7B) is able to bind a variety of transition metals with varying affinities. Whether other metals in addition to copper can be transported by the protein remains to be determined. In particular, we have found that the domain has a very strong affinity for copper regardless of whether it is given in the +1 or +2 oxidation state. Experimental evidence suggests that copper bound to the domain exists in the +1 oxidation state, suggesting that upon binding, Cu(II) atoms are reduced to Cu(I) by some mechanism. Furthermore, blotting experiments suggest that the binding of copper by the domain is cooperative in nature, however, this cannot be stated conclusively until further experiments are performed. If present, cooperative binding of copper may be a very important function of the protein *in vivo*.

REFERENCES

Bull, P. C., Thomas, G. R., Rommens, J. M., Forbes, J. R., and Cox, D. W. (1993). The Wilson disease gene is a putative copper transporting P-type ATPase similar to the Menkes gene. Nat. Genet. 5, 327–37.

Chelly, J., Tumer, Z., Tonnesen, T., Petterson, A., Ishikawa-Brush, Y., Tommerup, N., Horn, N., and Monaco, A. P. (1993). Isolation of a candidate gene for Menkes disease that encodes a potential heavy metal binding protein. Nat. Genet. *3*, 14–9.

Chen, P., Onana, P., Shaw, C. F. R., and Petering, D. H. (1996). Characterization of calf liver Cu,Zn-metallothionein: naturally variable Cu and Zn stoichiometries. Biochem. J. *317*, 389–94.

Conte, D., Narindrasorasak, S., and Sarkar, B. (1996). In vivo and in vitro iron-replaced zinc finger generates free radicals and causes DNA damage. J. Biol. Chem. *271*, 5125–30.

DiDonato, M., Narindrasorasak, S., Forbes, J. R., Cox, D. W., and Sarkar, B. (1997). Expression, purification, and metal binding properties of the N-terminal domain from the Wilson disease putative copper-transporting ATPase (ATP7B). J. Biol. Chem. *272*, 33279.

DiDonato, M., and Sarkar, B. (1997). Copper transport and its alterations in Menkes and Wilson diseases. BBA Mol. Basis. Dis. *1360*, 3–16.

Jin, J. P., and Smillie, L. B. (1994). An unusual metal-binding cluster found exclusively in the avian breast muscle troponin T of Galliformes and Craciformes. FEBS Lett. *341*, 135–40.

Mercer, J. F., Livingston, J., Hall, B., Paynter, J. A., Begy, C., Chandrasekharappa, S., Lockhart, P., Grimes, A., Bhave, M., Siemieniak, D., and et al. (1993). Isolation of a partial candidate gene for Menkes disease by positional cloning. Nat. Genet. *3*, 20–5.

O'Halloran, T. V. (1993). Transition metals in control of gene expression. Science *261*, 715–25.

Odermatt, A., Krapf, R., and Solioz, M. (1994). Induction of the putative copper ATPases, CopA and CopB, of Enterococcus hirae by Ag+ and Cu2+, and Ag+ extrusion by CopB. Biochem. Biophys. Res. Commu. *202*, 44–8.

Petris, M. J., Mercer, J. F., Culvenor, J. G., Lockhart, P., Gleeson, P. A., and Camakaris, J. (1996). Ligand-regulated transport of the Menkes copper P-type ATPase efflux pump from the Golgi apparatus to the plasma membrane: a novel mechanism of regulated trafficking. EMBO J. *15*, 6084–95.

Phung, L. T., Ajlani, G., and Haselkorn, R. (1994). P-type ATPase from the cyanobacterium Synechococcus 7942 related to the human Menkes and Wilson disease gene products. Pro. Natl. Acad. of Sci. *91*, 9651–4.

Pufahl, R. A., Singer, C. P., Peariso, K. L., Lin, S. J., Schmidt, P. J., Fahrni, C. J., Culotta, V. C., PennerHahn, J. E., and O'halloran, T. V. (1997). Metal ion chaperone function of the soluble Cu(I) receptor Atx1. Science *278*, 853–856.

Schiff, L. A., Nibert, M. L., and Fields, B. N. (1988). Characterization of a zinc blotting technique: Evidence that a retroviral gag protein binds zinc. Proc. Natl. Acad. Sci. *85*, 4195–4199.

Scopes, R. K. (1994). Protein Purification: Principles and Practice, 2nd Edition (Berlin: Springer-Verlag). pp. 317–324.

Silver, S., Nucifora, G., Chu, L., and Misra, T. K. (1989). Bacterial resistance ATPases: primary pumps for exporting toxic cations and anions. T.I.B.S. *14*, 76–80.

Tanzi, R. E., Petrukhin, K., Chernov, I., Pellequer, J. L., Wasco, W., Ross, B., Romano, D. M., Parano, E., Pavone, L., Brzustowicz, L. M., and et al. (1993). The Wilson disease gene is a copper transporting ATPase with homology to the Menkes disease gene. Nat. Genet. *5*, 344–50.

Tombs, M. P. (1985). Stability of enzymes. J. Appl. Biochem. *7*, 3–24.

Vulpe, C., Levinson, B., Whitney, S., Packman, S., and Gitschier, J. (1993). Isolation of a candidate gene for Menkes disease and evidence that it encodes a copper-transporting ATPase. Nat. Genet. *3*, 7–13.

15

STRUCTURE/FUNCTION RELATIONSHIPS IN CERULOPLASMIN

Giovanni Musci,[1] Fabio Polticelli,[2] and Lilia Calabrese[2,3]

[1]Department of Organic and Biological Chemistry
University of Messina
Salita Sperone, 31
98166 S. Agata, Messina Italy
[2]Department of Biology, University Roma Tre
viale Marconi, 446, 00146 Rome, Italy
[3]CNR Center of Molecular Biology
p.le A. Moro, 5, 00185 Rome, Italy

Ceruloplasmin belongs to the family of multinuclear copper oxidases, which also includes ascorbate oxidase and laccase (Solomon et al., 1996). These proteins are trinuclear copper cluster enzymes, as documented by spectroscopic and functional studies in solution for laccase (Allendorf et al., 1985; Spira-Solomon et al., 1986) and ceruloplasmin (Calabrese et al., 1988, 1989), and recently confirmed by X-ray crystallography for ascorbate oxidase (Messerschmidt et al., 1992) and ceruloplasmin (Zaitseva et al., 1996). Multiple copper sites are present, with a minimal functional unit constituted by a blue, or type 1, copper site, and three copper atoms clustered in a trinuclear arrangement. The blue copper site takes up an electron from a reducing substrate and transfers it, through long-range intramolecular electron transfer mediated by a Cys-His pathway, to the trinuclear cluster (Fig. 1), which provides four electrons to reduce oxygen to water. Among multicopper oxidases, ceruloplasmin is presently unique in that it possesses additional copper sites beside the four ions of the minimal unit. The three dimensional structure of ceruloplasmin solved at 3 Å resolution has shown that the enzyme is made up of six β-barrel domains arranged in a ternary simmetry (Zaitseva et al., 1996). Consistent with spectroscopic data (Wever et al., 1973, Musci et al., 1993), three of these domains (domains 2, 4 and 6) bind type 1 copper ions coordinated by nitrogen and sulphur ligands. Type 1 copper ions bound to domain 4 and 6 (hereafter referred to as Cu4 and Cu6) are coordinated by 2 histidines, a cysteine and a methionine. Type 1 copper bound to domain 2 (Cu2) should be preferably reduced (Musci et al., 1995; Solomon et al., 1996) since the axial methionine is replaced by leucine, in a geometry typical of a high redox potential copper site. The three copper ions of the trinuclear cluster lie at the interface between domain 1 and 6 and are coordi-

Copper Transport and Its Disorders, edited by Leone and Mercer.
Kluwer Academic / Plenum Publishers, New York, 1999.

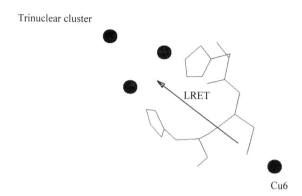

Figure 1. Scheme of long-range electron transfer (LRET) through the Cys-His pathway between type 1 copper and the trinuclear cluster in multicopper oxidases. The drawing refers to human ceruloplasmin.

nated by 8 histidine ligands and an oxygen ligand (OH or H_2O) bridging two of them. Labile metal binding sites have also been observed on domains 4 and 6, close to the respective prosthetic blue sites. Domain 2 appears different also in this respect, since it lacks the proper ligands for the corresponding labile site.

The high resolution of the three-dimensional structure of a metalloprotein provides the starting point for penetrating the complexity of its metal site(s), and is the beginning of any inquiry into its function. Thus, crystallographic data for ceruloplasmin, though at only 3 Å resolution, are now being used to elucidate the mechanism of its ferroxidase activity (Lindley et al., 1997), to identify low-affinity metal binding sites responsible for anti/pro-oxidant properties (Mukhopadhyay et al., 1997), and will hopefully be crucial for the understanding of the molecular mechanism of copper delivery to cells. When the electric charge of the protein residues is taken into account, crystallographic data allow to analyze the role of the charge distribution within the molecule and of the electric field(s) generated by the molecule. Thus, the relevance of electrostatic factors in the structure/function relationships of the protein, which have been shown to play an important role in protein-ligand interactions, can be investigated in deeper detail.

The electrostatic potential distribution around ceruloplasmin is depicted in Fig. 2. As already pointed out (Lindley et al., 1997), the region where blue copper binding sites are located is characterized by negative electrostatic potential values. In particular, wide areas of negative potential are observed on the internal side of the three tower-like structures which carry residues forming labile metal (Fe(II) or Cu(II)) binding sites (Fig. 2A). The opposite, flat region of the molecule is characterized by a slight prevalence of positive potential areas (Fig. 2B). Considering the recent finding of a mechanism of ceruloplasmin anchoring to the cell membrane through a GPI moiety attached to the C-terminus of the protein (Patel and David, 1997), this region could be involved in the interaction with negative groups on the membrane.

A more detailed analysis of the electrostatic potential distribution around the type 1 copper sites reveals interesting differences in the local environment of the copper ions (Fig. 3). In particular, two islands of negative potential, located on the sidechain carboxyl group of residues Glu971 and Glu272, are observed in the proximity of the labile metal binding sites, on domains 4 and 6, respectively. Such electrostatic potential minimum is not observed on copper site of domain 2. Thus, it can be concluded that unfavourable electrostatic factors add to the unfavourable ligand set (Lindley et al., 1997) in preventing

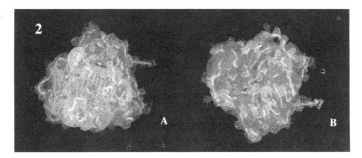

Figure 2. Mesh representation of electrostatic equipotential surfaces around human ceruloplasmin generated with the program GRASP (Nicholls et al., 1991), kindly provided by Prof. Barry Honig. Equipotential surfaces at -2 kT/e and +2 kT/e are represented in white and grey respectively. Protein backbone is represented by a white wireframe. A) Top view showing the negative equipotential surface enclosing the central area of the protein corresponding to type 1 copper binding sites. B) Bottom view of the molecule. The three dimensional structure of human ceruloplasmin has been used to model the potential around the protein. The two loop regions and the C-terminal portion of the polypeptide chain which are not clearly defined in the electron density map have been modelled using the "loop-search" procedure (Polticelli et al., 1994). The protein structure has been subsequently stereochemically regularized using the Maximin2 module of the program Sybyl by Tripos Associates.

metal binding on this site. Ferrous iron binding to the labile sites has been suggested as the initial step in the ferroxidase catalytic mechanism, followed by translocation of the resulting Fe(III) to a different region of the protein. If so, Cu2 would be excluded from the catalytic cycle, consistent with its -putatively too high- redox potential.

Early studies on the ferroxidase activity of ceruloplasmin demonstrated that in a reducing environment, which indeed the plasma is (Lane et al., 1987; Henry, 1984), enzymatic oxidation can account for the observed rate of iron turnover in the body, at variance with non catalyzed autoxidation rate of iron (Osaki et al., 1966). The ferroxidase activity of ceruloplasmin in the plasma can be considered as the mechanism of the antioxidant properties of ceruloplasmin, since it leads to clearance of Fe(II) from plasma without pro-

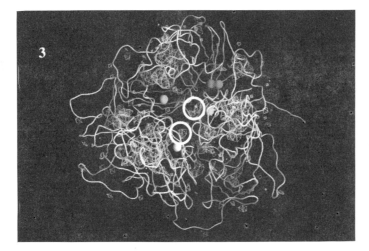

Figure 3. Equipotential surfaces at -30 kT/e around human ceruloplasmin. Protein backbone is represented by a white wireframe. White spheres represent copper atoms. The white circles on domain 4 and 6 labile metal sites evidentiate the presence of the negative equipotential surfaces which ar not observed in domain 2 copper site.

duction of the reactive oxygen species generated in the non-catalyzed autoxidation of Fe(II) (Gutteridge, 1978). The observation that ceruloplasmin can mediate iron release from liver cells (Osaki et al., 1971) and that individuals suffering of aceruloplasminemia due to a defective gene coding for ceruloplasmin, have a severely impaired iron metabolism, yet a normal copper homeostasis (Harris et al., 1995) has strongly reinforced the hypothesis that ceruloplasmin acts in vivo as a ferroxidase.

Kinetic study of iron oxidation by ceruloplasmin reveals biphasic curves in v versus v/Fe(II) plots, with two K_m values which differ by almost two orders of magnitude (K_{m1}=0.6 μM and K_{m2}=50 μM) (Osaki, 1966). This result can be interpreted in terms of two iron binding sites or alternatively by a rate-determining substrate activation mechanism. The latter mechanism is confirmed by loss of blue color of the enzyme with low iron concentrations and by the observed activation with other divalent metal ions (Huber and Frieden, 1970). Stopped flow techniques allow the determination of kinetic parameters of the reaction and the decomposition of the enzymatic mechanism in several steps: reduction of type 1 copper ions by iron would be the fastest step (1.2×10^6 $M^{-1}s^{-1}$), followed by a slow rate-determining step involving a conformational change of the enzyme (1.1 s^{-1}) and intramolecular electron transfer to the trinuclear copper cluster which would in turn transfer electrons to oxygen in a fast reaction (5.7×10^5 $M^{-1}s^{-1}$) (Osaki and Walaas, 1967). Kinetic evidence for the formation of an enzyme-substrate complex in the first phase of the reaction is consistent with the recent finding of the existence of labile metal binding sites within 10 Å from the type 1 copper ions in domain 4 and 6.

Two different redox potentials can be determined for the blue sites of human ceruloplasmin (Deinum and Vanngard, 1973). Since Cu4 and Cu6 share a very similar coordination geometry, differences in their redox potentials are expected to arise only from their different electrostatic environment. The study of the electrostatic effect of the protein moiety on the relative redox potentials of the type 1 copper ions[*] indicate that Cu6 should have a much higher redox potential than Cu4. The difference (ca. 200 mV) turns out to be of the same order of magnitude of that, ca. 100 mV, experimentally measured (Deinum and Vanngard, 1973). Thus, on this basis, Cu6 can be the preferential route of entry of electrons into the long-range electron transfer path to the trinuclear cluster, and can act as a sink which collects electrons when also Cu4 is reduced by the substrate.

A prominent role of domain 6 in the catalytic mechanism of ceruloplasmin is also indicated by Brownian dynamics simulations. This approach (Sergi et al., 1994) allows to quantitate the contribution of the electric field of the molecule on the ferroxidase activity. Simulating the diffusion of Fe(II) around the protein under the influence of brownian motion and of the protein-generated electrostatic field, the encounter probability between the substrate and either the type 1 copper site or the labile metal sites in domains 2, 4 and 6 can be evaluated. As summarized in Table 1, the electrostatic potential generated by the protein drives substrate diffusion preferentially towards domain 6 with a lower encounter probability with the metal sites in domains 2 and 4. There is not a significant difference when the prosthetic and the labile site of either domain 4 or 6 are taken as target for Fe(II), suggesting that they act as a functional unit. Addition of a divalent cation to simu-

---

[*] Electrostatic potential calculations allow to determine the contribution of the protein charge distribution to the redox potential of a metal according to the equation: $V = q \times \Phi \times 25.7$ where V is the redox potential in mV, q is the charge change at the metal site due to reduction, Φ is the electrostatic potential generated by the molecule at the metal site and 25.7 is a conversion factor from KT/e units to mV. Although such calculations do not provide absolute values of redox potentials, they are reliable in predicting relative figures when only electrostatic, and not geometric as for Cu4 and Cu6, factors are to be taken into account (Smith and Honig, 1994).

Table 1. Relative encounter probability of Fe(II) with type 1 copper or with the labile metal sites in human ceruloplasmin as determined by brownian dynamics simulations

Occupancy of	Type 1 copper			Labile site	
	Domain 2	Domain 4	Domain 6	Domain 4	Domain 6
-	0.093	0.055	0.118	0.055	0.118
Domain 4	0.095	-	0.154	-	0.159
Domain 6	0.143	0.129	-	0.100	-

late occupancy by Fe(II) of the labile metal sites on domains 4 and 6 results in an encounter probability between Fe(II) and Cu4 enhanced by occupancy of the labile site in domain 6, and vice versa (Table 1), and this can explain the mechanism underlying Fe(II) activation of the catalytic activity (Huber and Frieden, 1970).

To summarize, application of the electrostatic theory to the ferroxidase activity mechanism of ceruloplasmin seems to be in line with the available experimental data. In fact, even though diffusion of Fe(II) to the protein is not the rate-limiting step in the ferroxidase reaction catalyzed by human ceruloplasmin, nonetheless modelling enzyme substrate encounter according to a potential-driven diffusion mechanism is able to reproduce some properties of the system. This is somewhat surprising when we consider that in these analyses free Fe(II) ion is considered as the substrate, instead of some complexed form of ferrous ion. However, it should be kept in mind that all experimental data on the ferroxidase activity are obtained with iron salts as substrates, due to an absolute lack of knowledge of the physiological iron donor(s). Thus, it cannot be ruled out that free Fe(II) be indeed the true substrate, but it remains to be established which conditions, in terms of physiological environment, have to be met for this to happen.

The unique copper stoichiometry of ceruloplasmin, and the observation that Fe(II) is a very poor substrate for laccase and ascorbate oxidase could suggest that, for Fe(II) to be efficiently oxidized, the architecture of the copper sites of a multicopper oxidase must be more complex than a 4-copper catalytic unit. The yeast protein Fet3, belonging to the family of multicopper oxidases on the basis of primary structure homologies (Askwith et al., 1994; De Silva et al., 1995), has a stoichiometry of 4 copper ions/molecule (De Silva et al., 1997). Functional studies on yeast strains with a defective gene coding for Fet3 have unequivocally shown that Fet3 is involved in the high-affinity uptake of iron. Purified Fet3 has a ferroxidase activity with kinetic parameters comparable to those of ceruloplasmins. These data therefore indicate that no more than four copper ions are required to enzymatically oxidize Fe(II). It should be pointed out, however, that the classification of Fet3 as a multicopper oxidase awaits further, detailed investigation of its copper centers, especially concerning their redox state and reactivity. For this reason, we are currently investigating the functional properties of Fet3 in relation to its spectroscopic features which, according to preliminary data, appear more similar to those of laccase than of ceruloplasmin.

Why then has ceruloplasmin more than the minimum number of metal sites? As already mentioned, the labile metal sites in domains 4 and 6 have been shown to accomodate also copper, beside iron. Therefore, these sites which, on the basis of our previous considerations may have roles other than mere substrate binding in the protein activity when they interact with iron, could play a storage role when filled with copper. This can finally explain how ceruloplasmin can concomitantly act as an enzyme with its prosthetic sites, and as a copper transport protein, this latter function relying on numerous evidence

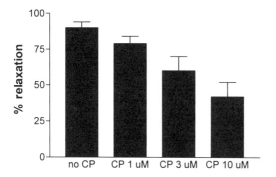

Figure 4. Effect of increasing concentrations of ceruloplasmin on the relaxation to 5×10^{-6} M acetylcholine of rabbit thoracic aorta preconstricted with PGF2a. Data are expressed as percent of relaxation under control conditions.

(Cousins, 1985). The participation of the prosthetic sites also in the copper transport cannot, however, be excluded and, as a matter of fact, these sites can be reversibly removed, at least in vitro. Both the loss and the gain of the copper ions are generally viewed as all-or-none processes (Musci et al., 1996), involving all six copper ions, although we have recent, preliminary evidence that one or two prosthetic copper ions can be selectively and reversibly removed from ceruloplasmin (Musci, Zumbahlen, Calabrese and McMillin, unpublished observations). This selective removal seems to be under kinetic, rather than thermodynamic, control, a hypothesis which is fully compatible with physiological events at the plasma membrane level.

Copper delivery from ceruloplasmin to tissues is generally interpreted as a mere transport function of the protein (Orena et al., 1986; Percival and Harris, 1990). The presence of specific receptors for ceruloplasmin on a number of cell types (Kataoka and Tavassoli, 1984, 1985; Dini et al., 1990), however, may be interpreted as a regulatory role of cellular functions by ceruloplasmin mediated by copper discharge. As a matter of fact, ceruloplasmin at physiological concentration is highly effective in reducing the endothelium-dependent relaxation of rabbit aorta stimulated with acetylcholine (Fig. 4) or ADP (Cappelli-Bigazzi et al., 1997), agonists known to activate the endothelial NO producing enzyme NO synthase (NOS). The effect of ceruloplasmin cannot be due to a scavenging action of the protein toward NO (Cappelli-Bigazzi et al., 1997). It is known that ceruloplasmin can donate copper to endothelial cells. This fact, along with experimental data including the observation that the enzymatic activity of NOS in vitro is impaired in the presence of copper leads to the hypothesis that copper be responsible for the ceruloplasmin-induced impairment of aorta relaxation (data to be published elsewhere).

REFERENCES

Allendorf, M.D., Spira, D. J., and Solomon, E.I. (1985). Low-temperature magnetic circular dichroism studies of native laccase: spectroscopic evidence for exogenous ligand bridging at a trinuclear copper active site. Proc Natl Acad Sci U S A 82, 3063–3067.

Askwith, C., Eide, D., Van Ho, A., Bernard, P.S., Li, L., Davis-Kaplan, S., Sipe, D.M., and Kaplan, J. (1994). The FET3 gene of S. cerevisiae encodes a multicopper oxidase required for ferrous iron uptake. Cell76, 403–410.

Calabrese, L., Carbonaro, M., and Musci, G. (1988). Chicken ceruloplasmin. Evidence in support of a trinuclear cluster involving type 2 and 3 copper centers. J. Biol. Chem. 263, 6480–6483.

Calabrese, L., Carbonaro, M., and Musci, G. (1989). Presence of coupled trinuclear copper cluster in mammalian ceruloplasmin is essential for efficient electron transfer to oxygen. J. Biol. Chem. 264, 6183–6187.

Cappelli-Bigazzi, M., Ambrosio, G., Musci, G., Battaglia, C., Bonaccorsi di Patti, M.C., Golino, P., ragni, M., Chiariello, M., and Calabrese, L. (1997) Ceruloplasmin impairs endothelium-dependent relaxation of rabbit aorta. Am. J. Physiol., in press.

Cousins, R.J. (1985). Absorption, transport, and hepatic metabolism of copper and zinc: special reference to metallothionein and ceruloplasmin. Physiol Rev 65, 238–309.

De Silva, D., Davis-Kaplan, S., Fergestad, J., and Kaplan, J. (1997). Purification and characterization of Fet3 protein, a yeast homologue of ceruloplasmin. J Biol Chem 272, 14208–14213.

De Silva, D.M., Askwith, C.C., Eide, D., and Kaplan, J. (1995). The FET3 gene product required for high affinity iron transport in yeast is a cell surface ferroxidase. J Biol Chem 270, 1098–1101.

Deinum, J., and Vanngard, T. (1973). The stoichiometry of the paramagnetic copper and the oxidation-reduction potentials of type I copper in human ceruloplasmin. Biochim Biophys Acta 310, 321–330.

Dini, L., Carbonaro, M., Musci, G., and Calabrese, L. (1990). The interaction of ceruloplasmin with Kupffer cells. Eur. J. Cell Biol. 52, 207–212.

Gutteridge, J.M.C. (1978). Ceruloplasmin: a plasma protein, enzyme and antioxidant. Ann. Clin. Biochem. 15, 293–296.

Harris Z.L., (1995). Aceruloplasminemia: molecular characterization of this disorder of iron metabolism. Proc. Natl. Acad. Sci. U.S.A.

Henry, J.B. (1984). Clinical diagnosis and management by laboratory methods. 7th ed. (Philadelphia, PA, W.B. Saunders).

Huber, C.T., and Frieden, E. (1970). Substrate activation and the kinetics of ferroxidase. J. Biol. Chem. 245, 3973–3978.

J Biol Chem 268(18), 13388–13395 (1993)

Kataoka, M., and Tavassoli, M. (1984). Ceruloplasmin receptors in liver cells suspensions are limited to endothelium. Exp. Cell Res. 155, 232–240.

Kataoka, M., and Tavassoli, M. (1985). Identification of ceruloplasmin receptors on the surface of human blood monocytes, granulocytes and limphocytes. Exp. Hematol. 13, 806–810.

Lane, E.E., and Walker, J.F. (1987). Clinical arterial blood gas analysis. (St. Louis, MS, C.V. Mosby).

Lindley, P.F., Card, G., Zaitseva, I., Zaitsev, V., Reinhammar, B., Selin-Lindgren, E., and Yoshida, K. (1997). An X-ray structural study of human ceruloplasmin in relation to ferroxidase activity. JBIC 2, 454–463.

Messerschmidt, A., Rossi, A., Ladenstein, R., Huber, R., Bolognesi, M., Gatti, G., Marchesini, A., Petruzzelli, and Finazzi-Agrò, A. (1989). X-ray crystal structure of ascorbate oxidase from zucchini: a preliminary analysis of the polypeptide fold and a model of the copper sites and ligands. J. Mol. Biol. 206, 513–529.

Mukhopadhyay, K.C., Mazumder, B., Lindley, P.F., and Fox, P.L. (1997). Identification of the prooxidant site of human ceruloplasmin: a model for oxidative damage by copper bound to protein surfaces. Proc. Natl. Acad. Sci. U.S.A. 94, 11546–11546.

Musci, G., Bonaccorsi di Patti, M.C., and Calabrese, L. (1993). The state of the copper sites in human ceruloplasmin. Arch. Biochem. Biophys. 306, 111–118.

Musci, G., Bonaccorsi di Patti, M.C., and Calabrese, L. (1995). Modulation of the redox state of the copper sites of human ceruloplasmin by chloride. J. Prot. Chem. 14, 611–619.

Orena, S.J., Goode, C.A., and Linder, M.C. (1986). Binding and uptake of copper from ceruloplasmin. Biochim. Biophys. Res. Commun. 139, 822–829.

Osaki, S. (1966). Kinetic studies of ferrous ion oxidation with crystallin human ferroxidase (ceruloplasmin). J. Biol. Chem. 241, 5053–5059.

Osaki, S., and Walaas, O. (1967). Kinetic studies of ferrous ion oxidation with crystallin human ferroxidase. II Rate constants at various steps and formation of a possible enzyme-substrate complex. J. Biol. Chem. 242, 2653–2657.

Osaki, S., Johnson, D.A., and Frieden, E. (1966). The possible significance of the ferrous oxidase activity of ceruloplasmin in normal human serum. J Biol Chem 241, 2746–2751.

Osaki, S., Johnson, D.A., and Frieden, E. (1971). The mobilization of iron from the perfused mammalian liver by a serum copper enzyme, ferroxidase I. J. Biol. Chem. 246, 3018–3023.

Patel, B.N., and David, S. (1997). A novel glycosylphosphatidylinositol-anchored form of ceruloplasmin is expressed by mammalian astrocytes. J. Biol. Chem. 272, 20185–20190.

Percival, S.S., and Harris, E.D. (1990) Copper transport from ceruloplasmin: characterization of the cellular uptake mechanism. Am. J. Physiol. 258, C140-C146.

Polticelli, F., Falconi, M., O'Neill, P., Petruzzelli, R., Galtieri, A., Lania, A., Calabrese, L., Rotilio, G., and Desideri, A. (1994). Molecular modeling and electrostatic potential calculations on chemically modified Cu,Zn superoxide dismutases from Bos taurus and shark Prionace glauca: role of Lys134 in electrostatically steering the substrate to the active site. Arch. Biochem. Biophys. 312, 22–30.

Sergi, A., Ferrario, M., Polticelli, F., P. O'Neill, and Desideri, A. (1994) Simulation of superoxide-superoxide dismutase assoction rate for six natural variants. Comparison with the experimental catalytic rate. J. Phys. Chem. 98, 10554–10557.

Smith, C.K., and Honig, B. (1994). Evaluation of the conformational free energies of loops in proteins. Proteins 18, 119–132.

Solomon, E.I., Sundaram, U.M., and Machonkin, T.E. (1996). Multicopper oxidases and oxygenases. Chem. Rev. 96, 2563–2605.

Spira-Solomon, D.J., Allendorf, M.D., and Solomon, E.I. (1986). Low temperature magnetic circular dichroism studies of native laccase: confirmation of a trinuclear copper active site. J. Amer. Chem. Soc. 108, 5318-.5328.

Wever, R., Van Leuween, F. X. R., and Van Gelder, B. F. (1973). The reaction of nitric oxide with ceruloplasmin. Biochim Biophys Acta 302, 236–239.

Zaitseva, I., Zaitsev, V., Card, G., Moshkov, K., Bax, B., Ralph, A., and Lindley, P. (1996). The X-ray structure of human serum ceruloplasmin at 3.1 Å: nature of the copper centres. JBIC 1, 15–23.

16

AUTOXIDATION OF AMYLOID PRECURSOR PROTEIN AND FORMATION OF REACTIVE OXYGEN SPECIES

G. Multhaup,* L. Hesse, T. Borchardt, Thomas Ruppert,[1] R. Cappai,[2]
C. L. Masters,[2] and K. Beyreuther

ZMBH-Center for Molecular Biology Heidelberg
University of Heidelberg
Im Neuenheimer Feld 282
D-69120 Heidelberg, Germany
[1]Max von Pettenkofer-Institute for Virology, Ludwig-Maximilians University
80336 Munich, Germany
[2]Department of Pathology, University of Melbourne
Parkville, Victoria, 3052, Australia and
Australia and Neuropathology Laboratory
Mental Health Research Institute of Victoria
Parkville, Victoria, 3052, Australia

INTRODUCTION

The major component of Alzheimer's disease (AD) amyloid, amyloid Aß protein is derived from the transmembrane amyloid precursor protein APP. The pathogenic mechanism leading to AD is not well understood. However, the central role of APP has emerged from the identification of genes that cosegregate with the disease. All of the identified susceptibilty genes linked to Alzheimer's disease appear to influence Aß amyloid formation (Beyreuther et al., 1996; Selkoe, 1996).

Although the normal cellular function of APP is unknown evidence has accumulated that APP has a function in repair processes. Work with different cell lines has shown that the secreted or membrane-associated forms of APP regulate cell growth, neurite length

---

* To whom correspondence should be addressed at the ZMBH-Center for Molecular Biology Heidelberg, University of Heidelberg, Im Neuenheimer Feld 282, D-69120 Heidelberg, Germany. Phone: +4-9622-1546849, Fax: +49-6221-545891, E-mail: g.multhaup@mail.zmbh.uni-heidelberg.de

Copper Transport and Its Disorders, edited by Leone and Mercer.
Kluwer Academic / Plenum Publishers, New York, 1999.

and participate in cell-cell and cell-matrix adhesion (Beher et al., 1996; Milward et al., 1992; Roch et al., 1992; Saitoh et al., 1989; Shivers et al., 1988). This idea is supported by the multi-ligand binding activity of APP to components of the extracellular matrix such as collagen, laminin and heparan sulfate side chains of proteoglycans (Breen, 1992; Multhaup, 1994; Multhaup et al., 1994; Narindrasorasak et al., 1992; Small et al., 1994). APP isoforms containing a region homologous to the Kunitz protease inhibitor (KPI) also form complexes with extracellular proteases which are internalized by the ApoE receptor LRP (Kounnas et al., 1995) (Fig. 1).

Regarding the binding of small, potentially neurotoxic ligands, APP interacts specifically with Zn(II) and Cu(II) at two distinct sites (Bush et al., 1993; Hesse et al., 1994). Of the two other known members of the APP gene family, APLP1 and APLP2, both binding sites are conserved in APLP2 whereas in APLP1 only the Zn(II) site is present (Bush et al., 1993; Hesse et al., 1994). Zn(II) and Cu(II) binding influence APP conformation, stability and homophilic interactions (Bush et al., 1993; Hesse et al., 1994; Multhaup et al., 1996).

Because transmembrane APP binds copper(II) our working hypothesis is that APP is involved in Cu homeostasis, and accumulation of APP in neurites as it occurs in AD (Cork et al., 1990; Giaccone et al., 1989; Wang and Munoz, 1995) may lead to disruption of Cu compartmentalization and Cu toxicity (Multhaup et al., 1997). APP not only binds copper ions but has the intrinsic activity to reduce specifically bound copper(II) to copper(I).

APP AND METAL-ION BINDING

Studies on metal-ion binding to APP were peformed to investigate an association between metals and APP metabolism. A novel Zn(II) binding motif was discovered in the cysteine rich N-terminal region of APP between residues 181–200 (Bush et al., 1993). This motif is distinct from the Zn(II) binding sites in the Aß region (Bush et al., 1994). The zinc binding site between residues 181–200 maps to exon 5 and is present at corresponding sites in all members of the APP superfamily. Zinc (II) binding has been shown to modulate the functional properties of APP, possibly by enhancing its macromolecular conformation. Incubation of APP with Zn(II) increases binding of APP to heparin (Multhaup, 1994; Multhaup et al., 1994) and it potentiates the inhibition of coagulation factor XIa by APP-KPI$^+$ isoforms (Komiyama et al., 1992; Van Nostrand, 1995). Zn(II) binding also influences the cleavage of the APP molecule, potentially via α-secretase activity (Bush et al., 1993; Li et al., 1995). The *in vivo* influences of zinc on APP metabolism were tested in rats given a dietary Zn(II) supply. This caused an increase in membrane associated forms and a reduction of soluble forms of APP (Whyte et al., 1997). This may be due either to altered APP trafficking or processing. An active transport of neuronal Zn(II) along cell processes or from cell to cell has been hypothesized (Simons et al., 1995).

Whereas Zn(II) is assumed to play a purely structural role, we found that APP binds Cu(II) with a dissociation constant of 10nM at pH 7.5 and reduces bound copper to Cu(I) (Hesse et al., 1994; Multhaup et al., 1996). Since Zn(II) exists exclusively in one oxidation state, only APP/Cu(II) complexes are sensitive to redox reactions. The APP copper binding motif corresponds to type II copper binding sites and is encompassed by residues APP135–155 within exon 4. The reduction of Cu(II) to Cu(I) by APP results in a corresponding oxidation of cysteines 144 and 158 in APP that involves an intramolecular oxidation reaction leading to a new disulfide bridge. This reaction was found to be very specific, as APP did not bind and reduce other metals such as Fe(III), Ni(II), Co(II) or Mg(II). Thus, Cu(II) binding leads to oxidative modification of APP, resulting in cystine

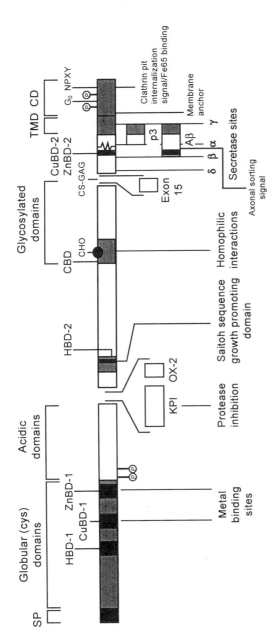

Figure 1. Schematic drawing of the structural domains and binding motifs of the APP molecule. Abbreviations used are: α, α-secretase site; "α", homologous to α-secretase cleavage site for release of the ectodomain; β, beta-secretase site; CD, cytoplasmic domain; CBD, collagen-binding domain; CHO, N-linked carbohydrate attachment site; CS-GAG, chondroitin sulfate glycosaminglycan; CuBD-1, copper-binding domain 1; CuBD-2, copper-binding domain 2; cys, cysteine; δ, delta-secretase site; γ, gamma-secretase site; GPD, growth promoting domain; HBD-1, heparin-binding domain 1; HBD-2, heparin-binding domain 2; KPI, Kunitz-type protease inhibitor; NPXY, asparagine-proline-any amino acid-tyrosine; OX-2, OX-2 homology domain; Poly-T, poly-threonine; P, phosphorylation site; SP, signal peptide; TMD, transmembrane domain; ZnBD-1, zinc-binding domain 1; ZnBD-2, zinc-binding domain 2.

and Cu(I) formation and the formation of an extra electron. This indicates that APP has an in vitro function in electron transfer to Cu(II) and possibly in radical formation.

APP REDUCES Cu(II) TO Cu(I)

To determine the requirements for copper reduction by APP, we incubated human APP, bacterial fusion proteins of APP (TP-APP$_{770}$, TP-APP$_{N262}$) and a synthetic peptide that represents the copper binding site of APP (APP135–156) with physiological concentrations of Cu(II) in the presence of the Cu(I)-indicator molecule bathocuproine disulfonate (BC) (Hesse et al., 1994; Multhaup et al., 1996). Incubation of Cu(II) with all forms of APP resulted in reduction to Cu(I) as indicated by formation of a peak with maximal absorbance at 480nm, characteristic of the BC-Cu(I) complex (Fig. 2).

In contrast, following incubation of other Cu(II) binding proteins such as Cu/Zn-SOD, keyhole limpet hemocyanin (KLH) or ovalbumin, with Cu(II) no increase in the absorbance at 480nm was observed (Hesse et al., 1994; Multhaup et al., 1996) (Fig. 2). The wavelength scans for the interaction of the longest APP isoform, TP-APP$_{770}$ and TP-APP$_{N262}$, comprising the N-terminal 262 residues of APP, showed the same reduction of Cu(II) to Cu(I). Carboxymethylation of TP-APP$_{770}$ prevented Cu(I) formation, suggesting that the oxidation of cysteines was necessary for the reduction of Cu(II). Thus, the NH$_2$-terminal 262 residues of APP are sufficient for this catalytic activity (Hesse et al., 1994; Multhaup et al., 1996).

APP not only binds and reduces copper, it thus also participates in electron transfer reactions. This transfer is Cu-ion mediated, has features of a redox-reaction and leads to intramolecular disulfide bond formation of APP, indicating that free sulfhydryl groups of APP are involved. The formation of APP-Cu(I) complexes suggests a fast electron transfer mechanism following direct binding of Cu(II) to APP. Since the oxidation of cysteines to cystine liberates two electrons and only one is required for Cu(II) reduction to Cu(I), this reaction enhances the production of hydroxyl radicals which could then attack sites near the location of the metal. The implication of oxygen radical-induced neuronal damage raises the possibility that Cu-mediated toxicity contributes to neurodegeneration in Alzheimer's disease.

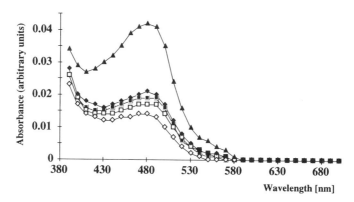

Figure 2. Reduction of Cu(II) to Cu(I) by APP. Incubations contained BC with Cu(II) (open diamonds); BC, Cu(II) and KLH (open squares); BC, Cu(II) and Cu/Zn-SOD (filled squares); BC, Cu(II) and ovalbumin (filled diamonds); BC, Cu(II) and human brain APP (filled triangles).

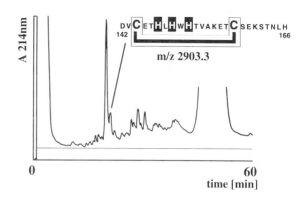

Figure 3. HPLC purification of copper binding peptides obtained by endoproteinase Asp-N digestion of APP_{N262} encoded by exons 1–6. Peptides were eluted from chelating Sepharose loaded with copper. Edman degradation and mass-spectrometry revealed the sequence of the indicated peak to contain an oxidized peptide with APP residues 142–166.

IDENTIFICATION OF THE OXIDIZED SULFHYDRYL GROUPS OF APP

To map the cystine formed during the reduction of Cu(II) to Cu(I) in APP, we used proteolytic fragments of the purified fusion protein APP (APP_{N262}) containing the previously identified Cu(II) binding site of APP residing within APP residues 135–155 (Multhaup et al., 1996). A digestion of the copper-oxidized and carboxymethylated APP_{N262} with endoprotease Asp-N yields one major peak followed by a minor one after affinity chromatography on copper(II) charged chelating Sepharose. To confirm the possible participation of Cys-144 and Cys-158 in the redox-reaction and cystine formation (Multhaup et al., 1996), ESI-MS was employed to obtain a complete analysis of the peptides eluting from the affinity column. These data indicate that the minor HPLC fraction contained the searched for singly oxidized peptide (APP residues 142–166), and thus permitted the assignment of a disulfide linkage between Cys-144 and Cys-158, differing in two mass units from the peptide in its reduced form (2905.6) (Fig. 3).

The complete analysis by electrospray ionisation mass-spectrometry (ESI-MS) of the fraction eluting from the affinity column failed to show dimerized or cross-linked APP142–166 peptides. A synthetic peptide, comprising Cys-144 and Cys-158, was oxidized on a Cu(II)-charged chelating Sepharose column and eluted with EDTA. ESI-MS revealed that HPLC fractions contained the peptide with an intramolecular disulfide linkage corresponding to a mass of 2903.6 compared to the non-oxidized peptide with a mass of 2905.6. Quantification using peak areas gives a 6-fold higher ratio for the formation of an intramolecular disulfide compared to an intermolecular disulfide cross-linked peptide APP142–166 (Fig. 4).

This provides convincing evidence for the intrinsic activity of APP residues 142–166 to form a site-directed intrachain disulfide bond between Cys-144 and Cys-158 (Fig. 5).

APP, COPPER TRANSPORT AND NEURONAL DEGENERATION

Copper is an essential trace element in the brain that is required for a number of enzyme activities including cytochrome c oxidase, Cu/Zn-superoxide dismutase (Cu/Zn-SOD) and dopamine-ß-hydroxylase with the latter being involved in the catecholamine biosynthetic pathway of the central nervous system (Linder and Hazegh Azam, 1996; Stewart and Klinman, 1988). Zinc and copper ions are needed for thermally stable native

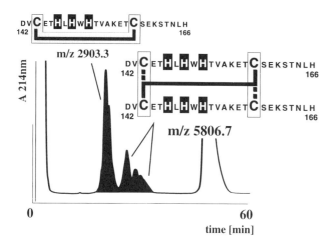

Figure 4. HPLC purification of synthetic peptide APP142–166 eluted from chelating Sepharose loaded with copper in three separate fractions. Sequence and mass-spectrometry analysis identified the majority of oxidized peptides with an intramolecular disulfide bridge and the minority to be composed of a mixture of disulfide cross-linked dimers.

Cu/Zn-SOD and to restore full catalytic activity. Whereas zinc can be substituted with cadmium, mercury and cobalt ions (Beem et al., 1974), no metal can replace copper in restoring catalytic function.

Despite the essential role in electron transfer, free copper is highly reactive and potentially toxic. Thus specialized pathways have evolved for the trafficking and compartmentation of this metal within cells (Hung et al., 1997). The following summarizes evidence that APP participates in copper transport. In neurons, APP is first delivered from the cell body to the axonal surface and then to the dendritic plasma membrane (Simons et al., 1995; Yamazaki et al., 1995). The function of APP transport in epithelial and neuronal cells is not known, but we assume that APP has an important physiological role in the cellular transport of the metal ions Zn(II) and Cu(II)/(I) in the periphery and in the central nervous system. Indirect evidence indicates that APP may also function in copper uptake during APP internalization. In non-neuronal cells, APP molecules can be reinternalized from the cell surface via endocytosis signals in the cytoplasmic tail (Haass et al., 1992; Koo and Squazzo, 1994) and then processed to Aß. The safe sequestration of transition metal ions is probably an important antioxidant defense in its own right. This is compatible with the presence of a single Cu(II)/Cu(I)

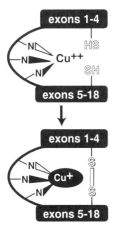

Figure 5. Cu(II) binding to the His-Xxx-His motif of APP (residues 142 to 166) which is characteristic for type II copper proteins leads to oxidative modification of APP, giving rise to cystine and Cu(I) formation.

binding site in APP (Multhaup et al., 1996) and the fact that organisms take great care in the handling of transition metal ions to minimize availability and hence reactivity.

APP mediated reduction of bound Cu(II) to Cu(I) suggests APP is acting as an extracellular copper reductase. The finding that an ion-mediated redox reaction leads to disulfide bridge formation in APP would indicate that free sulfhydryl groups in APP are involved (Multhaup et al., 1996). This is compatible with our knowledge on copper uptake and copper chaperon proteins that act as soluble cytoplasmic Cu(I) receptors in eucaryotic cells suggesting that copper is taken up and transported as Cu(I) rather than Cu(II) (Culotta et al., 1997; Pufahl et al., 1997). Furthermore, sulfhydryl-modifying reagents inhibit the uptake of copper, implicating that sulfur amino acids are required for the transport process in vivo (Harris, 1991; Percival and Harris, 1989). A regulatory mechanism for copper uptake has been discovered in the copper-dependent turnover of Ctr1p protein (Ooi et al., 1996). When cells are grown in low concentrations of copper, Ctr1p is a stable protein. But an increase in copper concentrations up to 10µM induces cleavage of Ctr1p at the cell surface (Ooi et al., 1996). This cleavage is believed to release the extracellular domain of Ctr1p that contains the copper binding site and thus inhibits further copper uptake. This may represent a general mechanism for other plasma proteins to regulate the uptake of ligands. This would be comparable with α-secretase cleavage of APP which occurs on or close to the cell surface (Sisodia et al., 1990) to release a large amino-terminal fragment of APP and precludes formation of Aß. This proteolytic pathway may inhibit copper uptake since the N-terminal domain of APP which contains the copper binding site is extracellular. The alternative and non-amyloidogenic α-secretase pathway indicates that APP may serve a function as "mammalian Ctr1p" in brain. APP cleavage by α-secretase may be the rate-limiting component responsible for the delivery of copper across the membrane to the cytosolic copper chaperone proteins of mammalian cells (Zhou and Gitschier, 1997). If cells are grown in low concentrations of copper the internalization of cell surface APP could be favored and the intracellular processing would be shifted to β- and γ-secretase cleavage of APP, i.e. the production of the amyloid protein Aß of Alzheimer's disease.

Two human inherited disorders illustrate the importance of copper transport and copper metabolism. The genes causing the recessive disorders Wilson disease and Menkes disease encode membrane Cu-transporting P-type ATPases. A failure to express functional protein results in decreased copper efflux, whilst overexpression will increase copper efflux and confer a stronger resistance to the toxic effects of copper (Camakaris et al., 1995). Neuronal degeneration as consequence of copper toxicity is also reported for familial amyotrophic lateral sclerosis (FALS). In FALS point mutations in Cu/Zn-superoxide dismutase (Cu/Zn-SOD) exhibit enhanced free radical-generating activity, while its dismutation activity is identical to that of the wild-type enzyme (Yim et al., 1996). It has been suggested that the disease is not due to loss of Cu/Zn-SOD function but rather due to an adverse or toxic gain of function of the mutant Cu/Zn-SOD molecule (Brown, 1995; Wong et al., 1995). Transgenic mice that overexpress the human Cu/Zn-SOD gene develop a clinical form of motor neuron disease which resembles human FALS (Gurney et al., 1994). A first indication for the underlying mechanism was the discovery that Cu/Zn-SOD carrying a FALS mutation is inactivated by H_2O_2 which rapidly reduces Cu(II) at the active site (Hodgson and Fridovich, 1975; Sato et al., 1992), and that Cu/Zn-was found to generate free °OH radicals from H_2O_2 (Yim et al., 1996). In an in vitro system, FALS-associated mutant Cu/Zn-SOD enzyme catalyzes the reduction of H_2O_2, thereby acting as a peroxidase (Wiedau-Pazos et al., 1996). This shows that Cu/Zn-SOD has a peroxidative activity that utilizes its own dismutation product, H_2O_2, as a substrate. This was found to be more pronounced for mutant than wild-type Cu/Zn-SOD, with the mutants being at least twice as reactive as the wild-type enzyme. The strong association found between those forms of familial amyotrophic lateral sclerosis (FALS) which carry mutations in the Cu/Zn superoxide dismutase (Cu/Zn-SOD) gene and oxygen radical formation provided the most striking evidence for a link between neurological disorders and reactive oxygen species (ROS) formation. How oxygen radicals lead to the selective degeneration of motor neurons in FALS remains to be elucidated (Wiedau-Pazos et al., 1996; Yim et al., 1997; Yim et al., 1996). Cu/Zn-SOD mutations may primarily affect Cu homeostasis of highly active neurons such as motor neurons or the anti-oxidant protection of these neurons is less efficient leading to dysfunction and death of neurons (Hottinger et al., 1997).

On the surface of neurons, APP-Cu(I) complexes may be particularly vulnerable to peroxides generated by extracellular forms of Cu/Zn-SOD. Such complexes are spontaneously formed since APP itself reduces bound Cu(II) to Cu(I). The Cu(II) ion-mediated redox reaction leads to disulfide formation in APP and to the formation of APP-Cu(I) complexes, even in the absence of hydrogen peroxide (Multhaup et al., 1996). A perturbation of free radical homeostasis may contribute to Alzheimer's disease because neuronal degeneration has been shown to be linked to the production of reactive oxygen species in FALS.

Aging coupled to environmental insults or genetic defects could exacerbate the consequences of APP-mediated copper toxicity. The most prevalent risk factor associated with late onset AD has also been suggested to be linked to cytotoxicity modulated by the varying antioxidant activity of the apoE isoforms (Miyata and Smith, 1996). It was shown that the E4 allele which is present at higher frequency in AD patients than in age matched controls possesses the lowest activity in protecting cells from hydrogen peroxide cytotoxicity (Miyata and Smith, 1996). Our hypothesis of an APP-mediated and radical-based neurotoxicity offers several novel strategies for the treatment of AD, such as protection by antioxidants, metal chelators which are specific for copper, antagonists for copper binding of APP and modulators of APP metabolism.

ACKNOWLEDGMENTS

We thank the Deutsche Forschungsgemeinschaft (DFG) for support through SFB317 and the BMBF through grant 030666A

REFERENCES

Beem, K. M., Rich, W. E., and Rajagopalan, K. V. (1974). Total reconstitution of copper-zinc superoxide dismutase. J. Biol. Chem. *249*, 7298–7305.

Beher, D., Hesse, L., Masters, C. L., and Multhaup, G. (1996). Regulation of amyloid protein precursor (APP) binding to collagen and mapping of the binding sites on APP and collagen type I. J Biol Chem *271*, 1613–1620.

Beyreuther, K., Multhaup, G., and Masters, C. L. (1996). Alzheimer's disease: genesis of amyloid. Ciba Found Symp *199*, 119–27; discussion 127–31.

Breen, K. C. (1992). APP-collagen interaction is mediated by a heparin bridge mechanism. Mol. Chem. Neuropathol. *16*, 109–121.

Brown, R. H., Jr. (1995). Amyotrophic lateral sclerosis: recent insights from genetics and transgenic mice. Cell *80*, 687–92.

Bush, A. I., Multhaup, G., Moir, R. D., Williamson, T. G., Small, D. H., Rumble, B., Pollwein, P., Beyreuther, K., and Masters, C. L. (1993). A novel zinc(II) binding site modulates the function of the beta A4 amyloid protein precursor of Alzheimer's disease. J. Biol. Chem. *268*, 16109–16112.

Bush, A. I., Pettingell, W. H., Multhaup, G., d Paradis, M., Vonsattel, J. P., Gusella, J. F., Beyreuther, K., Masters, C. L., and Tanzi, R. E. (1994). Rapid induction of Alzheimer A beta amyloid formation by zinc. Science *265*, 1464–7.

Camakaris, J., Petris, M. J., Bailey, L., Shen, P. Y., Lockhart, P., Glover, T. W., Barcroft, C. L., Patton, J., and Mercer, J. F. B. (1995). Gene amplification of the Menkes (Mnk ATP7a) P Type ATPase gene of CHO cells is associated with copper resistance and enhanced copper efflux. Hum. Mol. Genet. *4*, 2117–2123.

Cork, L. C., Masters, C., Beyreuther, K., and Price, D. L. (1990). Development of senile plaques. Relationships of neuronal abnormalities and amyloid deposits. Am J Pathol *137*, 1383–92.

Culotta, V. C., Klomp, L. W., Strain, J., Casareno, R. L., Krems, B., and Gitlin, J. D. (1997). The copper chaperone for superoxide dismutase. J Biol Chem *272*, 23469–72.

Giaccone, G., Tagliavini, F., Linoli, G., Bouras, C., Frigerio, L., Frangione, B., and Bugiani, O. (1989). Down patients: extracellular preamyloid deposits precede neuritic degeneration and senile plaques. Neurosci Lett *97*, 232–8.

Gurney, M. E., Pu, H., Chiu, A. Y., Dal Canto, M. C., Polchow, C. Y., Alexander, D. D., Caliendo, J., Hentati, A., Kwon, Y. W., Deng, H. X., Chen, W., Zhai, P., Sufit, R. L., and Siddique, T. (1994). Motor neuron degeneration in mice that express a human Cu,Zn superoxide dismutase mutation. Science *264*, 1772–5.

Haass, C., Schlossmacher, M. G., Hung, A. Y., Vigo-Pelfrey, C., Mellon, A., Ostaszewski, B. L., Lieberburg, I., Koo, E. H., Schenk, D., Teplow, D. B., and Selkoe, D. J. (1992). Amyloid beta-peptide is produced by cultured cells during normal metabolism •. Nature *359*, 322–5.

Harris, E. D. (1991). Copper transport: an overview. Proc. Soc. Exp. Biol. Med. *196*, 130–40.

Hesse, L., Beher, D., Masters, C. L., and Multhaup, G. (1994). The beta A4 amyloid precursor protein binding to copper. FEBS. Lett. *349*, 109–116.

Hodgson, E. K., and Fridovich, I. (1975). The interaction of bovine erythrocyte superoxide dismutase with hydrogen peroxide: inactivation of the enzyme. Biochemistry *14*, 5294–9.

Hottinger, A. F., Fine, E. G., Gurney, M. E., Zurn, A. D., and Aebischer, P. (1997). The copper chelator d-penicillamine delays onset of disease and extends survival in a transgenic mouse model of familial amyotrophic lateral sclerosis. Eur J Neurosci *9*, 1548–51.

Hung, I. H., Suzuki, M., Yamaguchi, Y., Yuan, D. S., Klausner, R. D., and Gitlin, J. D. (1997). Biochemical characterization of the wilson disease protein and functional expression in the yeast saccharomyces cerevisiae. J. Biol. Chem. *272*, 21461–2146.

Komiyama, Y., Murakami, T., Egawa, H., Okubo, S., Yasunaga, K., and Murata, K. (1992). Purification of factor XIa inhibitor from human platelets. Thromb Res *66*, 397–408.

Koo, E. H., and Squazzo, S. L. (1994). Evidence that production and release of amyloid beta-protein involves the endocytic pathway. J. Biol. Chem. *269*, 17386–17389.

Kounnas, M. Z., Moir, R. D., Rebeck, G. W., Bush, A. I., Argraves, W. S., Tanzi, R. E., Hyman, B. T., and Strickland, D. K. (1995). LDL receptor-related protein, a multifunctional ApoE receptor, binds secreted beta-amyloid precursor protein and mediates its degradation. Cell *82*, 331–340.

Li, Q. X., Evin, G., Small, D. H., Multhaup, G., Beyreuther, K., and Masters, C. L. (1995). Proteolytic processing of Alzheimer's disease beta A4 amyloid precursor protein in human platelets. J. Biol. Chem. *270*, 14140–14147.

Linder, M. C., and Hazegh Azam, M. (1996). Copper biochemistry and molecular biology. Am. J. Clin. Nutr. *63*, 797s–811s.

Milward, E. A., Papadopoulos, R., Fuller, S. J., Moir, R. D., Small, D., Beyreuther, K., and Masters, C. L. (1992). The amyloid protein precursor of Alzheimer's disease is a mediator of the effects of nerve growth factor on neurite outgrowth. Neuron *9*, 129–137.

Miyata, M., and Smith, J. D. (1996). Apolipoprotein E allele-specific antioxidant activity and effects on cytotoxicity by oxidative insults and beta-amyloid peptides. Nat. Genet. *14*, 55–61.

Multhaup, G. (1994). Identification and regulation of the high affinity binding site of the Alzheimer's disease amyloid protein precursor (APP) to glycosaminoglycans. Biochimie *76*, 304–311.

Multhaup, G., Bush, A. I., Pollwein, P., and Masters, C. L. (1994). Interaction between the zinc (II) and the heparin binding site of the Alzheimer's disease beta A4 amyloid precursor protein (APP). FEBS Lett *355*, 151–154.

Multhaup, G., Ruppert, T., Schlicksupp, A., Hesse, L., Beher, D., Masters, C. L., and Beyreuther, K. (1997). Reactive oxygen species and Alzheimer's disease. Biochem Pharmacol *54*, 533–9.

Multhaup, G., Schlicksupp, A., Hesse, L., Beher, D., Ruppert, T., Masters, C. L., and Beyreuther, K. (1996). The amyloid precursor protein of Alzheimer's disease in the reduction of copper(II) to copper(I). Science *271*, 1406–1409.

Narindrasorasak, S., Lowery, D. E., Altman, R. A., Gonzalez DeWhitt, P. A., Greenberg, B. D., and Kisilevsky, R. (1992). Characterization of high affinity binding between laminin and Alzheimer's disease amyloid precursor proteins. Lab. Invest. *67*, 643–652.

Ooi, C. E., Rabinovich, E., Dancis, A., Bonifacino, J. S., and Klausner, R. D. (1996). Copper-dependent degradation of the Saccharomyces cerevisiae plasma membrane copper transporter Ctr1p in the apparent absence of endocytosis. EMBO J *15*, 3515–3523.

Percival, S. S., and Harris, E. D. (1989). Ascorbate enhances copper transport from ceruloplasmin into human K562 cells. J. Nutr. *119*, 779–784.

Pufahl, R. A., Singer, C. P., Peariso, K. L., Lin, S., Schmidt, P. J., Fahrni, C. J., Culotta, V. C., Penner-Hahn, J. E., and O'Halloran, T. V. (1997). Metal ion chaperone function of the soluble Cu(I) receptor atx1. Science *278*, 853–6.

Roch, J. M., Shapiro, I. P., Sundsmo, M. P., Otero, D. A., Refolo, L. M., Robakis, N. K., and Saitoh, T. (1992). Bacterial expression, purification, and functional mapping of the amyloid beta/A4 protein precursor. J. Biol. Chem. *267*, 2214–2221.

Saitoh, T., Sundsmo, M., Roch, J. M., Kimura, N., Cole, G., Schubert, D., Oltersdorf, T., and Schenk, D. B. (1989). Secreted form of amyloid beta protein precursor is involved in the growth regulation of fibroblasts. Cell *58*, 615–622.

Sato, K., Akaike, T., Kohno, M., Ando, M., and Maeda, H. (1992). Hydroxyl radical production by H2O2 plus Cu,Zn-superoxide dismutase reflects the activity of free copper released from the oxidatively damaged enzyme. J Biol Chem *267*, 25371–7.

Selkoe, D. J. (1996). Amyloid beta-protein and the genetics of Alzheimer's disease. J Biol Chem *271*, 18295–18298.

Shivers, B. D., Hilbich, C., Multhaup, G., Salbaum, M., Beyreuther, K., and Seeburg, P. H. (1988). Alzheimer's disease amyloidogenic glycoprotein: expression pattern in rat brain suggests a role in cell contact. Embo J *7*, 1365–70.

Simons, M., Ikonen, E., Tienari, P. J., Cidarregui, A., Monning, U., Beyreuther, K., and Dotti, C. G. (1995). Intracellular routing of human amyloid protein precursor: Axonal delivery followed by transport to the dendrites. J. Neurosci. Res. *41*, 121–128.

Sisodia, S. S., Koo, E. H., Beyreuther, K., Unterbeck, A., and Price, D. L. (.1990). Evidence that beta-amyloid protein in Alzheimer's disease is not derived by normal processing. Science *248*, 492–495.

Small, D. H., Nurcombe, V., Reed, G., Clarris, H., Moir, R., Beyreuther, K., and Masters, C. L. (1994). A heparin-binding domain in the amyloid protein precursor of Alzheimer's disease is involved in the regulation of neurite outgrowth. J. Neurosci. *14*, 2117–2127.

Stewart, L. C., and Klinman, J. P. (1988). Dopamine beta-hydroxylase of adrenal chromaffin granules: structure and function. Annu. Rev. Biochem. *57*, 551–592.

Van Nostrand, W. E. (1995). Zinc (II) selectively enhances the inhibition of coagulation factor XIa by protease nexin-2/amyloid beta-protein precursor. Thromb Res *78*, 43–53.

Wang, D., and Munoz, D. G. (1995). Qualitative and quantitative differences in senile plaque dystrophic neurites of Alzheimer's disease and normal aged brain. J Neuropathol Exp Neurol *54*, 548–556.

Whyte, S., Jones, L., Coulson, E. J., Moir, R. D., Bush, A. I., Beyreuther, K., and Masters, C. L. (1997). The metabolism of the amyloid precursor protein of Alzheimer's disease and dietary zinc. In Alzheimer's Disease: Biology, Diagnosis and Therapeutics, K. Iqbal, B. Winblad, T. Nishimura, M. Takeda and H. M. Wisniewski, eds. (Chichester: John Wiley & Sons Ltd), pp. 417–422.

Wiedau-Pazos, M., Goto, J. J., Rabizadeh, S., Gralla, E. B., Roe, J. A., Lee, M. K., Valentine, J. S., and Bredesen, D. E. (1996). Altered reactivity of superoxide dismutase in familial amyotrophic lateral sclerosis. Science *271*, 515–518.

Wong, P. C., Pardo, C. A., Borchelt, D. R., Lee, M. K., Copeland, N. G., Jenkins, N. A., Sisodia, S. S., Cleveland, D. W., and Price, D. L. (1995). An adverse property of a familial ALS-linked Cu/Zn-SOD mutation causes motor neuron disease characterized by vacuolar degeneration of mitochondria. Neuron *14*, 1105–16.

Yamazaki, T., Selkoe, D. J., and Koo, E. H. (1995). Trafficking of cell surface beta-amyloid precursor protein: retrograde and transcytotic transport in cultured neurons. J. Cell. Biol. *129*, 431–442.

Yim, H. S., Kang, J. H., Chock, P. B., Stadtman, E. R., and Yim, M. B. (1997). A familial amyotrophic lateral sclerosis-associated A4V Cu,Zn-Superoxide dismutase mutant has a lower K-m for hydrogen peroxide. J Biol Chem *272*, 8861–8863.

Yim, M. B., Kang, J. H., Yim, H. S., Kwak, H. S., Chock, P. B., and Stadtman, E. R. (1996). A gain-of-function of an amyotrophic lateral sclerosis- associated Cu,Zn-superoxide dismutase mutant: An enhancement of free radical formation due to a decrease in K-m for hydrogen peroxide. Proc Natl Acad Sci USA *93*, 5709–5714.

Zhou, B., and Gitschier, J. (1997). hCTR1: a human gene for copper uptake identified by complementation in yeast. Proc Natl Acad Sci U S A *94*, 7481–6.

17

COPPER-ZINC SUPEROXIDE DISMUTASE AND ALS

Joan Selverstone Valentine, P. John Hart,[*] and Edith Butler Gralla

Department of Chemistry and Biochemistry, UCLA
Los Angeles, California 90095-1569

1. INTRODUCTION

1.1. Amyotrophic Lateral Sclerosis

Amyotrophic lateral sclerosis (ALS, Lou Gehrig's disease) is a neurodegenerative disorder characterized by the slow loss of large motor neurons in the spinal cord and brain. The mean age of onset is 55 years. After the disease begins, it progresses relentlessly to a lethal paralysis and culminates in the death of the afflicted person, usually within two to five years of symptom onset (Brown, 1997). The disease is inherited in approximately 10 % of cases and about one-fifth of those familial ALS (FALS) cases are associated with dominantly inherited, single-site mutations in *SOD1,* the gene that encodes human copper-zinc superoxide dismutase (CuZnSOD) (Cudkowicz and Brown, 1996). SOD-associated FALS thus represents only a fraction of the total cases of ALS, but it is the only form of the disease for which a cause is known. Many laboratories are actively seeking to understand how mutations in CuZnSOD cause this form of ALS in the hope that such knowledge will hasten the discovery of causes of this disease as well as potential therapeutic agents.

1.2. Copper-Zinc Superoxide Dismutase

CuZnSOD is a 32 kD homodimeric metalloenzyme which is present in the cytoplasm of most cells and is particularly abundant in red blood cells and in neurons, where it is ~1% of the mass of spinal tissue protein (Pardo et al., 1995). It lowers superoxide concentrations by catalyzing the disproportionation of that anionic radical to give hydrogen

---

[*] Present address: Department of Biochemistry, University of Texas Health Science Center at San Antonio, 7703 Floyd Curl Drive, San Antonio, Texas 78284-7660

Copper Transport and Its Disorders, edited by Leone and Mercer.
Kluwer Academic / Plenum Publishers, New York, 1999.

peroxide and dioxygen (Eqs. 1 and 2) and thus functions as an antioxidant enzyme (Fridovich, 1997).

$$O_2^- + Cu(II)ZnSOD \rightarrow O_2 + Cu(I)ZnSOD \qquad (1)$$

$$O_2^- + Cu(I)ZnSOD + 2H^+ \rightarrow H_2O_2 + Cu(II)ZnSOD \qquad (2)$$

Each of the two equivalent subunits of the enzyme binds one copper ion and one zinc ion in close proximity. In the oxidized Cu(II) form of the enzyme, crystallographic and spectroscopic studies have shown that a histidyl residue (His63 in human CuZnSOD) simultaneously coordinates both metal ions (Bertini et al., 1990; Valentine, 1994; Valentine and Pantoliano, 1981). This type of metal ion binding to a histidine side chain, termed the "histidine bridge" or "bridging imidazolate", is unique to CuZnSOD. The catalytic copper ion sits at the bottom of a channel that narrows from a shallow depression approximately 24 Å across at the surface of the molecule to a deeper channel about 10 Å wide and finally to an opening of less than 4 Å just above the copper ion. Access to this channel is limited to small anions (Tainer et al., 1983).

The copper-binding geometry in the oxidized protein is five-coordinate with the four histidine ligands (His46, 48, 63 and 120) in a highly distorted square planar arrangement and an axial water ligand (Bannister et al., 1987; Bertini et al., 1990; Tainer et al., 1982; Valentine and Pantoliano, 1981). In the Cu(I) form of the enzyme, crystallographic and spectroscopic analyses indicated that the bridging imidazolate is protonated on the copper-binding side at its NE2 atom. The loss of the His63-Cu bond upon copper reduction results in a nearly trigonal-planar copper coordination geometry with histidine residues 46, 48 and 120 acting as ligands (Bertini et al., 1990; Ogihara et al., 1996).

2. MUTATIONS IN CuZnSOD CAUSE ALS BY A GAIN–NOT A LOSS–OF FUNCTION

The initial report that mutations in the CuZnSOD gene could cause ALS (Rosen et al., 1993) and the observation of reduced SOD activity in red blood cells of some FALS patients (Deng et al., 1993) led to suggestions that a cause of the disease might be oxidative damage resulting from a lowering of the activity, i.e., loss of function, of this antioxidant enzyme. However, subsequent studies of transgenic mice demonstrated that overexpression of FALS mutant CuZnSODs caused a motor neuron degenerative syndrome despite normal or supranormal SOD enzymatic activities in these mice (Dal Canto and Gurney, 1997; Gurney, 1997; Gurney et al., 1994). By contrast, transgenic mice overexpressing wild type human CuZnSOD (Dal Canto and Gurney, 1997; Gurney, 1997; Gurney et al., 1994) as well as mice that were lacking CuZnSOD (Reaume et al., 1996) did not develop this syndrome. These experimental results gave strong support to a model in which the FALS mutations in CuZnSOD are dominant and exert their effects due to a (toxic) gain of function. This model is also supported by the findings that many FALS mutant human CuZnSOD genes rescue the oxygen-sensitive phenotype of *sod1*⁻ yeast (see section 3 below), that FALS mutations convert CuZnSOD from an anti-apoptotic gene to a pro-apoptotic gene in cultured neuronal cells (Rabizadeh et al., 1995), and that transfection of a human neuroblastoma cell line with plasmids directing constitutive expression of the FALS mutant G93A but not wild type human CuZnSOD induced a loss of mitochondrial membrane potential and an increase in cytosolic calcium concentration (Carri et al., 1997). Copper is strongly implicated in the toxicity of FALS mutant CuZnSODs based on the beneficial effects of copper chelators in cell culture model (Rabizadeh et al., 1995) and in the FALS SOD-expressing transgenic mice (Hottinger et al., 1997).

It should be noted that all of the published studies on FALS mutant CuZnSODs have been performed using one, or at most a few, of the more than 50 different mutations that have been identified. The validity of extrapolating from these limited data to the whole disease is not yet certain. If the disease has one cause, the conjunction of a single effect in all mutations may help to pinpoint that cause. If, on the other hand, a variety of different events can cause initiation of ALS, understanding the mechanisms of action of many different mutations will be essential. Therefore, until more is known, it will be important to widen the scope of these studies and to study as many mutations as possible.

3. MANY, BUT NOT ALL, HUMAN FALS MUTANT CuZnSODS RESCUE YEAST THAT LACK WILD TYPE CuZnSOD

In order to test the SOD activity of human FALS CuZnSOD mutants in an *in vivo* situation, they were expressed in yeast that lack the native CuZnSOD (*sod1Δ*). When grown in air, these *sod1Δ* yeast mutants exhibit a severe phenotype which includes slow growth, a requirement for certain amino acids, and extreme sensitivity to redox cycling drugs such as paraquat (PQ) (Gralla, 1997; Gralla and Kosman, 1992). As little as 0.01 mM PQ kills the *sod1Δ* cells, while wild type yeast are resistant to more than 1 mM PQ.

The wild type human CuZnSOD and nine FALS mutant human CuZnSODs were subcloned in the yeast multicopy vector YEP351 under the control of the yeast CuZnSOD promoter, and tested for the *sod1Δ* phenotype. Full rescue of the *sod1Δ* yeast was observed for most human FALS CuZnSODs tested, including A4V, L38V, G93A, G93C, G37R, G41D, and G85R. (The results for the first four were previously reported (Rabizadeh et al., 1995); the last three are newly reported here.) The only mutant proteins we have tested that did not rescue the *sod1Δ* yeast strain were H46R and H48Q, which have mutations in active site metal-binding ligands. So far as we have been able to observe, there is no partial rescue--either the enzyme is fully active in this assay or it is completely inactive. Assays of SOD activity performed on extracts of these cells indicated that levels of SOD activity provided by the ALS-SOD proteins that rescue the *sod1Δ* yeast strain are at least as high as the activity in normal wild type yeast, while non-rescuing mutants exhibit no activity. Figure 1 shows a test of paraquat sensitivity for some of the new mutants.

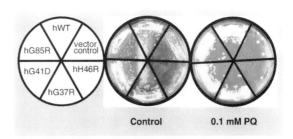

Figure 1. Paraquat sensitivity of *sod1Δ* yeast expressing some human FALS mutant CuZn-SODs. Cells from an overnight culture of the indicated strains (see text) were diluted to similar densities and streaked on freshly prepared YPD (rich medium) plates containing or lacking the indicated amount of paraquat (PQ) and incubated at 30 degrees C. for three days. Similar results were observed for paraquat concentrations ranging from 0.01 to 1.0 mM.

The rescue of the sod1Δ yeast by most of the human FALS mutant CuZnSODs tested demonstrates that these proteins are enzymatically active and capable of functioning in vivo. The resistance of S. cerevisiae expressing FALS mutant CuZnSODs to whatever property causes harm to motor neurons may be due to the fact that these yeast are considerably more resistant to oxidative stress than mammalian cells. This difference in vulnerability to oxidative stress may be due to the fact that these yeast, unlike mammalian cells, contain no polyunsaturated fatty acids and thus their lipids are highly resistant to peroxidation. It is also very possible that the target(s) of the toxic action of the FALS mutant CuZnSODs in the motor neurons, e.g., neurofilament proteins, is not present in yeast.

4. HYPOTHESES EXPLAINING THE GAIN OF A TOXIC FUNCTION

Several different hypotheses have been suggested to explain the one or more molecular mechanisms by which the FALS mutant human CuZnSODs exert a deleterious effect that causes motor neurons to die.

4.1. Neurofilament Abnormalities, Protein-Protein Interactions, and Aggregation of CuZnSOD in Vivo

Neurofilament abnormalities have been implicated in ALS for some time and have been observed in the transgenic mice expressing FALS mutant CuZnSODs (Bruijn and Cleveland, 1996; Tu et al., 1997; Williamson et al., 1996). Mechanisms that might lead to damage to neurofilaments have therefore been particularly favored. Recent studies of transgenic mice expressing the human G93A CuZnSOD provide additional support for the hypothesis that the deleterious effect of the FALS mutant CuZnSODs cause neurofilament abnormalities with the demonstration of impaired axonal transport in the ventral roots of these mice and the appearance of neurofilament inclusions and vacuoles in vulnerable motor neurons (Zhang et al., 1997). There is also evidence for aggregation of human FALS mutant CuZnSODs, but not of wild type human CuZnSOD, when expressed in cultured spinal motor neurons (Durham et al., 1997) and two proteins, lysyl-tRNA synthetase and translocon-associated protein delta, were found in a yeast two-hybrid assay to bind to FALS mutant but not wild type CuZnSOD (Kunst et al., 1997).

4.2. Studies of Isolated FALS Mutant CuZnSODs

Several laboratories have expressed, purified, and characterized FALS mutant CuZn-SODs in an effort to learn in what respects they differ from the wild type enzyme and what properties they have in common that might explain their pathogenicity. These studies have led to several gain of function hypotheses: (1) altered metal ion binding leading to abnormal metal ion metabolism, (2) increased non-specific peroxidative activity, or (3) increased generation of peroxynitrite; either of these last two hypotheses would lead to increased oxidative stress.

4.2.1. Metal Binding Properties. The ALS mutations were found to alter the characteristic metal ion binding properties of wild type CuZnSODs first in the case of analogous mutants of yeast CuZnSOD (Nishida et al., 1994) and then for the mutant human enzymes

(Crow et al., 1997; Lyons et al., 1996). These results suggest that metal ion metabolism, particularly of copper or zinc, might be altered in SOD-associated FALS.

4.2.2. Non-Specific Peroxidative Activity. FALS mutant CuZnSODs have also been reported to function as nonspecific peroxidases, using hydrogen peroxide as a substrate to generate hydroxyl radical ($^\bullet$OH), a very reactive species that could damage cellular constituents. This mode of reactivity of CuZnSOD was originally discovered by Yim and Stadtman who found that bovine wild type CuZnSOD would catalyze the oxidation of a model substrate, the spin trap DMPO (5,5'-dimethyl-1-pyrroline N-oxide) by hydrogen peroxide (Yim et al., 1990; Yim et al., 1993). Studies of two human FALS mutants, A4V and G93A using the same techniques have demonstrated that this peroxidative reaction occurs at higher rates for these FALS mutant CuZnSODs than for wild-type human CuZn-SOD (Wiedau-Pazos et al., 1996; Yim et al., 1997; Yim et al., 1996). These results suggest the possibility that FALS mutant CuZnSODs might catalyze the oxidation of unknown intracellular substrates by hydrogen peroxide *in vivo*.

The mechanism of this reaction is believed to involve first reduction of the Cu(II) form of the enzyme by hydrogen peroxide (Eq. 3). The Cu(I) enzyme that is formed then reacts with a second hydrogen peroxide in a Fenton-type reaction to produce a species capable of transferring the elements of a hydroxyl radical, $^\bullet$OH, to a substrate, in this case DMPO (Eq. 4). Inside the cell, other reducing agents such as ascorbate may take the place of hydrogen peroxide in the first step of reaction (Eq. 3) (Lyons et al., 1996). A major unknown in evaluating the possibility that this mechanism acts in motor neurons is the concentration of hydrogen peroxide that might be present in the cell.

$$H_2O_2 + Cu(II)ZnSOD \rightarrow O_2^- + H^+ + Cu(I)ZnSOD \quad (3)$$

$$H_2O_2 + DMPO + Cu(I)ZnSOD \rightarrow DMPO\text{-}OH + OH^- + Cu(II)ZnSOD \quad (4)$$

An interesting related finding is the observation of a superoxide-dependent peroxidase activity for the H48Q FALS CuZnSOD mutant enzyme. The same activity was not observed for the G93A, G93R, or E100G mutants (Liochev et al., 1997).

Whether or not this mechanism accounts for the toxic gain of function of the FALS mutant CuZnSODs, it is relevant here to note here that there is an increasing body of evidence of oxidative damage to tissues in ALS. For example, elevated levels of characteristic markers of oxidative damage to nucleic acids, proteins, and lipids have recently been reported in tissue samples obtained from both sporadic ALS (SALS) and SOD-associated FALS patients (Ferrante et al., 1997). Moreover, studies of the transgenic mice expressing FALS mutant CuZnSODs have shown beneficial effects of the lipid antioxidant vitamin E (Gurney et al., 1996).

4.2.3. Tyrosine Nitration Catalyst. The suggestion has also been made that FALS mutant CuZnSODs might have an enhanced capacity to catalyze nitration of tyrosine residues by peroxynitrite (ONOO$^-$), formed from the reaction of superoxide (O_2^-) with nitric oxide (NO) *in vivo* (Beckman et al., 1993).

$$SOD\text{-}Cu^{2+} + {}^-OONO \rightarrow SOD\text{-}CuO\cdots NO_2^+ \quad (5)$$

$$SOD\text{-}CuO\cdots NO_2^+ + H\text{-}Tyr \rightarrow SOD\text{-}Cu^{2+} + HO^- + NO_2\text{-}Tyr \quad (6)$$

In support of this hypothesis, mice expressing human FALS mutant G37R CuZnSOD exhibit 2- to 3-fold elevation in free nitrotyrosine levels in spinal cord tissue relative to normal mice or mice expressing high levels of wild type human enzyme (Bruijn et al., 1997), and increased 3-nitrotyrosine immunoreactivity was observed in motor neurons of both sporadic and familial ALS patients (Beal et al., 1997). In addition, isolated FALS mutant CuZnSODs have been found to lose zinc ion readily and to have an enhanced ability to catalyze peroxynitrite-mediated tyrosine nitration when they were zinc-deficient (Crow et al., 1997).

5. STRUCTURAL PROPERTIES OF FALS MUTANT CuZnSODS

5.1. Locations of FALS Mutations in the Three-Dimensional Structure of CuZnSOD

Eleven different missense mutations in CuZnSOD were initially identified in thirteen different FALS families (Rosen et al., 1993). The locations of these mutations were clustered at the dimer interface and in loop regions at the ends of β-strands (Deng et al., 1993). By 1998, the number of distinct FALS CuZnSOD mutations identified has risen to about 50 (Juneja et al., 1997; Siddique and Deng, 1996). Figure 2 illustrates the distribution of the FALS mutated residues on the known structure of the wild-type human CuZn-

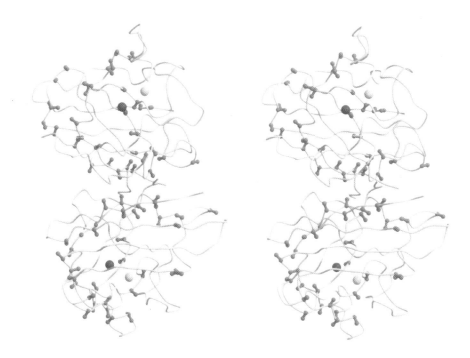

Figure 2. Stereo view of the human CuZnSOD homodimer highlighting the three-dimensional distribution of the FALS mutations. The molecular two-fold axis is horizontal in the plane of the paper. The protein backbone is represented as a light gray coil, the positions of the currently known FALS mutations as dark gray ball-and-stick β-carbons, and the copper and zinc ions as dark and light gray spheres respectively.

SOD molecule (pdb1spd) (Deng et al., 1993; Hart et al., 1998). FALS mutations in *SOD1* are found in all exons. A clustering of mutant residues at the dimer interface and at the ends of β-strands is still evident in the three-dimensional structure, but several recently identified mutations map to other regions as well. Examination of the locations of the different FALS mutations suggests that the enzyme molecule may be affected in a global sense by the FALS mutations.

5.2. X-Ray Structure of Human G37R CuZnSOD

The human G37R CuZnSOD mutant enzyme has been studied extensively both *in vivo* and *in vitro*. When expressed in transgenic mice, it causes motor neuron disease (Cleveland et al., 1996) and a 2- to 3-fold elevation in free nitrotyrosine levels in spinal cord tissue, relative to normal mice or mice expressing high levels of wild type human enzyme (Bruijn et al., 1997). It has been reported to have full specific activity (Borchelt et al., 1994), an approximate 2-fold reduction in polypeptide stability relative to wild type (Borchelt et al., 1994), and to enhance apoptosis in a dominant fashion (Rabizadeh et al., 1995).

The x-ray structure of the human G37R FALS mutant was determined and analyzed to 1.9 Å resolution (Hart et al., 1998). The structure of G37R CuZnSOD shows the Greek β-barrel topology typical of known CuZnSOD structures, and there are no gross deviations in backbone positions relative to the wild-type and thermostable mutant human CuZnSOD protein coordinates available in the Protein Data Bank (pdb1spd, pdb1sos) (Deng et al., 1993; Parge et al., 1992). The subunits are observed to have been properly N-acetylated by the yeast expression system despite the fact that wild-type yeast CuZnSOD is not similarly modified (Hallewell et al., 1987; Wiedau-Pazos et al., 1996). (Human CuZnSOD expressed in *E. coli* is not N-acetylated.) The glycine to arginine mutation causes little rearrangement of the protein backbone relative to wild type.

The two SOD subunits have distinct environments in the crystal and are different in structure at their copper binding sites. In one subunit, the bridging imidazolate (His63) coordinates both metal ions at distances of 2.70 Å (Cu site) and 1.87 Å (Zn site) and shows strong continuous electron density between it and the metal ions. The metal ion in the copper site has a four-coordinate ligand geometry suggestive of Cu(II). The other subunit shows a distorted trigonal planar geometry around the metal ion in the copper site. The electron density is continuous between the zinc site ion and His63 but is broken between the copper site ion and that same residue. Thus His63 coordinates only the metal ion in the zinc site at a distance of 1.81 Å. The distance between the copper site metal ion and NE2 of His63 is 2.95 Å. The three-coordinate geometry is suggestive of Cu(I) (see Figure 3).

Another indication of structural differences between the two subunits is the higher atomic displacement parameters for the copper ion (40 Å^2 vs. 28 Å^2) and backbone atoms (<B> = 30 ± 10 Å^2) vs. <B> = 24 ± 11 Å^2) for one subunit relative to the other subunit. Such large differences between the two subunits has not been reported for other CuZnSOD structures. These results imply that the G37R molecule is more flexible than wild type and suggest a possible link of a looser structure to FALS (Hart et al., 1998).

6. CONCLUSIONS

Copper ions and copper coordination complexes are frequently quite toxic, presumably due to their ability to promote adverse oxidation reactions. This mode of toxicity is

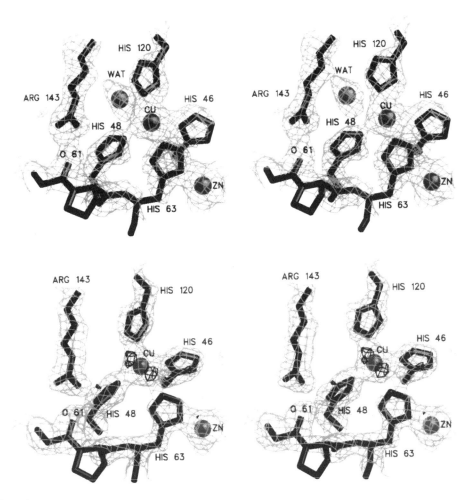

Figure 3. Stereo views of the bridge-intact and bridge-broken G37R FALS mutant copper sites superimposed on 1.9 Å electron density. The electron density (light gray) is an annealed omit map contoured at 1 σ. Copper ligands (His46, His48, His63, His120) and Arg143 are shown as black tubes. The copper and zinc ions are represented by dark gray spheres. A) Bridge-intact subunit. Note the continuous electron density between the copper atom and the bridging imidazolate, His63. The water molecule, represented by a dark gray sphere is ~2.9 Å from the copper ion. B) Bridge-broken subunit. Note the lack of continuous electron density between the copper atom and His63. The copper is coordinated by a roughly trigonal-planar arrangement of histidine residues. The copper atom exhibits anisotropy represented by the black cages of difference electron density contoured at 3 σ. (Hart et al., 1998.)

repressed in normal copper proteins such as CuZnSOD presumably by limiting access of substrates to the copper site and modulating the reactivity of the copper ion by adjusting its coordination environment. A loosening of the protein structure of CuZnSOD in FALS mutants may de-repress this inherent toxicity of the copper center, converting the enzyme-bound copper ion into a "wolf in sheep's clothing". It is interesting in this regard to note that there is substantial evidence implicating oxidative damage catalyzed by copper or zinc in several other neurodegenerative diseases (Valentine et al., 1998) as well.

ACKNOWLEDGMENTS

Support for this research from the USPHS (GM28222) and the ALS Association is gratefully acknowledged.

REFERENCES

Bannister, J. V., Bannister, W. H., and Rotilio, G. (1987). Aspects of the structure, function, and applications of superoxide dismutase. CRC Crit. Rev. Biochem. *22*, 111–180.
Beal, M. F., Ferrante, R. J., Browne, S. E., Matthews, R. T., Kowall, N. W., and Brown, R. H., Jr. (1997). Increased 3-nitrotyrosine in both sporadic and familial amyotrophic lateral sclerosis. Ann. Neurol. *42*, 644–654.
Beckman, J. S., Carson, M., Smith, C. D., and Koppenol, W. H. (1993). ALS, SOD and peroxynitrite. Nature *364*, 584.
Bertini, I., Banci, L., and Piccioli, M. (1990). Spectroscopic studies on Cu_2Zn_2SOD: A continuous advancement of investigation tools. Coord. Chem. Rev. *100*, 67–103.
Borchelt, D. R., Lee, M. K., Slunt, H. S., Guarnieri, M., Xu, Z. S., Wong, P. C., Brown, R. H., Jr., Price, D. L., Sisodia, S. S., and Cleveland, D. W. (1994). Superoxide dismutase 1 with mutations linked to familial amyotrophic lateral sclerosis possesses significant activity. Proc. Natl. Acad. Sci. (USA) *91*, 8292–8296.
Brown, R. H., Jr. (1997). Amyotrophic lateral sclerosis. Arch. Neurol. *54*, 1246–1250.
Bruijn, L. I., Beal, M. F., Becher, M. W., Schulz, J. B., Wong, P. C., Price, D. L., and Cleveland, D. W. (1997). Elevated free nitrotyrosine levels, but not protein-bound nitrotyrosine or hydroxyl radicals, throughout amyotrophic lateral sclerosis (ALS)-like disease, implicate tyrosine nitration as an aberrant *in vivo* property of one familial ALS-linked superoxide dismutase 1 mutant. Proc. Natl. Acad. Sci. (USA) *94*, 7606–7611.
Bruijn, L. I., and Cleveland, D. W. (1996). Mechanisms of selective motor neuron death in ALS: Insights from transgenic mouse models of motor neuron disease. Neuropathol. Appl. Neurobiol. *22*, 373–387.
Carri, M. T., Ferri, A., Battistoni, A., Famhy, L., Gabbianelli, R., Poccia, F., and Rotilio, G. (1997). Expression of a Cu, Zn superoxide dismutase typical of familial amyotrophic lateral sclerosis induces mitochondrial alteration and increase of cytosolic Ca^{2+} concentration in transfected neuroblastoma SH-SY5Y cells. FEBS Lett. *414*, 365–368.
Cleveland, D. W., Bruijn, L. I., Wong, P. C., Marszalek, J. R., Vechio, J. D., Lee, M. K., Xu, X. S., Borchelt, D. R., Sisodia, S. S., and Price, D. L. (1996). Mechanisms of selective motor neuron death in transgenic mouse models of motor neuron disease. Neurology *47*, 54–62.
Crow, J. P., Sampson, J. B., Zhuang, Y., Thompson, J. A., and Beckman, J. S. (1997). Decreased zinc affinity of amyotrophic lateral sclerosis-associated superoxide dismutase mutants leads to enhanced catalysis of tyrosine nitration by peroxynitrite. J. Neurochem. *69*, 1936–1944.
Cudkowicz, M. E., and Brown, R. H., Jr. (1996). An update on superoxide dismutase 1 in familial amyotrophic lateral sclerosis. J. Neurol. Sci. *139*, 10–15.
Dal Canto, M. C., and Gurney, M. E. (1997). A low expressor line of transgenic mice carrying a mutant human Cu, Zn superoxide dismutase (*SOD1*) gene develops pathological changes that most closely resemble those in human amyotrophic lateral sclerosis. Acta Neuropathol. (Berlin) *93*, 537–550.
Deng, H. X., Hentati, A., Tainer, J. A., Iqbal, Z., Cayabyab, A., Hung, W. Y., Getzoff, E. D., Hu, P., Herzfeldt, B., Roos, R. P., and et al. (1993). Amyotrophic lateral sclerosis and structural defects in Cu, Zn superoxide dismutase. Science *261*, 1047–1051.
Durham, H. D., Roy, J., Dong, L., and Figlewicz, D. A. (1997). Aggregation of mutant Cu/Zn superoxide dismutase proteins in a culture model of ALS. J. Neuropathol. Exp. Neurol. *56*, 523–530.
Ferrante, R. J., Browne, S. E., Shinobu, L. A., Bowling, A. C., Baik, M. J., MacGarvey, U., Kowall, N. W., Brown, R. H., Jr., and Beal, M. F. (1997). Evidence of increased oxidative damage in both sporadic and familial amyotrophic lateral sclerosis. J. Neurochem. *69*, 2064–2074.
Fridovich, I. (1997). Superoxide anion radical (O_2^-), superoxide dismutases, and related matters. J. Biol. Chem. *272*, 18515–18517.
Gralla, E. B. (1997). Superoxide Dismutase: Studies in the Yeast Saccharomyces Cerevisiae. In Oxidative Stress and the Molecular Biology of Antioxidant Defenses, J. Scandalios, ed. (Cold Spring Harbor, NY: Cold Spring Harbor Laboratory Press), pp. 495–525.
Gralla, E. B., and Kosman, D. (1992). Molecular genetics of superoxide dismutases in yeasts and related fungi. Adv. Genet. *30*, 251–319.

Gurney, M. E. (1997). Transgenic animal models of familial amyotrophic lateral sclerosis. J. Neurol. *244 Suppl 2*, S15–20.

Gurney, M. E., Cutting, F. B., Zhai, P., Doble, A., Taylor, C. P., Andrus, P. K., and Hall, E. D. (1996). Benefit of vitamin E, riluzole, and gabapentin in a transgenic model of familial amyotrophic lateral sclerosis. Ann. Neurol. *39*, 147–157.

Gurney, M. E., Pu, H., Chiu, A. Y., Dal Canto, M. C., Polchow, C. Y., Alexander, D. D., Caliendo, J., Hentati, A., Kwon, Y. W., Deng, H. X., and et al. (1994). Motor neuron degeneration in mice that express a human Cu, Zn superoxide dismutase mutation. Science *264*, 1772–1775.

Hallewell, R. A., Mills, R., Tekamp-Olson, P., Blacher, R., Rosenberg, S., Otting, F., Masiarz, F. R., and Scandella, C. J. (1987). Amino terminal acylation of authentic human Cu, Zn superoxide dismutase produced in yeast. Bio/Technology *5*, 363–366.

Hart, P. J., Hart, P. J., Liu, H., Pellegrini, M., Nersissian, A. M., Gralla, E. B., Valentine, J. S., and Eisenberg, D. (1998). Subunit asymmetry in the three-dimensional structure of a human CuZnSOD mutant found in familial amyotrophic lateral sclerosis. Protein Science (in press).

Hottinger, A. F., Fine, E. G., Gurney, M. E., Zurn, A. D., and Aebischer, P. (1997). The copper chelator d-penicillamine delays onset of disease and extends survival in a transgenic mouse model of familial amyotrophic lateral sclerosis. Eur. J. Neurosci. *9*, 1548–1551.

Juneja, T., Pericak-Vance, M. A., Laing, N. G., Dave, S., and Siddique, T. (1997). Prognosis in familial amyotrophic lateral sclerosis: progression and survival in patients with glu100gly and ala4val mutations in Cu, Zn superoxide dismutase. Neurology *48*, 55–57.

Kunst, C. B., Mezey, E., Brownstein, M. J., and Patterson, D. (1997). Mutations in *SOD1* associated with amyotrophic lateral sclerosis cause novel protein interactions. Nat. Genet. *15*, 91–94.

Liochev, S. I., Chen, L. L., Hallewell, R. A., and Fridovich, I. (1997). Superoxide-dependent peroxidase activity of H48Q: a superoxide dismutase variant associated with familial amyotrophic lateral sclerosis. Arch. Biochem. Biophys. *346*, 263–268.

Lyons, T. J., Liu, H., Goto, J. J., Nersissian, A., Roe, J. A., Graden, J. A., Cafe, C., Ellerby, L. M., Bredesen, D. E., Gralla, E. B., and Valentine, J. S. (1996). Mutations in copper-zinc superoxide dismutase that cause amyotrophic lateral sclerosis alter the zinc binding site and the redox behavior of the protein. Proc. Natl. Acad. Sci. (USA) *93*, 12240–12244.

Nishida, C. R., Gralla, E. B., and Valentine, J. S. (1994). Characterization of three yeast copper-zinc superoxide dismutase mutants analogous to those coded for in familial amyotrophic lateral sclerosis. Proc. Natl. Acad. Sci. (USA) *91*, 9906–9910.

Ogihara, N. L., Parge, H. E., Hart, P. J., Weiss, M. S., Goto, J. J., Crane, B. R., Tsang, J., Slater, K., Roe, J. A., Valentine, J. S., Eisenberg, D., and Tainer, J. A. (1996). Unusual trigonal-planar copper configuration revealed in the atomic structure of yeast copper-zinc superoxide dismutase. Biochemistry *35*, 2316–2321.

Pardo, C. A., Xu, Z., Borchelt, D. R., Price, D. L., Sisodia, S. S., and Cleveland, D. W. (1995). Superoxide dismutase is an abundant component in cell bodies, dendrites, and axons of motor neurons and in a subset of other neurons. Proc. Natl. Acad. Sci. (USA) *92*, 954–958.

Parge, H. E., Hallewell, R. A., and Tainer, J. A. (1992). Atomic structures of wild-type and thermostable mutant recombinant human Cu, Zn superoxide dismutase. Proc. Acad. Sci. (USA) *89*, 6109–6113.

Rabizadeh, S., Gralla, E. B., Borchelt, D. R., Gwinn, R., Valentine, J. S., Sisodia, S., Wong, P., Lee, M., Hahn, H., and Bredesen, D. E. (1995). Mutations associated with amyotrophic lateral sclerosis convert superoxide dismutase from an antiapoptotic gene to a proapoptotic gene: studies in yeast and neural cells. Proc. Natl. Acad. Sci. (USA) *92*, 3024–3028.

Reaume, A. G., Elliot, J. L., Hoffman, E. K., Kowall, N. W., Ferrante, R. J., Siwek, D. F., Wilcox, H. M., Flood, D. G., Beal, M. F., Jr., R. H. B., Scott, R. W., and Snider, W. D. (1996). Motor neurons in Cu/Zn superoxide dismutase-deficient mice develop normally but exhibit enhanced cell death after axonal injury. Nat. Genet. *13*, 43–47.

Rosen, D. R., Siddique, T., Patterson, D., Figlewicz, D. A., Sapp, P., Hentati, A., Donaldson, D., Goto, J., O'Regan, J. P., Deng, H.-X., Rahmani, Z., Krizus, A., McKenna-Yasek, D., Cayabyab, A., Gaston, S. M., Berger, R., Tanzi, R. E., Halperin, J. J., Herzfeldt, B., Van den Bergh, R., Hung, W.-Y., Bird, T., Deng, G., Mulder, D. W., Smyth, C., Laing, N. G., Soriano, E., Pericak-Vance, M. A., Haines, J., Rouleau, G. A., Gusella, J. S., Horvitz, H. R., and Brown Jr., R. H. (1993). Mutations in Cu/Zn superoxide dismutase gene are associated with familial amyotrophic lateral sclerosis. Nature *362*, 59–62.

Siddique, T., and Deng, H. X. (1996). Genetics of amyotrophic lateral sclerosis. Hum. Mol. Genet. *5 Spec No*, 1465–1470.

Tainer, J. A., Getzoff, E. D., Beem, K. M., Richardson, J. S., and Richardson, D. C. (1982). Determination and analysis of the 2 A structure of copper, zinc superoxide dismutase. J. Mol. Biol. *160*, 181.

Tainer, J. A., Getzoff, E. D., Richardson, J. S., and Richardson, D. C. (1983). Structure and mechanism of copper, zinc superoxide dismutase. Nature *306*, 284–287.

Tu, P. H., Gurney, M. E., Julien, J. P., Lee, V. M., and Trojanowski, J. Q. (1997). Oxidative stress, mutant SOD1, and neurofilament pathology in transgenic mouse models of human motor neuron disease. Lab. Invest. *76*, 441–456.

Valentine, J. S. (1994). Dioxygen Reactions. In Bioinorganic Chemistry, I. Bertini, H. B. Gray, S. J. Lippard and J. S. Valentine, eds. (Mill Valley, California: University Science Books) pp. 253–314.

Valentine, J. S., and Pantoliano, M. W. (1981). Protein-metal ion interactions in cuprozinc protein (superoxide dismutase): A major intracellular repository for copper and zinc in the eukaryotic cell. In Copper Proteins, T. G. Spiro, ed.: John Wiley and Sons, Inc.), pp. 292–358.

Valentine, J. S., Wertz, D. L., Lyons, T. J., Liou, L.-L., Goto, J. J., and Gralla, E. B. (1998). The dark side of dioxygen biochemistry. Curr. Op. Chem. Biol. (in press).

Wiedau-Pazos, M., Goto, J. J., Rabizadeh, S., Gralla, E. B., Roe, J. A., Lee, M. K., Valentine, J. S., and Bredesen, D. E. (1996). Altered reactivity of superoxide dismutase in familial amyotrophic lateral sclerosis. Science *271*, 515–518.

Williamson, T. L., Marszalek, J. R., Vechio, J. D., Bruijn, L. I., Lee, M. K., Xu, Z., Brown, R. H., Jr., and Cleveland, D. W. (1996). Neurofilaments, radial growth of axons, and mechanisms of motor neuron disease. Cold Spring Harb. Symp. Quant. Biol. *61*, 709–723.

Yim, H. S., Kang, J. H., Chock, P. B., Stadtman, E. R., and Yim, M. B. (1997). A familial amyotrophic lateral sclerosis-associated A4V Cu, Zn-superoxide dismutase mutant has a lower Km for hydrogen peroxide. Correlation between clinical severity and the Km value. J. Biol. Chem. *272*, 8861–8863.

Yim, M. B., Chock, P. B., and Stadtman, E. R. (1990). Copper, zinc superoxide dismutase catalyzes hydroxyl radical production from hydrogen peroxide. Proc. Natl. Acad. Sci. (USA) *87*, 5006–5010.

Yim, M. B., Chock, P. B., and Stadtman, E. R. (1993). Enzyme function of copper, zinc superoxide dismutase as a free radical generator. J. Biol. Chem. *268*, 4099–4105.

Yim, M. B., Kang, J. H., Yim, H. S., Kwak, H. S., Chock, P. B., and Stadtman, E. R. (1996). A gain-of-function of an amyotrophic lateral sclerosis-associated Cu, Zn-superoxide dismutase mutant: An enhancement of free radical formation due to a decrease in Km for hydrogen peroxide. Proc. Natl. Acad. Sci. (USA) *93*, 5709–5714.

Zhang, B., Tu, P., Abtahian, F., Trojanowski, J. Q., and Lee, V. M. (1997). Neurofilaments and orthograde transport are reduced in ventral root axons of transgenic mice that express human SOD1 with a G93A mutation. J. Cell. Biol. *139*, 1307–1315.

A STUDY OF THE DUAL ROLE OF COPPER IN SUPEROXIDE DISMUTASE AS ANTIOXIDANT AND PRO-OXIDANT IN CELLULAR MODELS OF AMYOTROPHIC LATERAL SCLEROSIS

M. T. Carrì, A. Battistoni, A. Ferri, R. Gabbianelli, and G. Rotilio

Department of Biology, University of Rome "Tor Vergata"
Centro di Neurobiologia Sperimentale "Mondino-Tor Vergata-S.Lucia"
Rome, Italy

1. DUAL ROLE OF COPPER IN THE PROCESS OF OXYGEN ACTIVATION

All living organisms require dietary copper for continued growth and development. This nutritional requirement arises from the essential role of the metal in the function of cuproenzymes, which play a critical role in the biochemistry of oxygen activation, not only for energy transduction (e.g. ATP synthesis by cytochrome oxidase or dopamine β-hydroxilation), but also in futile oxygen activation for chemical transformation of metabolic intermediates (e.g. H_2O_2-producing amine oxidase) (Linder and Hazegh-Azam, 1996). However, excess copper - as exemplified by non-physiological conditions such as experimental copper overload or by clinical conditions related to impaired metal transport as Wilson disease - mediates free radical production and direct oxidation of cellular components. Reduction of copper by physiological reductants triggers a series of radical reactions. Autoxidation of Cu^+ yields O_2^-, which dismutes to H_2O_2, reacting in turn with Cu^+ with the ultimate production of OH^-, the actual agent of oxidative damage (Fridovich, 1995). In order to prevent and counterbalance such reactions, living organisms have evolved various mechanisms of defense. These involve, besides the direct sequestering of copper and other metal ions by metallothioneins, so that they never exist in the free form, a complex system of oxyradicals interception including several enzymatic and non-enzymatic scavengers. Among the species involved in oxyradicals interception, a key role is played by Cu,Zn superoxide dismutase (SOD) where copper scavenges the primary source of damage (see above) at diffusion-limited rate.

2. Cu,Zn SUPEROXIDE DISMUTASE AND HUMAN DISEASES

Both a defect and an excess of SOD activity seem to be involved in important human pathologies, including neurodegenerative disorders. Overexpression of SOD1 has

Copper Transport and Its Disorders, edited by Leone and Mercer.
Kluwer Academic / Plenum Publishers, New York, 1999.

been proposed to be responsible for some of the clinical symptoms of Down's syndrome (Elroy-Stein and Groner, 1988), possibly because excess SOD could be more easily inactivated by H_2O_2 and then react with it as a peroxidase. On the other hand, decreased Cu,Zn superoxide dismutase is known to be associated with Fanconi's anaemia and with increased susceptibility to chromosomal aberrations (Mavelli et al., 1981); also, superoxide is known to be responsible for severe cellular damage in a variety of physiological and pathological conditions, including ageing (Marcocci et al.,1989; Stadtman, 1992).

Genetic evidence has recently led to the association of Cu,ZnSOD with the familial form of amyotrophic lateral sclerosis (ALS). ALS is a progressive, lethal disease characterized by degeneration of cortical and spinal motoneuros. ALS occurs both sporadically and as a familial, age-dependent autosomal dominant disorder; about 20% of familial ALS patients possess point mutations in the gene coding for Cu,ZnSOD. Up to now, more than 50 different mutations have been reported as responsible for familial amyotrophic lateral sclerosis (FALS) (Rosen et al., 1993; Deng et al., 1993, Siddique et al., 1996). Most autosomal-dominant FALS cases are indistinguishable from sporadic ALS on the basis of clinical and pathological criteria (Bowling et al.,1993). This similarity suggests that sporadic and familial forms share similar pathogenetic mechanisms. However, little is known about those mechanisms which would explain how mutations in Cu,ZnSOD might result in motoneuron injury.

In early reports (Ogasawara et al., 1993) it has been suggested that the severity of FALS is to be related to the residual Cu,ZnSOD activity. However, no direct correlation resulted, for instance, in the case of mutation of copper-binding residue His46, which caused a complete inactivation of the enzyme (Carrì et al., 1994) although this mutation is reportedly typical of a "mild" Japanese form of FALS (Ogasawara et al., 1993). Furthermore, transgenic mice carrying FALS-typical SOD1 mutations showed a FALS phenotype even in the absence of reduction of total Cu,ZnSOD activity (Gurney et al., 1994).

More recently, several lines of evidence have pointed out to the possibility that mutated Cu,ZnSOD are responsible for ALS through the acquisition of an as-yet-unidentified toxic property. Two proposed hypotheses are that toxicity may arise *i.* from imperfectly folded mutant SOD catalyzing the nitration of tyrosines through use of peroxynitrite as a substrate (Beckman et al., 1993) or *ii.* from peroxidation arising from elevated production of hydroxyl radicals through use of hydrogen peroxide as a substrate (Wiedau-Pazos et al., 1996).

The presence of elevated free nitrotyrosine levels has been reported in transgenic mice expressing SOD mutants (Bruijn et al., 1997). Although nitration of specific, critical targets cannot be ruled out on the basis of that work, bulk nitration of proteins is probably not the main mechanism involved in human disease, since protein-bound nitrotyrosine was not increased in this model.

It has been demonstrated (Yim et al., 1990) that wild-type Cu,ZnSOD has a secondary, one-electron peroxidase activity having as substrates H_2O_2 and anionic molecules, including some neurotransmitters such as glutamate and taurine, which become free radicals by the peroxidase reaction. This might be relevant to FALS, since at least two of the ALS-linked mutant SOD (A4V and G93A) can increase the rate of formation of hydroxyl radical *in vitro* compared to wild-type SOD1 in this reaction (Wideau-Pazos et al., 1996, Yim et al., 1996; Yim et al., 1997). However, neither hydroxyl radicals were increased nor the peroxidation product malondialdehyde was detected in transgenic mice expressing G37R (Bruijn et al., 1997).

Both hypotheses point to a different, increased availability of active-site copper to substrate as a prerequisite for the gain of function of Cu,ZnSODs typical of FALS patients. For instance, enhanced peroxidase activity of FALS-SOD is copper-dependent in that it can be blocked by stoichiometric copper chelating agents as diethyldithiocarbamate and penicillamine *in vitro*, while the peroxidase activity of the wild-type enzyme is unaffected unless copper is removed from the active site by an excess of chelators (Wiedau-Pazos et al., 1996, Bredesen et al., 1996). Furthermore, the metal-binding properties of the ALS mutant proteins and their redox behaviour are altered; Lyons and co-workers have proposed that alteration in the properties of the zinc site will alter the metal binding affinity of the copper site as well (Lyons et al., 1996), possibly inducing a loosening of the protein structure.

3. EXPERIMENTAL MODELS FOR THE STUDY OF FALS

Most of the early studies on the function gained by FALS mutant SODs have been undertaken *in vitro* on recombinant enzymes. However, investigation on the relationship between mutation in Cu,Zn SOD and ALS obviously need a model system *in vivo* where the above mentioned hypotheses can be tested. Several lines of transgenic mice overexpressing mutated human Cu,Zn SOD have been established and used in studies aimed to the elucidation of FALS pathogenesis (Gurney et al., 1994; Bruijn et al., 1997). However, some criticism might arise on the use of this experimental system as a model for ALS, since *i.* those mice express 5 to 15 times more human enzyme than FALS patients plus the normal mouse SOD complement; the overall acceleration of the pathogenic process in mice resulting from the elevated levels of the toxic mutant protein might mask more subtle, crucial alterations occurring in man; *ii.* no neuronal alteration similar to ALS has ever been described for mice, casting some doubt on the possibility to transfer results obtained in such a model to the human disorder, at least without a direct verification of data on man; *iii.* differences in the anatomies of mouse and man restrict the pathology to lower motor neurons in mice; *iv.* the dramatic vacuolation observed in transgenic mice has not been reported in cytological studies on patients.

As a possibly better approximation to FALS, we deviced a human cell system where ratio of expression of mutant SOD to wild-type SOD approximated 1:1, a situation resembling that of heterozigous patients.

Several human neuroblastoma SH-SY5Y cell lines have been established, expressing either wild-type or one of the FALS-linked SOD1 mutants under control of constitutive promoter CMV (Carrì et al., 1997). SH-SY5Y cells constitutively express Bcl-2 (Itano et al., 1996), which is known to inhibit apoptosis without altering intracellular free Ca^{2+} (Kane et al., 1993). This fact could be of help in establishing a model where overexpression of the mutant enzyme might have led to rapid apoptosis and cell death. We were able to propagate a total of about 40 independent lines expressing different mutations (G37R, H46R, G85R, G93A, I113T). For further experiments we selected two indipendent monoclonal cell lines expressing mutant G93A (named 93B and 93C) and a monoclonal cell line expressing mutant H46R (line 46D). These particular variants were chosen since mutation of these two residues exert widely different effects in wild-type Cu,ZnSOD *in vitro*. H46R, which is typical of a "mild" japanese form of ALS, has been demonstrated to possess almost no superoxide dismutase activity because of its low copper content and distorted geometry of the active site (Carrì et al., 1994), while G93A is fully metallated and displays enzyme activity comparable to the wild-type (Borchelt et al., 1994).

By Western blot analysis we have determined the level of Cu,ZnSOD immunoreactive protein in cells transfected with the plasmid coding for wild-type human Cu,ZnSOD (wt) and in lines 93B, 93C and 46D : all lines possess a significantly higher content of im-

Table 1. SOD activity and immunoreactive protein in control lines (SH-SY5Y and wild-type), and in lines expressing FALS-SOD

Cell line	SOD activity (U/mg tot.protein)	SOD immunoreactive protein (arbitrary units)
SH-SY5Y	10.4 ± 1.2 n = 6	100
Wild-type	22.6 ± 1.2 n = 4	230
93 B	21.5 ± 3.3 n = 4	210
93 C	17.7 ± 1.9 n = 4	170
46 D	11.0 ± 0.7 n = 3	190

munoreactive enzyme as compared to untrasfected SH-SY5Y cells, and the "exogenous" enzymes are expressed in a ratio close to 1:1 with the endogenous wild-type protein (Table 1). Total superoxide dismutase activity does not match the immunoreactive protein as expected on the basis of the above mentioned reports on FALS-SODs (Table 1).

5. EFFECT OF MUTANT Cu,ZnSODS IN TRANSFECTED SH-SY5Y CELLS

As a first approach to the detection of putative peroxidative damage in cells expressing FALS-SODs, we studied the status of polarization of mitochondrial membranes in our transfected lines. The rationale lies on the fact that mitochondria are a major cellular source of H2O2 (Boveris and Cadenas, 1982), but they also represent a preferred target (Zhang et al., 1990) of peroxidase reaction due to the presence of sensitive lipid in their membranes. Peroxidative attack results in further production of H_2O_2 by the damaged respiratory chain. This vicious cycle ultimately yields in mitochondrial swelling, uncoupling of oxidative phosphorilation, lipoperoxidation and dissipation of the Ca^{2+} gradient (Bowling and Beal, 1995).

When analyzed by staining with fluorochrome JC-1, our lines expressing mutant Cu,ZnSODs showed a significant decrease in mitochondrial membrane potential compared with both parental line SH-SY5Y and with cells expressing wild-type SOD (Fig.1). When depolarization of membranes was induced by increasing doses of valinomycin, lines 93B, 93C and 46D showed a higher sensitivity to this drug than both control cell lines (SH-SY5Y and WT) (Fig.2). Indeed, transgenic mice carrying FALS mutation G37R were demonstrated to possess mitochondrial damage, selectively localized in neurones and uniformely absent in other non-neuronal cell types (i.e. glia) (Wong et al., 1995) despite significantly elevated free radical scavenging activity.

Table 2. Cystolic Ca^{2+} concentration in control lines (SH-SY5Y and wild-type) and in lines expressing FALS-SOD (93B, 93C, 46D)

Cell line	Cytosolic Ca^{2+} concentration (nM)	
SH-SY5Y	187 ± 7	n = 8
Wild-type	198 ± 13	n = 5
93 B	263 ± 22	n = 6
93 C	261 ± 23	n = 4
46 D	275 ± 13	n = 3

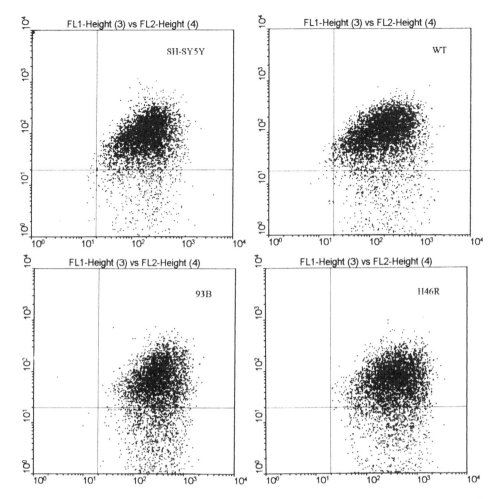

Figure 1. Mitochondrial electrochemical membrane potential was monitored by treating intact cells in suspension with fluorochrome JC-1 (10 μg/ml) and determining FL2 fluorescence shift in a Facscan cytometer (Becton Dickinson). A single, typical experiment is shown, consisting of 5 x 10^3 events measured in each sample.

Damage in mitochondria could lead to impairment of the long-term storage of calcium and this was indeed observed when monitoring cytosolic Ca^{2+} concentration determined by fluo3-AM: the expected alteration in cells carrying mutated Cu,ZnSODs was indicated by the net increase of staining (Fig. 3 and Table 2), which was found to be specific for cell lines expressing FALS-type mutants and was not related to total superoxide dismutase activity.

Both effects observed in our cell lines, damage of mitochondrial membrane and cytosolic Ca^{2+} increase, might offer a rationale for the highly selective vulnerability of the corticoneuronal system in ALS. There is compelling evidence that free radicals together with increases in cytosolic Ca^{2+} play a major role in neuronal death, although neither the source of these radicals nor the direct connection between Ca^{2+} mobilization and radical production has been conclusively identified (Dykens, 1994). Ca^{2+} buffering potential in the cytosol seems to be crucial in determining the selective vulnerability to insults of cer-

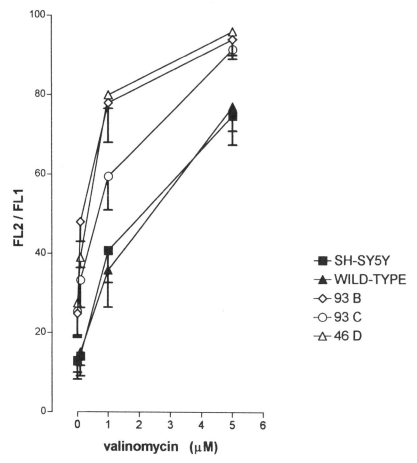

Figure 2. Depolarization of mitochondria membranes by increasing concentrations of valinomycin (0, 0.1, 1 and 5 µM), determined by JC-1 fluorescence shift. Mean values ± SD are represented; n = 3.

tain populations of cells. Only trace amounts of immunoreactivity to the calcium-buffering proteins calbindin-D28k and parvalbumin can be observed in human motoneurons (Krieger et al., 1996) and this immunoreactivity is absent in motor neuron damaged early or severely in human ALS (Alexianu et al. 1994). This process of neuronal injury may be self-sustaining. Elevation of cytosolic Ca^{2+} levels may compromise the structural integrity of mitochondria; it has been demonstrated that exposition to elevated Ca^{2+} induces production of free radicals in isolated mitochondria (Dykens 1994). This could lead, in turn, to enhanced mitochondrial release of OH· radicals and other reactive oxygen species.

The observation that motoneurons are deficient in Ca^{2+} binding proteins might be relevant also for copper-mediated mechanisms of pathogenesis. It has been reported that calcium-binding protein S100b from bovine brain is able to sequester copper ions *in vitro* and reduce oxidative cell damage induced by $CuCl_2$ plus H_2O_2 in *E.coli* (Nishikawa et al. 1997). Cytosolic superoxide dismutase is present in eukaryotic cells both in the *holo*-, Cu,Zn form and as *apo*-, copper-free enzyme and a role in the cell copper-buffering system has been proposed (Galiazzo et al. 1991, Steinkuhler et al. 1991, Steinkuhler et al. 1994, Culotta et al. 1995; Petrovic et al. 1997). Alteration of Cu-binding site in ALS-type

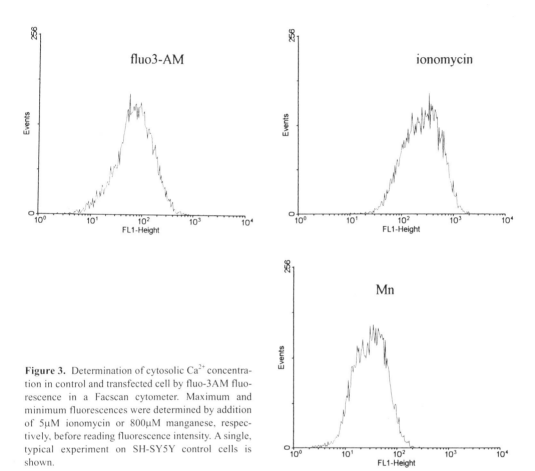

Figure 3. Determination of cytosolic Ca^{2+} concentration in control and transfected cell by fluo-3AM fluorescence in a Facscan cytometer. Maximum and minimum fluorescences were determined by addition of 5μM ionomycin or 800μM manganese, respectively, before reading fluorescence intensity. A single, typical experiment on SH-SY5Y control cells is shown.

SODs might : *i.* cause the release of the metal ion increasing both cellular copper availability and the ratio of *apo-* to *holo-* SOD; *ii.* impair copper buffering by SOD, therefore causing metal-mediated oxidative stress; *iii.* lead to a novel peroxidative function of the enzyme. These effects might be not adequately counterbalanced by the relatively ineffective Ca-binding system in motoneurons (Krieger, 1996) possibly involved in the buffering of copper as well.

ACKNOWLEDGMENTS

The financial support of Telethon - Italy (Grant no.700) is gratefully acknowledged. A.F. is a fellow of the Fondazione "C. Mondino".

REFERENCES

Alexianu, M.E., Ho, B.-K., Mohamed, H., LaBella, V., Smith, R.G. and Appel, S.H. (1994) The role of calcium-binding proteins in selective motoneuron vulnerability in amyotrophic lateral sclerosis. Ann.Neurol. 36, 846–858.

Beckman, J.S., Carson, M., Smith, C.D., Koppenol, W.H. (1993). ALS, SOD and peroxynitrite. Nature 364, 584.

Borchelt, D.R., Lee, M.K., Slunt, H.S., Guarnieri, M., Xu, Z.S., Wong, P.C., Brown, R.H., Jr., Price, D.L., Sisodia, S.S. and Cleveland, D.W. (1994). Superoxide dismutase 1 with mutations linked to familial amyotrophic lateral sclerosis possesses significant activity. Proc. Natl. Acad. Sci. 91, 8292–8296.

Boveris, A. and Cadenas, E. (1982). Production of superoxide radicals and hydrogen peroxide in mitochondria, in Superoxide dismutase. Vol.II (Oberley L.W. ed.), pp.15–30. CRC Press, Boca Raton, Florida.

Bowling, A.C., Schulz, J.B., Brown, R.H., Jr. and Beal, M.F. (1993). Superoxide dismutase activity, oxidative damage and mithocondrial energy metabolism in familial and sporadic amyotrophic lateral sclerosis. J. Neurochem. 61, 2322–2325.

Bowling, A.C. and Beal, M.F. (1995). Bioenergetic and oxidative stress in neurodegenerative diseases. Life Sci. 56, 1151–1171.

Bredesen, D.E., Wiedau-Pazos, M., Goto, J.J., Rabizadeh, S., Roe, J.A., Gralla, E:B., Ellerby, L.M. and Valentine, J.S. (1996). Cell death mechanisms in ALS. Neurology 47, (Suppl 2) S36-S39.

Bruijn, L.I., Beal, M.F., Becher, M.W., Schulz, J.B., Wong, P.C., Price, D.L. and Cleveland, D.W. (1997). Elevated free nitrotyrosine levels, but no protein-bound nitrotyrosine or hydroxyl radicals, throughout amyotrophic lateral sclerosis (ALS)-like disease implicate tyrosine nitration as an aberrant *in vivo* property of one familial ALS-linked superoxide dismutase 1 mutant. Proc. Natl. Acad. Sci USA 94, 7606–7611.

Carri', M.T., Battistoni, A., Polizio, F., Desideri, A. and Rotilio G. (1994). Impaired copper binding by the H46R mutant of human Cu,Zn superoxide dismutase, involved in amyotrophic lateral sclerosis. FEBS letters 356, 314–316.

Carri, M.T., Ferri, A., Battistoni, A., Famhy, L., Gabbianelli, R., Poccia, F. and Rotilio, G. (1997) Expression of a Cu,Zn superoxide dismutase typical of familial amyotrophic lateral sclerosis induces mitochondrial alteration and increase of cytosolic Ca^{2+} concentration in transfected neuroblastoma SH-SY5Y cells. FEBS letters 414, 365–368.

Culotta, V.C., Joh, H.-D., Lin, S.-J., Slekar, K.H. and Strain, J. (1995). A physiological role for Saccharomyces cerevisiae Copper/zinc superoxide dismutase in copper buffering. J.Biol.Chem., 270, 29991–29997.

Deng, H.X., Hentati, A., Tainer, J.A., Iqbal, Z., Cayabyab, A., Hung, W.Y., Getzoff, E.D., Hu, P., Herzfeldt, B., Roos, R.P., Warner, C., Deng, G., Soriano, E., Smith, C., Parge, H.E., Ahmed, A., Roses, A.D., Hallewell, R.A., Pericak-vance, M.A. and Siddique, T. (1993). Amyotrophic lateral sclerosis and structural defects in Cu,Zn superoxide dismutase. Science 261, 1047–1051.

Dykens, J.A. (1994) Isolated cerebral and cerebellar mitochondria produce free radicals when exposed to elevated Ca^{2+} and Na^+: Implications for neurodegeneration. J.of Neurochem. 63, 584–591.

Elroy-Stein, O., Bernstein, Y. and Groner, Y (1986). Overproduction of human Cu/Zn-superoxide dismutase in transfected cells: extenuation of paraquat-mediated cytotoxicity and enhancement of lipid peroxidation. Embo J. 5, 615–622.

Fridovich, I. (1995). Superoxide radical and superoxide dismutase. Ann. Rev. Biochem. 64, 97–112.

Galiazzo,F.,Ciriolo, M.R., Carri, M.T., Civitareale, P., Marcocci, L., Marmocchi, F., and Rotilio, G. (1991). Activation and induction by copper of Cu,Zn superoxide dismutase in *Saccharomyces cerevisiae*. Eur. J. Biochem 196, 545–549.

Gurney, M.E., Haifeng, P., Chiu, A.Y. Dal Canto, M.C., Polchow, C.Y., Alexander, D.D., Caliendo, J., Hentati, A., Kwon, Y.W., Deng, H.X., Chen, W., Zhai, P., Sufit, R.L., Siddique, T. (1994). Motor neuron degeneration in mice that express a human cu, Zn superoxide dismutase mutation. Science 264, 1772–1775.

Itano, Y., Ito, A., Uehara, T. and Nomura, Y. (1996) Regulation of Bcl-2 Protein expression in human neuroblastoma SH-SY5Y Cells : Positive and negative effects of protein kinase C and A, respectively. J.Neurochem. 67, 131–137.

Kane, D.J. Sarafian, T.A., Anton, R., Hahn, H., Gralla, E.B., Valentine, J.S., Ord, T. and Bredesen, D.E. (1993). Bcl-2 inhibition of neural death: decreased generation of reactive oxygen species. Science, 262, 1274–1277

Krieger, C., Lanius, R.A., Pelech, S.L. and Shaw, C.A. (1996) Amyotrophic lateral sclerosis : the involvement of intracellular Ca^{2+} and protein kinase C. TiPs, 17, 114–120.

Linder,M.C.and Hazegh-Azan, M. (1996). Copper biochemistry and molecular biology. Am.J.Clin. Nutr. 63, 797–811.

Lyons, T.J., Liu, H., Goto, J.J., Nersissian, A., Roe, J.A., Graden, J.A., Cafè, C., Ellerby, L.M., Bredesen, D.E., Gralla, E.B. and Valentine, J.S. (1996) Proc.Natl.Acad.Sci.U.S.A. 93, 12240–12244.

Marcocci L., Carri, M.T., Battistoni, A. e Rotilio, G. (1989). Bioengineering of superoxide dismutase and related enzymes. Basic and clinical aspects. In Bioengineered molecules : basic and clinical aspects, Verna, R., Blumenthal, R. e Frati, L. eds. Serono Symposia Series Adv. Exp. Med. 1 pp.11–27 Raven Press.

Mavelli, I., Ciriolo, M.R., Rotilio, G., DeSole, P., Castorino, M. and Stabile, A. (1982) Superoxide dismutase, glutathione peroxidase and catalase in oxidative hemolysis. A study of Fanconi's anemia erythrocytes. Biochem.Biophys.Res.Commun., 106, 286–290.

Nishikawa, T., Lee, S.M., Shiraishi, N., Ishikawa, T., Ohta, Y.and Nishikimi, M. (1997). Identification of S100b protein as copper-binding protein and its suppression of copper-induced cell damage. J. Biol. Chem. 272, 23037–23041.

Ogasawara, M., Matsubara, Y. and Narisawa, K. (1993). Mild ALS in Japan associated with novel SOD mutation. Nature Genetics 5, 323–324.

Petrovic, N., Comi, A., Ettinger, M.J. (1996). Identification of an apo-superoxide dismutase pool in human lymphoblasts. J. Biol. Chem. 271, 28331–28334.

Rosen, D.R., Siddique, T., Patterson, D., Figlewicz, D.A., Sapp, P., Hentati, A., Donaldson, D., Goto, J., O'Regan, J.P., Deng, H.-X., Rahamani, Z., Krizus, A., McKenna-Yasek, D., Cayabyab, A., Gaston, S.M., Berger, R., Tanzi, R.E., Halperin, J.J., Herzfeldt, B., Van den Bergh, R., Hung, W.-Y., Bird, T., Deng, G., Molder, D.W., Smyth, C., Laing, N.G., Soriano, E., Pericak-Vance, M.A., Haines, J., Rouleau, G.A., Gusella, J.S., Hovitz, H.R. and Brown jr, R.H. (1993) Mutations in Cu/Zn superoxide dismutase gene are associated with familial amyotrophic lateral sclerosis. Nature 362, 59–62.

Siddique, T., Nijhawan, D. and Hantati, A. (1996) Molecular genetic basis of familial ALS. Neurol. 47 (Suppl.2) S27.

Stadman, E.R. (1992) Protein oxidation. Science 257, 1220–1224

Steinkhuler, C., Sapora, O., Carri, M.T., Nagel, W., Marcocci, L., Ciriolo, M.R., Weser, U. e Rotilio, G. (1991). Increase of Cu/Zn superoxide dismutase activity during differentiation of K562 cells involves activation by copper of a constantly expressed copper-free protein. J. Biol. Chem. 266, 24580–24587 (1991).

Steinkhuler, C., Carri, M.T., Micheli, G., Knoepfel, L., Weser, U. and Rotilio, G. (1994). Copper-dependent metabolism of Cu,Zn-superoxide dismutase in human K562 cells. Biochem.J. 302, 687–694.

Wideau-Pazos,M., Goto, J.J., Rabizadeh, S., Gralla, E.B., Roe, J.A., Lee, M.K., Valentine, J.S. and Bredesen, D.E. (1996) Altered reactivity of superoxide dismutase in familial amyotrophic lateral sclerosis. Science, 271, 515–518.

Wong, P.C., Pardo, C.A., Borchelt, D.R., Lee, M.K., Copeland, N.G., Jenkins, N.A., Sisodia, S.S., Cleveland, D.W. and Price, D.L. (1995) An adverse property of familial ALS-linked SOD1 mutation causes motor neuron disease characterized by vacuolar degeneration of mitochondria. Neuron 14, 1105–1116.

Yim, M.B., Chock, P.B., Stadtman, E.R. (1990). Copper,zinc superoxide dismutase catalyzes hydroxyl radical production from hydrogen peroxide. Proc.Natl.Acad.Sci. USA 87, 5006–5010.

Yim, M.B., Kang, J.-H., Yim, H.-S., Kwak, H.-S., Chock, P.B. and Stadtman, E.R. (1996) A gain of function of an amyotrophic lateral sclerosis-associated Cu,Zn superoxide dismutase mutant : an enhancement of free radical formation due to a decrease in Km for hydrogen peroxide. Proc.Natl.Acad.Sci.USA 93, 5709–5714.

Yim, H.-S., Kang, J.-H., Chock, P.B., Stadtman, E.R. and Yim, M.B. (1997) A familial amyotrophic lateral sclerosis-associated A4V Cu,Zn-superoxide dismutase mutant has a lower K_m for hydrogen peroxide. J.Biol.Chem. 272, 8861–8863.

Zhang, Y., Marcillat, O., Giulivi, C., Ernster, L. and Davies, K.J.A. (1990) The oxidative inactivation of mitochondrial electron transport chain components and ATPase. J.Biol.Chem. 265, 16330–16336.

19

THE EFFECT OF COPPER ON TIGHT JUNCTIONAL PERMEABILITY IN A HUMAN INTESTINAL CELL LINE (Caco-2)

Simonetta Ferruzza,[1] Yula Sambuy,[1] Giuseppe Rotilio,[1,2] Maria Rosa Ciriolo,[3] and Maria Laura Scarino[1]

[1]Istituto Nazionale della Nutrizione
Roma, Italy
[2]Dipartimento di Biologia, Università di Roma "Tor Vergata"
Roma, Italy
[3]Dipartimento di Scienze Biomediche, Università di Chieti
Chieti, Italy

1. INTRODUCTION

1.1 Copper Nutrition and Toxicity

Copper is a trace element essential to life, participating to a wide number of biochemical reactions as cofactor of many enzymes (Linder, 1991). However, it also exhibits toxic effects and can impair cellular functions (Goyer, 1994). For this reason it is crucial to understand the relashionship between requirements and toxicity of this heavy metal.

The vast majority of copper is efficiently absorbed in the duodenum. The intestinal epithelial cells are therefore exposed to variable amounts of copper ingested with food or as mineral integrator. Early, tissue-specific signs of copper toxicity are of great importance in tracing the bordeline between copper nutrition and toxicity at the cellular lever and, ultimately to help in the assessment of the safe and adequate requirements of this metal for the human population.

1.2 Intestinal Barrier Function and the Tight Junctions

The intestinal mucosa represents both a target and a site of entry for toxic compounds in the intestinal lumen. An early sign of toxicity at the intestinal level can in fact result in the alteration of the barrier function of the mucosa, leading to an uncontrolled influx of substances into the blood. The barrier function is guaranteed by the presence of junctional complexes, highly specialized structures joining adjacent epithelial cells. Tight junctions (TJ), the most apical component of the junctional complex, form a selective permeability barrier along the paracellular pathway of epithelial cells (Anderson et al., 1993; Ballard et al., 1995; Schneeberger and Lynch, 1992). In response to different stimuli the TJ rapidly change their permeability and functional properties, allowing dynamic fluxes of ions and solutes as well as the trans-epithelial passage of whole cells. At least seven proteins have been identified as structural or regulatory components of the TJ (Anderson et al., 1993; Ballard et al., 1995; Keon et al., 1996). Among these proteins, occludin is the only integral membrane protein localized at the points of membrane-membrane interaction of the TJ, and appears to be responsible for the barrier function of the junction (Chen et al., 1997; Furuse et al., 1993). The intracellular domain of occludin interacts with other TJ-associated proteins, that in turn perform bridging and regulatory functions. In the epithelial cells an extensive actin network supports the microvilli and is linked to the TJ. Immediately below the TJ the adherens junctions are tightly coupled to a circumferential actin-myosin ring that also attaches directly onto the cytoplasmic surface of TJ contacts and may transmit cytoskeletal changes to the junctional complex. Recent studies have established that TJ assembly and barrier functions are influenced by all the classic second messangers and signalling pathways including tyrosine kinases, intracellular calcium, protein kinase C, heterotrimeric G proteins, calmodulin, cAMP and phospholipase C. In addition, since TJ are linked to the cytoskeleton, almost any perturbation of perijunctional actin, will disrupt the paracellular barrier (Anderson et al., 1993; Ballard et al., 1995; Schneeberger et al., 1992).

TJ integrity can be measured by either assessing the permeability to hydrophylic solutes that are not absorbed by the cells, varying in molecular size and charge, or by determining the trans-epithelial electrical resistance (TEER) that depends upon trans-junctional ion fluxes. These two parameters are usually well correlated although perturbation to the TJ integrity can induce ionic fluxes producing a drop in TEER even before changes in the permeability to larger solutes can be detected (Adson et al., 1994; Schneeberger et al., 1992).

1.3 Caco-2 Cells as an in Vitro Model for the Study of Intestinal Absorption and Toxicity

The Caco-2 cell line, isolated from a human colon adenocarcinoma (Fogh et al., 1977), undergoes in culture a process of spontaneous differentiation starting at confluency and leading, in two to three weeks, to the formation of a monolayer of highly polarized cells, joined by functional TJ, with well developed and organized microvilli on the apical (AP) membrane and expressing many enzyme activities (disaccharidases and peptidases) and transport proteins typical of the small intestinal absorptive enterocyte (Neutra and Louvard, 1989; Zweibaum et al., 1991). Conversely, electrical properties, ionic conductivity and permeability characteristics of the differentiated Caco-2 cells resemble more those of the colon crypt cells than those of the small intestinal enterocytes (Grasset et al., 1984).

The Caco-2 cell line grown and differentiated on microporous filter supports, forms a two compartment system where the cell monolayer separates the AP compartment, corresponding in vivo to the intestinal lumen, from the basolateral (BL) compartment, in vivo the blood capillaries space (Figure 1).

This system allows full accessibility to both sides of the cells and has extensively been used as an in vitro model of the small intestinal epithelium to study the transport of molecules across the monolayer and the toxicity of substances added to either plasma membrane domain.

The Effect of Copper on Tight Junctional Permeability in a Human Intestinal Cell Line (Caco-2)

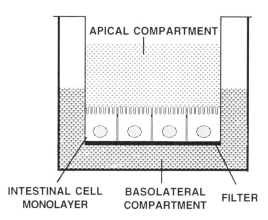

Figure 1. Diagram of Caco-2 cell monolayers grown on filter substrate for transport and toxicity studies. In this system the Caco-2 cells seeded on the filter form a monolayer of cells coupled by tight junctions that divides the apical (AP) from the basolateral (BL) compartment, allowing separate control of the composition of the medium in the two compartments for transport and toxicity studies.

During the establishment of Caco-2 confluent cell monolayers and the subsequent differentiation process TEER values reach a peak before achieving steady state values, corresponding to the formation of fully developed and functional junctions (Ferruzza et al., 1995; Ranaldi et al., 1994). Determinations of TEER in Caco-2 cells have been used in several studies as an indicator of early sub-lethal epithelial toxicity of different compounds, including heavy metals (Hashimoto et al., 1994; Hecht et al., 1988; Narai et al., 1997; Rossi et al., 1996; Scarino et al., 1997).

The aim of this study was to determine the effects of ionic copper on the integrity of Caco-2 tight junctions by measuring TEER values and permeability to the extracellular fluid marker, mannitol. Ionic copper added from the AP side to fully differentiated Caco-2 cells was shown to reversibly alter the permeability of tight junctions.

2. MATERIALS AND METHODS

Caco-2 cells were grown and maintained as previously described (Ferruzza et al., 1995) in Dulbecco Modified Minimum Essential Medium containing 25 mmol/L glucose, 3.7 g/L $NaHCO_3$ and supplemented with 4 mmol/L L-glutamine, 10% heat inactivated fetal calf serum, 1% non-essential amino acids, 1×10^5 U/L penicillin, 100 µg/L streptomycin. For the experiments the cells were seeded on polycarbonate filter cell culture chamber inserts (Transwell, 12 mm diameter, 1.13 cm^2 area, 0.45 µm pore diameter; Costar Europe, Badhoevedorp, The Netherlands,) at a density of 4×10^5 cells / cm^2 and were left to differentiate for 15–17 days after confluency; the medium was regularly changed three times a week.

To investigate the effects of copper ions on the permeability of TJ, Caco-2 cells were treated for 3 to 4 h at 37°C with increasing concentrations of $CuCl_2$ in Hank's balanced saline solution (HBSS) at pH 6.0, unless otherwise stated, in the AP compartment. The BL compartment contained HBSS at pH 7.4 with 0.4 % copper-free bovine serum albumin (BSA) and reduced glutathione (GSH) in a 1:2 molar ratio. The pH gradient between the AP and BL compartment reproduces the pH conditions of the microenvironment in proximity of the small intestinal villi and in the submucosal compartment (Wilson, 1989). At the end of the experiment, the cell monolayer was washed with HBSS and the permeability of the TJ was determined by measuring the TEER of filter-grown cell monolayers in HBSS using a commercial apparatus (Millicell ERS; Millipore Co., Bedford, MA) as previously described (Ferruzza et al., 1995). TEER was expressed as ½ $\times$ cm^2. After TEER measurements, the trans-epithelial passage of the radiolabelled extracellular marker D-1[^{3}H(N)]-mannitol (specific activity 706.7 Gbq/mmol) (Life Science Products, Bruxelles, Belgium) across the cell monolayers was determined as previously described (Ranaldi et al., 1994). Briefly, the radioactive compound in complete growth medium was added to the AP compartment and, after 2 h of incubation at 37°C, the radioactivity in the BL medium was measured in a liquid scintillation counter (LS1801; Beckmann Instruments Inc., Irvine, California) and expressed as a percentage of the AP applied radioactivity passing in 1 h.

In recovery experiments, after the treatment and the monitoring of TJ permeability (TEER and mannitol passage), the cells were transferred in complete medium and kept at 37°C in the incubator for 24 h before measuring again TEER values.

3. RESULTS AND DISCUSSION

When Caco-2 cells were treated with concentrations of $CuCl_2$ ranging between 5 and 100 μM there was a dose-dependent decrease in TEER accompanied by a parallel increase in the % mannitol passage/h (Figure 2).

Although the effects of copper on TJ permeability have, to the best of our knowledge, never been reported, other heavy metals such as cadmium and methylmercury have been shown to alter the permeability of TJ in epithelial cells (Bohme et al., 1992; Rossi et al., 1996). Although for cadmium this effect may be due to its ability to compete calcium

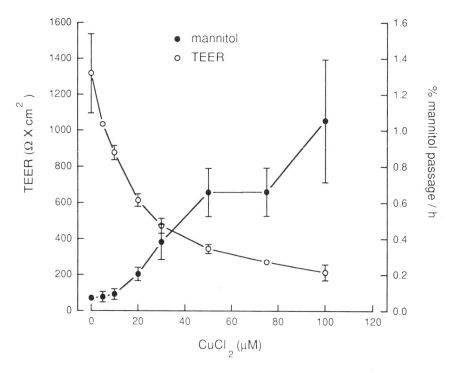

Figure 2. Permeability of Caco-2 tight junctions in the presence of $CuCl_2$. Differentiated Caco-2 cells grown on filters were treated with 5 to 100 μM $CuCl_2$ in HBSS at pH 6.0 in the AP compartment. After 3 h of treatment the permeability of the cell monolayer was monitored by measuring the trans-epithelial electrical resistance (TEER) and ^3H mannitol passage. The results are the mean ±SD of triplicate filters in a representative experiment.

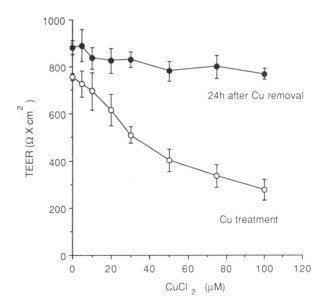

Figure 3. Recovery of TJ integrity after copper treatment for 3 hours. Cells were treated as in Figure 2 with 5–100 µM $CuCl_2$. After the treatment the AP medium was removed, cells were washed and complete culture medium was added to the cells for 24 h at 37°C. The TEER values measured after the treatment and after the 24 h of recovery are shown for each $CuCl_2$ concentration. The results are the mean ±SD of triplicate filters in a representative experiment.

binding to calmodulin (Goering et al., 1995), for other metals the mechanisms of action on the TJ are largely unknown.

The permeability of TJ can be modulated by different physiological factors (Ballard et al., 1995; Schneeberger et al., 1992). In addition substances, such as calcium chelators or phorbol esters, induce a disassemby of TJ that is fully reversible (Cereijido et al., 1993; Hecht et al., 1994). Determining the reversibility of permeability changes of TJ can be an important indicator of the type of damage to the TJ and hence of the mechanism involved.

Figure 3 shows the maximum decrease in TEER obtained at different $CuCl_2$ concentrations after 3 h of treatment and, at each concentration, the TEER values measured after 24 h of recovery in complete culture medium. In all cases the concentration-dependent decrease in TEER observed after the treatment was fully reversible after 24 h of recovery. However, at concentrations above 100 µM the TEER did not fully recover to the control values (data not shown).

It therefore appears that ionic copper between 5 and 100 µM interferes with the TJ withouth irreversibly altering their structure, while at higher concentrations the effect may well extend beyond the TJ, involving a more general cytotoxic effect.

A proton gradient across the AP membrane of intestinal epithelial cells, including Caco-2 cells, is known to drive the transport of nutrients such as dipeptides, amino acids and metal ions (Ganapathy and Leibach, 1991; Gunshin et al., 1997; Thwaites et al., 1994; Thwaites et al., 1995). Changing the AP extracellular pH between 6.0 and 7.0 reduces this trans-membrane proton gradient and therefore affects any rheogenic transport. The effect of increasing $CuCl_2$ concentrations on TEER, expressed as % of control, was determined at pH 6.0 and at pH 7.0. As shown in Figure 4 at pH 6.0 the TEER decreased with increasing copper concentration (as also shown in Figure 2), while at pH 7.0 $CuCl_2$ only moderately affected the TEER values with respect to control, between 5 and 100 µM.

Since, at these two pH values the TEER of control Caco-2 cells was not significantly different, namely $704 \pm 2 \frac{1}{2} X cm^2$ at pH 6.0 and $693 \pm 5 \frac{1}{2} X cm^2$ at pH 7.0, the copper-dependent changes in TEER observed at pH 6.0 and 7.0 could indirectly result from changes in copper uptake at these two pH values. It has in fact recently been reported that a metal ion transporter, DCT1, expressed in the intestine and exhibiting a broad substrate specificity, in-

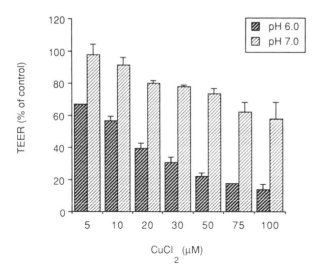

Figure 4. Effect of pH on copper-dependent TEER decrease. Caco-2 cells were treated with 5–100 μM $CuCl_2$ in HBSS at pH 6.0 or pH 7.0 for 3 h. The TEER values at the end of the experiment are expressed as a percentage of their copper-free control at the appropriate pH. No significant differences in the TEER values of control cells at pH 6.0 and pH 7.0 were observed over the time of the experiment. The results are the mean ± SD of two experiments performed in triplicate.

cluding copper, is driven by an inwardly directed proton gradient (Gunshin et al., 1997). Conditions that favour copper uptake (such as pH 6.0) may, due to the high copper reactivity with reduced sulfhydryl groups of proteins (Dawson and Ballatori, 1995), result in perturbations of TJ, either directly or through interactions with the actin cytoskeleton.

Among other possible toxic effects of copper is the production of reactive oxygen species in aqueous solution, that can exhert a variety of cytotoxic effects (Stohs and Bagchi, 1995). Different antioxidants were therefore used to prevent the effect of copper on the TJ integrity. Ascorbic acid in a 1:1 molar ratio with $CuCl_2$ did not prevent the decrease in TEER produced by 25 and 50 μM $CuCl_2$ (Figure 5).

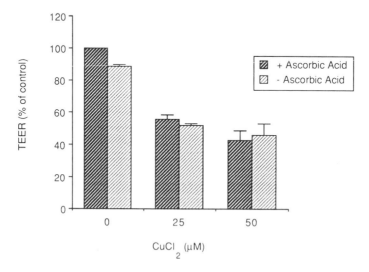

Figure 5. Effect of ascorbic acid on TEER in copper-treated Caco-2 cells. Caco-2 cells were treated for 4 h with 25 and 50 μM $CuCl_2$ in HBSS at pH 6.0, with or without ascorbic acid in a 1:1 molar ratio with copper. The histogram shows the TEER values at the end of the experiment of cells treated with copper in the presence or absence of ascorbic acid. The results are expressed as percentage of the TEER value of the control at time zero, and represent the mean ± SD of two experiments performed in triplicate.

Other reactive oxygen species scavengers such as the vitamin E soluble derivative, Trolox (1:2 molar ratio), mannitol (500 µM) and dimethylsulfoxide (2 mM) were employed to counteract copper effects on TJ. In analogy to what shown for ascorbic acid, these antioxidants in the conditions used, did not prevent the decrease in TEER produced by copper between 10 and 100 µM (data not shown).

Previous studies had shown that reactive oxygen metabolites could cause an increase in TJ permeability of Caco-2 cells (Baker et al., 1995; Manna et al. 1997). The copper-induced increase in TJ permeability observed in the present work does not, however, appear to be the direct consequence of oxidative damage.

4. CONCLUSIONS

Ionic copper in the form of $CuCl_2$ produced a dose-dependent alteration in the integrity of TJ as monitored by changes in TEER and in trans-epithelial mannitol passage. This effect was evident at concentrations as low as 5 µM and was fully reversible up to a concentration of 100 µM. This reversibility suggests, though it does not prove, that the effect of copper in this concentration range is specifically aimed at the TJ and their regulation.

The fact that the effect of $CuCl_2$ on TJ was observed at pH 6.0 but not at pH 7.0 raises the possibility that toxicity to TJ is related to differences in copper uptake at the two pH values. At least one metal ion transporter that is likely to carry copper across the intestinal mucosa is proton-dependent and does operates better at pH 6.0 than at pH 7.0. The intracellular copper may therefore bind to TJ or cytoskeletal proteins leading to perturbation of TJ integrity.

The alternative possibility that intracellular copper may cause alterations to the TJ through the generation of reactive oxygen species has not been confirmed. The antioxidants employed, at least under the conditions used in our experiments, have not in fact been able to counteract the copper effects on the TJ.

Differentiated Caco-2 cells on permeable filter supports provide a powerful sensitive and discriminating tool to investigate the mechanisms of action of metal ions exherting toxic effects on the barrier function of the intestinal mucosa.

ACKNOWLEDGMENTS

This work was supported by the European Community, FOODCUE: Contract n.FAIR-CT95-0813

REFERENCES

Adson, A., Raub, T., Burton, P., Barshun, C., Hilgers, A., Audus, K., and Ho, N. (1994). Quantitative approaches to delineate paracellular diffusion in cultured epithelial cell monolayers. J Pharm Sci 83, 1529–1536.
Anderson, J., Balda, M., and Fanning, A. (1993). The structure and regulation of tight junctions. Curr. Opin. Cell. Biol. 5, 772–778.
Baker, R., Baker, S., and La Rosa, K. (1995). Polarized Caco-2 cells. Effect of reactive oxygen metabolites on enterocyte barrier function. Dig. Dis. Sci. 40, 510–518.
Ballard, S., Hunter, J., and Taylor, A. (1995). Regulation of tight junction permeability during nutrient absorption across intestinal epithelium. Annu. Rev. Nutr. 15, 35–55.
Bohme, M., Diener, M., Mestres, P., and Rummel, W. (1992). Direct and indirect actions of $HgCl_2$ and methyl mercury chloride on permeability and chloride secretion across the rat colonic mucosa. Toxicol. Appl. Pharmacol. 114, 285–294.
Cereijido, M., Gonzalez-Mariscal, R., Contreras, R., Gallardo, J., Garcia-Villegas, R., and Valdes, J. (1993). The making of a tight junction. J. Cell Sci. Suppl. 17, 127–132.

Chen, Y., Merzdorf, C., Paul, D., and Goodenough, D. (1997). COOH-terminus of occludin is required for tight junction barrier function in early Xenopus embryos. J Cell Biol 891–899.

Dawson, D., and Ballatori, N. (1995). Membrane transporters as sites of action and routes of entry for toxic metals. In Toxicology of metals. Biochemical aspects., R. GoyerandM. Cherian, eds. (Berlin: Springer-Verlag), pp. 53–76.

Ferruzza, S., Ranaldi, G., Di Girolamo, M., and Sambuy, Y. (1995). The transport of lysine across monolayers of human cultured intestinal cells (Caco-2) depends on Na^+-dependent and Na^+-independent mechanisms on different plasma membrane domains. J. Nutr. 125, 2577–2585.

Fogh, J., Fogh, J.M., and Orfeo, T. (1977). One hundred and twenty seven cultured human tumor cell lines producing tumors in nude mice. J. Natl. Cancer Inst. 59, 221–226.

Furuse, M., Hirase, T., Itoh, M., Nagafuchi, A., Yonemura, S., Tsukita, S., and Tsukita, S. (1993). Occludin: a novel integral membrane protein localizing at tight junctions. J Cell Biol 123, 1777–1788.

Ganapathy, V., and Leibach, F.H. (1991). Proton-coupled solute transport in the animal cell plasma membrane. Curr. Opin. Cell Biol 3, 695–701.

Goering, P., Waalkes, M., and Klaassen, C. (1995). Toxicology of cadmium. In Toxicology of metals. Biochemical aspects., R. Goyer and M. Cherian, eds. (Berlin: Springer-Verlag), pp. 189–214.

Goyer, R. (1994). Biology and nutrition of essential elements. In Risk assessment of essential elements., C. Abernathy, A. Mertz and S. Olin, eds. (Washington DC: International Life Sciences Institute Press).

Grasset, E., Pinto, M., Dussaulx, E., Zweibaum, A., and Desjeux, J.F. (1984). Epithelial properties of human colonic carcinoma cell line Caco 2: electrical parameters. Am. J. Physiol. 247, C26-C267.

Gunshin, H., Mackenzie, B., Berger, U., Gunshin, Y., Romero, M., Boron, W., Nussberger, S., Gollan, J., and Hediger, M. (1997). Cloning and characterization of a mammalian proton-coupled metal-ion transporter. Nature 338, 482–488.

Hashimoto, K., Matsunaga, N., and Shimizu, M. (1994). Effect of vegetable extracts on the transepithelial permeability of the human intestinal Caco-2 cell monolayer. Biosci. Biotech. Biochem. 58, 1345 - 1346.

Hecht, G., Pothoulakis, C., La Mont, J.T., and Madara, J.L. (1988). Clostridium difficile toxin A perturbs cytoskeletal structure and tight junction permeability of cultured human intestinal epithelial monolayers. J. Clin. Invest. 82, 1516–1524.

Hecht, G., Robinson, B., and Koutsouris, A. (1994). Reversible disassembly of an intestinal epithelial monolayer by prolonged exposure to phorbol esters. Am. J. Physiol. 266, G214-G221.

Keon, B.H., Schafer, C., Kuhn, C., Grund, C. and Franke, W. (1996). Symplekin, a novel type of tight junction plaque protein. J Cell Biol 134, 1003–1018.

Linder, M. (1991). The biochemistry of copper. Plenum Press, New York.

Manna, C., Galletti, P., Cucciolla, V., Moltedo, O., Leone, A., and Zappia, V. (1997). The protective effect of olive oil polyphenol (3,4- dihydroxyphenyl)-ethanol counteracts reactive oxygen metabolite-induced cytotoxicity in Caco-2 cells.

Narai, A., Arai, S., and Shimizu, M. (1997). Rapid decrease in transepithelial electrical resistance of human intestinal Caco-2 cell monolayers by cytotoxic membrane perturbants. Toxicol. in Vitro 11, 347 -354.

Neutra, M., and Louvard, D. (1989). Differentiation of intestinal cells in vitro. Mod. Cell Biol. 8, 363–398.

Ranaldi, G., Islam, K., and Sambuy, Y. (1994). D-cycloserine uses an active transport mechanism in the human intestinal cell line Caco 2. Antimicrob. Agents Chemother. 38, 1239–1245.

Rossi, A., Poverini, R., Di Lullo, G., Modesti, A., Modica, A., and Scarino, M.L. (1996). Heavy metal toxicity following apical and basolateral exposure in the human intestinal cell line Caco-2. Toxicol. In Vitro 10, 27–36.

Schneeberger, E., and Lynch, R. (1992). Structure, function, and regulation of cellular tight junctions. Am. J. Physiol. 262, L647-L661.

Stohs, S., and Bagchi, D. (1995). Oxidative mechanisms in th etoxicity of metal ions. Free Radical Biol. Med. 18, 321–336.

Thwaites, D., Hirst, B., and Simmons, N. (1994). Substrate specificity of the di/tripeptide transporter in human intestinal epithelia (Caco-2): identification of substrates that undergo H^+-coupled absorption. Br. J. Pharmacol. 113, 1050–1056.

Thwaites, D., McEwan, G., and Simmons, N. (1995). The role of the proton electrochemical gradient in the transepithelial absorption of amino acids by human intestinal Caco-2 cell monolayers. J. Membr. Biol. 145, 245–256.

Wilson, G. (1989). Cell culture techniques for the study of drug transport. Eur. J. Drug Metab. Pharmacokinet. 15, 159–163.

Zweibaum, A., Laburthe, M., Grasset, E., and Louvard, D (1991). Use of cultured cell lines in studies of intestinal cell differentiation and function. In Handbook of Physiology: The Gastrointestinal System, M. Field and R.A. Frizzel, eds. (Bethesda, MD: Am. Physiol. Soc.), pp. 223–255.

20

METAL REGULATION OF METALLOTHIONEIN GENE TRANSCRIPTION IN MAMMALS

P. Remondelli,[1] O. Moltedo,[1] M. C. Pascale,[2] and Arturo Leone[2,*]

[1]Dipartimento di Biochimica e Biotecnologie Mediche
Università degli Studi di Napoli "Federico II"
[2]Dipartimento di Scienze Farmaceutiche
Università degli Studi di Salerno

INTRODUCTION

Heavy metals play their essential role as nutrients and as cofactors for a variety of enzymes and metallo-proteins (O'Halloran, 1989). Metals are normally present in trace amount in the cell but these levels can increase consistently following environmental or nutritional changes. To avoid toxic effects and death due to metal overload, cells have developed during evolution several biochemical and molecular mechanisms which regulate the metal uptake, its intracellular distribution and elimination from the intracellular compartments. Therefore, it appears that two main processes control intracellular metal homeostasis, the first based on the regulation of the enzymatic activities of metal pumps and transporters, the second activating gene transcription.

Metal ions are able to induce the expression of a number of genes (Thiele, 1992). The synthesis of two different families of proteins is particularly increased in the presence of high concentrations of metal ions: the metallothioneins (MTs), which directly bind the excess of metal in the cytosol through their cysteine-rich domains α and β (Bremner and Beattie, 1990) and the heat shock proteins (Hsps), which preserve from molecular damage the native folding of cell polypeptides (Remondelli at al., 1988; Morimoto, 1993). Regulation of their expression by metal ions occurs mainly at the level of transcription, via the activation of specific gene transcription factors (Morimoto, 1993). In this chapter we will focus on the metal-mediated transcriptional control of MT genes.

---

[*] To whom all correspondence should be addressed: Prof. Arturo Leone, Dipartimento di Scienze Farmaceutiche, Università degli Studi di Salerno, Piazza V. Emanuele, 9–84080 Penta di Fisciano (Salerno) Italy, Tel. +39+89-9689. 16, Fax. +39+89-9689. 10, e-mail: leone@ponza. dia. unisa. it

Copper Transport and Its Disorders, edited by Leone and Mercer.
Kluwer Academic / Plenum Publishers, New York, 1999.

LOW EUKARYOTES MT TRANSCRIPTIONAL ACTIVATORS

The MT gene CUP1 of the baker yeast *Saccharomyces cerevisiae* was first isolated for its ability to confer resistance to high copper concentrations through the binding of its protein product to the excess of metal in the cytoplasm (Karin at al., 1984a). The transcription factor ace1 (**A**ctivator **C**opper **E**lement 1) regulates CUP1 expression in response to copper (Thiele, 1988; Szczypka and Thiele, 1989). A similar transcriptional activator, amt1 (**A**ctivator of **M**etallothionein **G**ene 1), has been cloned and characterized in *Candida glabrata* (Zhou and Thiele, 1991; Winge, 1998).

Both *ace1* and *amt1* transcription factors are constituted by two functional domains: a metal-binding domain and a DNA-binding domain, both requiring copper for their functional activity. Increases in the intracellular concentrations of copper allows the interaction of the metal in excess to the apo-factor and the formation of the active transcription factor, which can bind properly to the regulatory sequences present in the promoter of the yeast MT genes, thus initiating transcription (Fürst et al., 1988; Thiele, 1992).

THE MAMMALIAN MT TRANSCRIPTIONAL ACTIVATORS

In mammals metal regulation of MT transcription is a more complex phenomenon. Despite the copper specificity of yeast MT regulation, a wide range of both essential or toxic metals are able to induce MT gene expression in mammals. Moreover, a number of non metallic stimuli have also been shown to stimulate MT gene transcription (Kägi, 1991). Analyses of the structures of promoter regions of MT genes show a complex array of cis-acting elements corresponding to the binding sites of defined transcription factors (Karin and Richards, 1982). Metal activation is restricted to short metal regulating elements, MREs, present in multiple imperfect copies (from 2 to 6) with a consensus sequence TGCRCNC (R, purine; N, any, nucleotide) found in all the MT promoter regions analyzed so far. MREs require the positive interaction with metal regulated transcription factors (MRTFs) (Karin et al.,1987) and are also able to confer metal induction to non metal responsive minimal promoters (Stuart et al., 1984). MRE-binding factors are therefore key proteins to understand the mechanisms of the metal-mediated transcription activation in mammals.

Analysis of nuclear extracts from mammalian cells revealed the presence of distinct MRE-binding proteins of various mr (Sèguin and Prevost, 1988; Searle et al., 1990; Czupryn et al., 1992; Westin and Shaffner, 1988; Koizumi et al., 1992; Minichiello et al., 1994). A common feature of MRE binding factors is the requirement of zinc for their DNA-interaction. Likewise the *in vitro* assays, *in vivo* genomic foot-printing performed on nuclei from metal-treated cells revealed that the interactions between proteins and MREs were improved by exposure of the cells to zinc, suggesting that such proteins may act as a positive metal-dependent transactivators (Karin et al., 1984b; Mueller et al., 1988). The cloning or purification of MRTFs from higher organisms has been performed by different groups with the result of the isolation of different molecules whose role in the control of the MT gene expression is still under investigation.

The **M**etal **B**inding **F**actor I or MBF1, was first purified by affinity chromatography (Imbert et al., 1989) for its potential to bind an MRE element from the trout MT-B gene (Zafarullah et al., 1988). Although MBF1 was purified from mouse L cells there has been no evidences that MBF1 was not able to interact with functional mammalian MREs.

A 108kd mouse protein, designated as the **M**etal-regulating **E**lement **P**rotein 1 or MEP1, was originally detected in mouse L cells (Sèguin and Prévost, 1988). First evidences showed that MEP1 acts as a zinc-regulated factor and displays MRE-binding properties distinct to the basal factor SP1 whose binding site overlaps with the MRE core sequence in the MRE_d region of the mouse MTI gene (Labbè et al., 1991). Moreover, purified MEP1 was able to protect the MREs present on the mouse MTIa promoter in the presence of exonuclease III (Labbè et al., 1991).

The cDNA clone encoding for a "zinc finger" protein named mMTF-1 has been isolated from a mouse cDNA expression library using the synthetic MRE-S as probe (Radtke et al., 1993). The overexpression of mMTF-1 leaded to an increased constitutive expression of MRE regulated reporter plasmids, suggesting a role of mMTF1 in the regulation of basal mMT expression. Interestingly, the "knock out" of mMTF-1 gene prevented the induction of MT genes in mouse Embryonic Stem cells exposed to different heavy metals (Palmiter, 1994).

Another mouse cDNA encoding for the zinc finger protein MafY, has also been isolated by the screening of an expression library in yeast cells (Xu, 1993). MafY is able to activate an MRE_d-driven reporter gene in yeast cells, but no analysis of its DNA binding activities or metal dependency has been reported.

CLONING OF THE ZINC REGULATED FACTOR: ZiRF1

A distinct mammalian MRTF was isolated by the screening in yeast cells of a mouse cDNA library for sequences encoding for DNA binding domains having the ability to recognize one of the MREs of the mMT-I gene: the MRE_d (Inouye et al., 1994) (Fig.1). This region was shown to be a good MRE binding site *in vivo* (Culotta and Hamer, 1989) and it is protected in a zinc-dependent manner both *in vivo* (Mueller et al.,1988) and *in vitro* (Labbè et al., 1991).The insertion of the MRE_d sequence in not metal-dependent minimal

```
MREd/c     5'-AATTCTC TGCACTCCG CCCG AAAAG TGCGCTCGG -3'
                           SP1

MRE3/4     5'-AATTCGG TGCGCCCGG CCC AG TGCGCGCGG CCG -3'
                           SP1

MRE-S      5'-CGAGGGAGCTC TGCACACGG CCCG AAAAGT -3'
                           SP1
                                         *   *
mutG2      5'-AATTCTC TGCACTCCG TCT GAAAAG TGCGCTCGG -3'


SP-HSV     5'-CCGGCCCCGCCCATCCCCGGCCCCGCCCATCC -3'


MRE consensus sequence    TGCRCNC   (R, purine; N, any nt)
```

Figure 1. Sequences of the cited oligonucleotides.

promoters non responsive to metals was also able to confer them a strong metal regulation (Culotta and Hamer, 1989). The strategy we used allows the direct cloning of a gene; these conditions are more physiological compared with the screening of expression libraries with synthetic ologonucleotides having appropriate binding sites (Elledge et al., 1991).

Yeast were used as recipient cells to select for proteins containing the activation domain of GAL4 fused to fragments of mammalian proteins with a functional MRE-binding domain (Fig. 2A). The interaction of these fusion proteins with the MRE_d target sequence allows the transcription of a reporter gene to which the target sequence is linked (Fig.2A and 2B). Using this genetic selection Inouye et al. isolated M96, a mouse cDNA clone which codes for a protein of the zinc-finger type (Inouye at al.,1994). Analysis of the M96 cDNA coding sequence revealed the presence of regions particularly rich in cysteine and histidine residues (22 cysteine and 12 histidine) (Fig.3). These residues could enable the

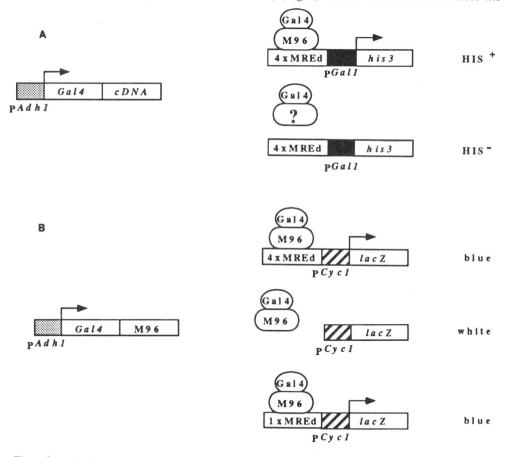

Figure 2. A. Isolation of M96 cDNA. The multimerized MRE sequences, $4XMRE_d$, are linked to the GAL1 promoter and drive the expression of the HIS3 gene. In the indicator strains there is no transcription from the reporter constructs and cells are not able to grow in absence of histidine (his⁻). If the cDNA encodes a protein able to recognize the MREs, then the binding of this factor will lead to the activation of the GAL1 promoter via the Gal4 activation domain, and cells will become (his⁺). B. Activation of reporter genes by GAL4/M96 is MRE dependent. The hybrid protein Gal4/M96 is under the control of the ADH1 promoter. This fusion protein is able to activate the CYC1 promoter driven by multimerized MRE_d or single copy MRE_d. M96 cDNA encodes a protein able to recognize the MREs and therefore the binding of this factor leads to the activation of the lacZ reporter gene via the Gal4 activation domain, with a bleu staining of the cells in presence of X-gal. Control cells transformed with reporter plasmids containing no MRE sequences in the promoter region present white staining in presence of X-Gal.

Metal Regulation of Metallothionein Gene Transcription in Mammals

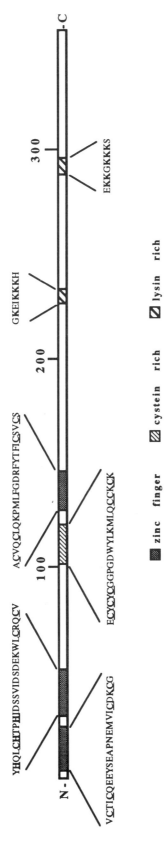

Figure 3. Structure and functional sequences of the zinc regulated factor, ZiRF1. Cysteine and histidine residues are underlined. Domains encoding for zinc finger structures of putative zinc binding domains (cystein rich) are indicated. Lysine residues in the two lysine rich domains are in bold.

factor both to bind bivalent cations and to form "zinc finger" structures (Gibson et al., 1988). The zinc-fingers domains of M96 are similar to the DNA binding domains formerly reported for the Drosophila trithorax and the human analog transcriptional activators (Tkachuk et al., 1993).

We also observed that M96 is constitutively expressed at the mRNA level, in agreement with previous findings indicating that the metal induction of MT genes does not require *de novo* protein synthesis (Karin and Hershman, 1980).

CHARACTERIZATION OF THE ZiRF1 MRE BINDING ACTIVITY

We first showed in gel retardation experiments the ability of the M96-encoded protein to specifically recognize the synthetic $MRE_{3/4}$ oligonucleotide of the human metallothionein-IIa promoter (Karin et al., 1984b) and the single MREd region of the mouse MTIa promoter. In addition, this complex could be competed by the wild-type MRE_d but not by a mutant MRE_d oligonucleotide containing a single mutation in the core sequence which inactivates the metal regulating properties of this region *in vivo* (Culotta and Hamer, 1989).

According to models proposed for yeast MT activation (Fürst et al., 1988; Thiele, 1992) a putative MRTF must be a molecular sensor of the variations in the metal concentrations. Several experiments were therefore performed in order to estimate the potential of metal ions to modulate the MRE binding activity of the M96 product.

Former experiments were conducted on *Escherichia coli* cells harboring the GST-M96 plasmid and grown in culture medium in the absence or presence of zinc. The recombinant protein was able to bind the $MRE_{3/4}$ oligonucleotide - a functional metal regulating region present in the human MTIIa promoter (Fig.1) - only when produced from cells grown in the presence of the metal (Inouye et al., 1994). The requirement of zinc for the binding activities of the M96 encoded protein prompted us to name this transcriptional activator **Z**inc **R**egulated **F**actor 1, or ZiRF1.

We also showed that the interaction of the GST-ZiRF1 fusion protein with the MREs was metal-dependent. The native GST-ZiRF1 fusion protein was metal-depleted by the use of chelating agents with different affinities toward heavy metals and the effect of metal depletion was measured by the use of mobility shift assays. In this way, we obtained a dose dependent binding inhibition of the purified GST-ZiRF1 fusion protein to functional MREs (Remondelli et al.,1997). The need of a specific metal ion in regulating the ZiRF1/MRE binding activity was examined by pre-incubating the EDTA-dialyzed GST-ZiRF1 protein with disparate metal ions prior to the addition of the $MRE_{3/4}$ binding site (Fig.4, lanes 3–12). In the presence of zinc the DNA binding activity was reestablished at levels equivalent to those displayed by the untreated protein (Fig.4, compare lane 2 with lanes 4–6). Similar assays showed that two other metals, cadmium and copper, severely suppressed the binding activity of the EDTA-treated protein (Fig.4, lanes 7–12). This finding could be explained by the possible interference of those metals with the appropriate conformation of the protein.

The methylation interference method (Garner and Revzin, 1981) was used for a more precise definition of the sequence interaction of ZiRF1 with the mouse MTIa promoter. Promoter fragments corresponding to the coding and non coding strand of the MT Ia mouse promoter (Searle et al., 1984), spanning from nucleotides -150 to -123, were partially methylated with dimethyl sulfate (DMS) and incubated in the presence of the GST-ZiRF1 fusion protein (Remondelli and Leone, 1997). This method revealed two specific

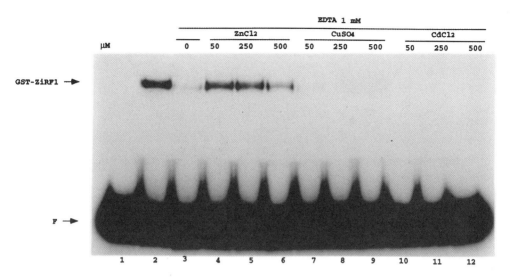

Figure 4. Zinc restores the DNA binding activity of the EDTA treated GST-ZiRF1 fusion protein. The binding of EDTA-treated recombinant GST-ZiRF1 protein with labeled $MRE_{3/4}$ was analyzed in mobility shift assays following its incubation with increasing amounts of metals salts $ZnCl_2$, $CdCl_2$ or $CuSO_4$. Lane 1, no protein; lane 2, native protein; lanes 3–12 EDTA-treated protein; EDTA-treated GST-ZiRF1 protein preincubated with: $ZnCl_2$, lanes 3–6; $CuSO_4$, lanes 6–8; $CdCl_2$, lanes 10–12.

contacts corresponding to guanines at positions -128 and -130 on the cis MRE_c region (Fig.5 A and C). We also mapped the contacts on the non coding strand which corresponded to three guanine residues located at positions -140, -139, -138 on the 3'-end of the MRE_d element and at positions -129, -127, -125 of the MRE_c element (Fig.4 B and C).

The presence of SP1 binding sites in the core sequence of several mammalian MREs and the isolation of the metal regulated factor mMTF1 suggested us to investigate the effect of competition of oligonucleotides containing the binding sites for these two transcriptional activators on the interaction of the GST-ZiRF1 fusion protein with the $MRE_{d/c}$ sequence (Fig.5A). Surprisingly, the multiple SP1 binding site, SP-HSV, (Fig.1) (Jones and Tjian, 1985; Radtke et al. 1993) exhibited in this assay levels of inhibition comparable to the wild type $MRE_{d/c}$, suggesting that ZiRF1 and SP1 have common sequence recognition properties (Fig. 5A, compare lanes 3–4 with lanes 9–10). The competition with the mutated version of the $MRE_{d/c}$ oligonucleotide mutG2, did not affect the MRE binding activity, in agreement with the alteration in the mutG2 sequence in the nucleotides interacting with ZiRF1 in methylation interference analysis (Fig.5A, lanes 7–8). With a lesser extent, the recognition site for the mMTF1 factor, the oligonucleotide MRE-S, gave rise to inhibition of the $ZiRF1/MRE_{d/c}$ interaction (Fig.5A, lanes 5–6).

The potential of the SP1 transcription factor to interact with the $MRE_{d/c}$ region was also investigated using human recombinant SP1 protein (Jones and Tjian, 1985). Competition experiments performed with purified SP1 showed that the specificity of SP1 toward the $MRE_{d/c}$ oligonucleotide was comparable to that displayed by ZiRF1 (compare Fig. 5A with Fig. 5B) and that the mutG2 oligonucleotide was not able to inhibit the binding of SP1 (Fig. 5B, lanes 7–8). Moreover, the $SP1/MRE_{d/c}$ complex displayed similar resistance to competition with the MRE-S, thus confirming the similarities in the interactions of SP1 and ZiRF1 with the functional $MRE_{d/c}$ region (Fig.5B, lanes 5–6).

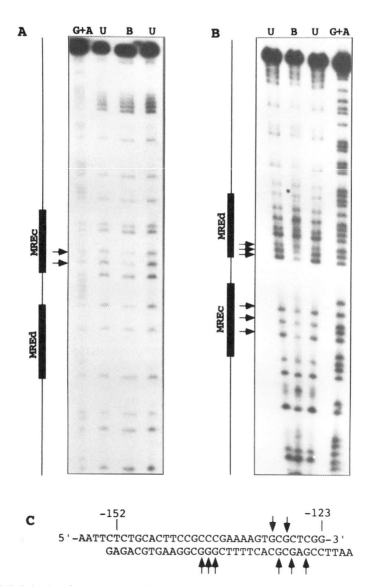

Figure 5. Methylation interference analysis. Methylation interference footprinting was performed by using partially purified GST-ZiRF1 and ^{32}P-labelled MRE$_{d/c}$ fragments labeled on either strands. A) Analysis of the pBS-MRE$_{d/c}$ coding strand fragment. B) Analysis of the pBS-MRE$_{d/c}$ non-coding strand. U represents unbound probe, showing cleavage at all guanines; B represents bound DNA-protein complexes; the G+A ladder shows cleavage at guanines and adenines residues. Arrows indicate guanines interacting with the GST-ZiRF1 fusion protein. C) Nucleotide sequence of the MRE$_{d/c}$ region spanning from nucleotides -151 to -123. Guanine residues interacting with ZiRF1 are represented by arrows.

These results confirm that basal SP1 transcription factor exhibits MRE recognition properties similar to those evidenced for the metal-regulated ZiRF1 and mMTF1 factors and that the basal SP1 activator could be therefore considered a bona fide MRE-binding protein.

DISCUSSION

Despite the functional and structural homology displayed by the yeast copper activated factors, mammalian MRTFs so far isolated are startling different. The isolation of various mammalian MRTFs indicate the possibility that the metal-dependent activation of MT genes may involve more than one protein with distinct affinities for MRE sequences or different mechanisms of activation during metal exposure.

The expression of the mouse MTF1, a "zinc finger" protein isolated by Radtke et al., triggers the activation of MRE-driven reporter genes in transient transfections experiments in the absence of metal induction (Radtke et al., 1993). Role of this factor in the regulation of MT gene expression has been established by the observation that both the basal and the inducible MT gene expression was silent in mouse ES cells when the mMTF1 gene was inactivated (Palmiter, 1994). A model for the activation of mMTF1 has been proposed in which an inhibitory protein which is the intracellular sensor of the metal concentration could modulate the activity of mMTF1 in a manner similar to that described for NFKb transcriptional complex (Palmiter, 1994).

The zinc-regulated MRE-binding factor ZiRF1 displays an open reading frame encoding for cysteine and histidine reach domains that could enable the factor both to bind bivalent cations and to form "zinc finger" structures. ZiRF1 specifically binds functional MREs present in the promoter regions of both the human MTIIa and the mouse MTIa genes.

The molecular structure of ZiRF1 suggests that this protein could become activated in a manner analogous to the yeast MT regulator ACE1/CUP2 (Fürst et al., 1988), i.e. zinc ions may interact with the metal-binding domain of ZiRF1 which preexists as an apo-protein. In this case, the "apo-ZiRF1" could act as a molecular sensor of the intracellular metal content. Apo-ZiRF1 obtained by the metal depletion with various metal chelating agents, 1,10 phenanthroline (1,10 PHE), ethylenediaminetetraacetic acid (EDTA) and tetrakis(2-pyridylmethyl-) ethylenediamine (TPEN), does not exhibit MRE binding activities *in vitro*. Only metal reconstitution in the presence of zinc ions allows the recovery of the MRE-binding properties, suggesting that this metal plays a key role in the activation of this factor. Interestingly, activation of the MRE-binding activity by zinc is a common feature of the different mammalian MRTFs so far identified (Sèguin and Prevost, 1988; Searle, 1990; Sèguin, 1991; Czupryn et al., 1992; Westin and Shaffner, 1988; Koizumi et al., 1992; Minichiello et al., 1994; Remondelli et al. 1997).

It has been demonstrated by both *in vivo* and *in vitro* footprinting analyses that this metal regulatory region which spans from nucleotides -150 to -123 of the mouse MTI promoter and interacts with ZiRF1 is protected by specific nuclear proteins *in vivo* following cell exposure to zinc (Mueller et al., 1988). Moreover, nucleotide changes in the sites of interaction with ZiRF1, which were not investigated in the earlier analysis of the point mutations of MRE regions (Culotta and Hamer, 1989), abolished the $MRE_{d/c}$ metal-dependent activation of a heterologous promoter (Remondelli and Leone, 1997).

Our data indicated the multiple interactions of the transcription factors ZiRF1, SP1 and mMTF1 with the $MRE_{d/c}$ region of the mouse MTI promoter (Remondelli and Leone,

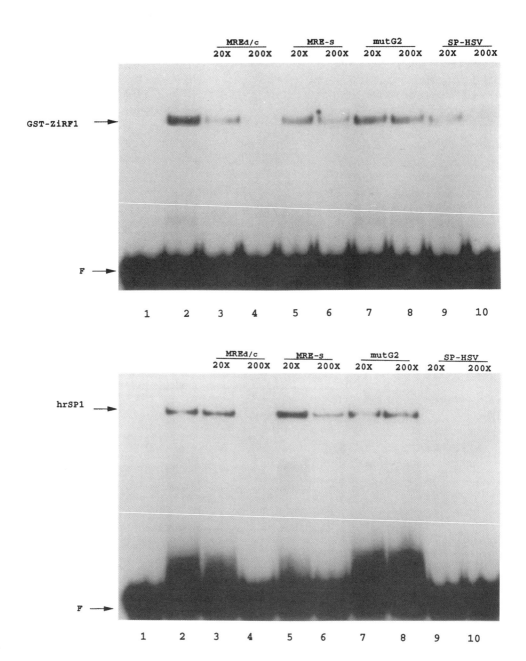

Figure 6. Binding activities of the recombinant ZiRF1 and SP1 factors to the $MRE_{d/c}$ region of the mMT-Ia promoter. A) Binding of the partially purified GST-ZiRF1 protein to the ^{32}P-labeled wild-type $MRE_{d/c}$. Lane 1, no added protein; lanes 2–10, 10 ng bacterial expressed GST-ZiRF1 protein with 20- and 200-fold molar excess of the cold competitors oligonucleotides: lanes 3–4, $MRE_{d/c}$; lanes 5–6, MRE-S; lanes 7–8, the mutant mutG2; lanes 9–10, SP-HSV. B) Binding of the recombinant human SP1 to the ^{32}P-labeled wild-type $MRE_{d/c}$. Lane 1, no added protein; lanes 2–10, 10ng of purified human SP1; lanes 3–10, same amounts of competitors as in A.

1997). Furthermore, analyses of the MRE-interactions with purified human SP1 indicated that this transcription factor displays properties of a MRE-binding protein. In fact, specificity of SP1 toward the $MRE_{d/c}$ region was highly similar to that displayed by ZiRF1 (Remondelli and Leone, 1997). These data supports the hypothesis of a competition between basal and metal-regulated transcription factors modulated by the increases or decreases in the intracellular concentrations of free metals. In agreement with a model of regulation by multiple factors of the MRE-regulated MT gene transcription is our observation that the down regulation of a hMTTIIa promoter-CAT fusion plasmid in both uninduced and zinc-induced transfected cells expressing antisense ZiRF1 cDNA was never complete (Fig.7). The consistent levels of hMTIIa promoter activity which are still detected following antisense expression could be explained by the possibility that different factors may compete for the same regulatory regions present in the MT promoters.

The observation that the activities of other nuclear factors are changed following exposure of the cell to metals strongly suggest that different mechanisms of regulation are involved in the metal -dependent gene expression. In response to various stresses, including metals, heat shock factors (HSFs) activate transcription of their related genes (Morimoto, 1993). It is known that the activation of the HSFs by heat is the result of different molecular events including site specific phosphorilation, protein-protein interactions and import in the nucleus (Morimoto, 1993). The effect of heavy metals at the different levels of HSFs activation has not been elucidated.

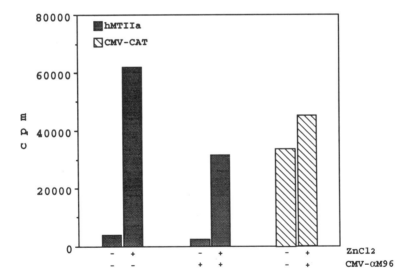

Figure 7. M96 cDNA antisense down regulation of hMTIIa promoter. Human 293 cells were transfected with a DNA mixture containing 5μg of the indicated CAT plasmids, 1μg CMV-αM96 antisense expression vector when indicated (+), and 0.5 μg of the CMV-β-gal plasmid. CMV-αM96 contains the entire M96 cDNA cloned in the antisense direction under the control of the citomegalovirus promoter (CMV); CMV-β-gal and CMV-CAT contained the CMV promoter directing the expression of the β-galactosidase (β-gal) and the chloramphenicol-acetyl-trasferase (CAT) bacterial genes, respectively. hMT-IIa-CAT reporter contained the entire human MT-IIa promoter region driving the expression of the CAT reporter gene. Cells were induced with 100 μM $ZnCl_2$ for 6 hours before cells harvesting. β-galactosidase equivalents of cell extracts were used to calculate CAT activities. Each data point represents the average of three transfection experiments.

Other evidences show that transcription factors related to the jun/AP1 family (i.e. fungal yAP1 and CAD1) could confer resistance to toxic concentrations of metals, presumably via the up-regulation of the expression of genes involved in the response to redox state changes during the oxidative stress (Wu et al., 1993; Shuzuke and Jones, 1994; Wemmie et al., 1994).

In mammals, some metal ions like Cd or Zn can also induce the expression of c-jun and c-fos, the cellular protooncogenes involved in the formation of the transcription complex AP1, a known target of tumor promoters (Jin and Rongertz, 1990; Abshire et al., 1996). Following the exposure to toxic levels of the heavy metals Cd, Zn or Cu , the AP1 complex is activated at a magnitude comparable to the stimulation obtained using the phorbol ester PMA (Remondelli et al., unpublished results). The phorbol ester-mediated activation of AP1 (Angel et al., 1987) is the product of a cascade of phosphorilation events mediated by the activity of the cellular protooncogene ras (ras pathway) which causes the phosporylation mediated by the stress activated protein kinase, JNK, that phosphorilates specific sites in the N-terminal domain of c-jun (Davis, 1995; Lim et al., 1996). A different signal transduction pathway of the AP1 activation has also been proposed, based on the increase of the generation of hydroxyl radicals. Such event would cause the subsequent decrease in the levels of reduced glutathione (GSH), and perturb the intracellular redox potential (Abshire et al., 1996). Therefore, the effect of heavy metal treatment on the AP1 activity could be explained with a similar mechanism, since metal ions show high affinity to reactive SH groups. Moreover, metal ions (zinc and sodium arsenate) are also powerful phosphatase inhibitors which have been shown to participate in the modulation of the signal transduction pathways (Cavigelli et al., 1996).

The overall of these results indicate that metals ions could modulate gene expression not only by direct activation of nuclear factors present in the cell in apo-protein form (i.e. CUP1), but also affecting distinct trasduction pathways. MT gene transcription in mammals could well be regulated through different signalling pathways (i.e. MRE- and/or AP1-mediated) in order to ensure the proper intracellular metal content during the basal and induced state. In this view, the functional studies of metal-specific transporters like the copper Menkes and the Wilson ATPases (Monaco and Chelly, 1995; Di Donato and Sarkar, 1997) or the zinc ZNT-2 and ZNT-3 proteins (Palmiter at al. 1996a, Palmiter at al. 1996b) can well represent the first step to link the transcriptional regulation of MT genes to the enzimatic control of intracellular metal homestasis.

ACKNOWLEDGMENTS

Telethon fellowship was awarded to O.M.. This work was supported by the following grants to A.L.: M.U.R.S.T. 40%; Comitato Scienze Biologiche e Mediche, C.N.R.; and Telethon grant n.E.373 .

REFERENCES

Abshire, M.K., Buzard, G.S., Shiraishi, N., and Waalkes, M.P. (1996). Induction of c-myc and c-jun proto-oncogene expression in rat L6 myoblasts by cadmium is inhibited by zinc preinduction of the metallothionein gene. J. Toxicol. Environ. Health 48, 359–377

Angel, P., M. Imagawa, R. Chiu, B. Stein, R. J. Imbra, H. J. Rahmsdorf, C. Jonat, P. Herrlich, and Karin, M. (1987). Phorbol ester inducible genes contain a common cis element recognized by a TPA-modulated trans-acting factor. Cell 49, 729–739.

Bremner, I., and J. H. Beattie. (1990). Metallothionein and the trace minerals. Ann. Rev. Nutr. 10, 63–83.
Cavigelli, M., Li, W.W., Lin, A., Su, B., Yoshioka, K., and Karin, M. (1996). The tumor promoter arsenite stimulates AP-1 activity by inhibiting a JNK phospatase. EMBO J. 15, 6269–6279.
Culotta, V. C., and Hamer, D. H.. (1989). Fine mapping of a mouse metallothionein gene metal response element. Mol. Cell. Biol. 9, 1376–1380.
Czupryn,M., Brown, W. E., and Vallee, B. L. (1992). Zinc rapidly induces a metal response element binding factor. Proc.Natl.Acad.Sci.USA 89, 10395–10399.
Di Donato, M., and Sarkar, B. (1997). Copper transport and its alterations in Menkes and Wilson deseases. Biochim. Biophys. Acta 1360(1), 3–16.
Davis, R.J. (1995). Transcriptional regulation by MAP kinases. Mol. Reprod. Dev. 42, 459–467.
Elledge, S. J., J. T. Mulligan, S. W. Ramer, M. Spottswood, and Davis,R. W. (1991). LambdaYES: a multifunctional cDNA expression vector for the isolation of genes by complementation of yeast and *Escherichia coli* mutations. Proc. Natl. Acad. Sci. USA 88, 1731–1735.
Fürst, P., S. Hu, R. Hackett, and Hamer, D. (1988). Copper activates metallothionein gene transcription by altering the conformation of a specific DNA binding protein. Cell 55, 705–717.
Garner, M., and Revzin, A. (1981). A gel electrophoresis method for quantifying the binding of proteins to specific DNA regions: application to components of the *E. coli* lactose regulatory system. Nucl. Acids Res. 9, 3047–3060.
Gibson, T. J., Postma, J. P. M.,Brown, R. S. and Argos, P. (1988) A model for tertiary structure of the 28 residue DNA-binding motif ('zinc finger') common to many eukaryotic transcriptional regulatory proteins. Protein Engineering 2, 209–218.
Imbert, J., Zafarullah, M., Culotta, V. C., Gedamu, L., and Hamer, D. (1989). Transcription factor MBF-I interacts with metal regulatory elements of higher eucaryotic metallothionein genes. Mol. Cell. Biol. 9, 5315–5323.
Inouye, C., Remondelli, P., Karin, M. and Elledge, S. (1994). Isolation of a metal response element binding protein using a novel expression cloning procedure: the one hybrid system. DNA Cell Biol. 13, 731–742.
Jin, P., and Ringertz, N.R. (1990). Cadmium induces transcription of proto-oncogenes c-jun and c-myc in rat L6 myoblasts. J. Biol.Chem. 265, 14061–14064.
Jones, K.A. and Tjian, R. (1985). Sp1 binds to promoter sequences and activates herpes simplex virus 'immediate-early' gene transcription in vitro. Nature (London) 317, 179–182.
Kägi, J. H. R. (1991). Overview of metallothionein. Methods Enzymol. 205, 613–626.
Karin, M., and Herschman, H. R. (1980). Characterization of metallothioneins induced in HeLa cells by dexamethasone and zinc. Eur. J. Biochem. 107, 395–401.
Karin, M., and Richards, R. I. (1982). Human metallothionein genes: primary structure of the metallothionein-II gene and a related processed gene. Nature 299, 797–802.
Karin, M., Najaran, R., Haslinger, A., Valenzuela, P., Welch, J. and Fogel, S. (1984a). Primary structure and transcription of an amplified genetic locus: the CUP1 locus in yeast. Proc. Nat. Acad. Sci. USA 81, 337–341.
Karin, M., Haslinger, A., Holtgreve, H., Cathala, G., Slater, E., and Baxter, J.D. (1984b). Activation of a heterologous promoter in response to dexamethasone and cadmium by metallothionein gene 5'-flanking DNA. Cell 36, 371–379.
Karin, M., Haslinger, A., Heguy, A., Dietlin, T., and Cooke, T. (1987). Metal-responsive elements act as positive modulators of human metallothionein-IIA enhancer activity. Mol. Cell. Biol. 7, 606–613.
Koizumi, S., Suzuki, K., and Otsuka, K. (1992). A nuclear factor that recognizes the metal-responsive element of the human metallothionein IIa gene. J. Biol. Chem. 267, 18659–18664.
Labbè, S., Prévost, J., Remondelli, P., Leone, A., and Sèguin, C. (1991). A nuclear factor binds to the metal regulatory elements of the mouse gene encoding metallothionein-I. Nucl.Ac.Res. 19, 4225–4231.
Lim, L., Manser, E., Leung, T., and Hall, C. (1996). Regulation of phosphorilation pathways by p21 GTPases. The p21 Ras-related Rho subfamily and its role in phosphorilation signalling pathways. Eur. J. Biochem. 242, 171–185.
Minichiello, L., Remondelli, P., Cigliano, S., Bonatti, S. and Leone, A. (1994). Interactions of nuclear proteins from uninduced,induced and superinduced HeLa cells with the metal regulatory elements MRE 3 and 4 of the human metallothionein IIa encoding gene. Gene 143, 289–294.
Monaco, A.P., and Chelly, J. (1995). Menkes and Wilson deseases. Adv. Genet. 33, 233–253.
Morimoto, R.I. (1993). Cells in stress: transcriptional activation of heat shock genes. Science 259, 1409–1410.
Mueller, P. R., Salser, S. J., and Wold, B. (1988). Constitutive and metal-inducible protein: DNA interactions at the mouse metallothionein I promoter examined by in vivo and in vitro footprinting. Genes Dev. 2, 412–427.
O'Halloran, T.V. (1989) in Metal Ion in Biological Systems (Siegel, H. and Sigel, A., eds.) Mercel Dekker, Inc. New York, pp. 105–146.

Palmiter, R. D. (1994). Regulation of metallothionein genes by heavy metals appears to be mediated by a zinc-sensitive inhibitor that interacts with a costitutively active transcription factor, MTF1. Proc. Natl. Acad. Sci. USA 91, 1219–1223. EMBO J. 15, 1784–1791

Palmiter, R.D., Cole, T.B., and Findley, S.D. (1996). ZnT-2, a mammalian protein that confers resistance to zinc by facilitating vesicular sequestration. EMBO J. 15, 1784–1791.

Palmiter, R.D., Cole, T.B., Quaife, C.F., and Findley, S.D. (1996). ZnT-3 a putative tansporter of zinc into synaptic vesicles. Proc. Natl. Acad. Sci. USA 93, 14934–14939.

Radtke, F., Heuchel, R., Georgiev, O., Hergersberg, M., Gariglio, M., Dembic, Z., and Schaffner, W. (1993). Cloned transcription factor MTF-I activates the mouse metallothionein I promoter. EMBO J. 12, 1355–1362.

Remondelli, P., Pascale, M.C., and Leone, A. (1988). Effects of zinc,copper and cadmium on protein biosynthesis of two differentiated human hepatoma cell lines. In Metal Ion Homeostasis: Molecular Biology and Chemistry (D. Winge and D.H. Hamer Eds.) UCLA Symposia on Molecular and Cellular Biology; Alan R.Liss, Inc., New York, NY, U.S.A. 98, 56–69.

Remondelli, P. and Leone, A. (1997). Interactions of the zinc regulated transcription factor (ZiRF1) with the the mouse MT Ia promoter. Biochem. J. 323, 79–80.

Remondelli, P., Moltedo, O. and A. Leone (1997). Regulation of ZiRF1 and basal SP1 transcription factors MRE-binding activity by transition metals. FEBS Letters 416, 254–258.

Searle, P. (1990). Zinc dependent binding of a liver nuclear factor to metal response element MREa of the mouse metallothionein-I gene and variant sequences. Nucl. Acids Res. 18, 4683–4690.

Searle, P. F., Davidson, B.L., Stuart, G.W., Wilkie, T.M., Norstedt, G., and Palmiter, R.D. (1984). Regulation, linkage, and sequence of mouse metallothionein I and II genes. Mol. Cell. Biol.4, 1221–1230.

Sèguin, C. (1991). A nuclear factor requires Zn^{2+} to bind a regulatory MRE element of the mouse gene encoding metallothionein-1. Gene 97, 295–300.

Sèguin, C., and Prevost, J. (1988). Detection of a nuclear protein that interacts with a metal regulatory element of the mouse metallothionein I gene. Nucl. Acıds Res. 16, 10547–10560.

Shuzuke, K. and Jones, N. (1994). YAP1 dependent activation of TRX2 is essential for the response of *Saccharomices cerevisiae* to oxidative stress by hydroperoxides. *EMBO J.* 13, 655–664.

Stuart, G. W., Searle, P. F., Chen, H. Y., Brinster, R. L., and Palmiter, R. D. (1984). A 12-base-pair DNA motif that is repeated several times in metallothionein gene promoters confers metal regulation to a heterologous gene. Proc. Natl. Acad. Sci. USA 81, 7318–7322.

Szczypka, M.S. and Thiele, D. J. (1989). A cysteine reach nuclear protein activates yeast metallothionein gene transcription. Mol. Cell. Biol. 9, 421–429.

Thiele, D.J. (1988). ACE1 regulates expression of the *Saccharomices cerevisiae* metallothionein gene. Mol. Cell. Biol. 8, 2745–2752.

Thiele, D. J. (1992). Metal regulated transcription in eukaryotes. Nucl. Acid. Res. 20, 1183–1191.

Tkachuk, D. C., Kohler, S., and Cleary, M. L..(1994). Involvement of a homolog of drosophila thrithorax by 11q23 chromosomal translocations in acute leukemias. Cell. 71, 691–700.

Wemmie, J.A., Wu, A.L., Harshman, K.D., Parker, C.S. and Moye-Rowley, W.S. (1994). Transcriptional activation mediated by yeast AP-1 protein is required for normal cadmium tolerance. J. Biol. Chem. 269, 14690–14697.

Westin, G., and Schaffner, W. (1988). A zinc-responsive factor interacts with a metal-regulated enhancer element (MRE) of the mouse metallothionein gene. EMBO J. 7, 3763–3770.

Winge, D.R. (1998). Copper-regulatory domain involved in gene expression. In: Copper transport and its disorders: molecular and cellular aspects (A. Leone and J.F. Mercer Eds.); Advances in Experimental Medicine and Biology; Plenum Press Inc., this volume.

Wu, A., Wemmie, J.A., Edginton, N.P., Goebl, M., Guevara, J.L., Moye-Rowley, W.S. (1993). Yeast bZIP proteins mediate pleiotropic drug and metal resistance. J. Biol. Chem. 268, 18850–18858

Xu, C. (1993). cDNA cloning of a mouse factor that activates transcription frm a metal response element of the mouse metallothionein-I gene. DNA Cell Biology 12, 517–525.

Zafarullah, M., Bonham, K. and Gedamu, L. (1988). Structure of the raimbow trout metallothionein B gene and characterization of its metal-responsive region. Mol. Cell. Biol. 8, 4469–4476.

Zhou, P. and Thiele, D. J. (1991). Isolation of metal activated transcription factor gene in *Candida glabrata* by complementation in *Saccharomyces cerevisiae* . Proc.Nat. Acad. Sci. USA 88, 6112–6116.

21

COPPER-REGULATORY DOMAIN INVOLVED IN GENE EXPRESSION

Dennis R. Winge

University of Utah Health Sciences Center
Departments of Medicine and Biochemistry
Salt Lake City, Utah 84132

Cells in the natural world experience a changing environment. Physiological responses to such changes enable cells to survive variation in nutrient concentrations or environmental factors. Sensory systems are required to detect nutrient or environmental changes and transduce signals into physiological responses. A common primary response is the transcriptional regulation of genes whose products have protective functions or participate in nutrient metabolism.

Cells experience changes in the availability of metal ions. Under conditions of limiting levels of a particular metal ion, genes encoding proteins involved in uptake of copper, iron and zinc ions are derepressed in yeast (Dancis et al., 1994; Labbe et al., 1997; Stearman et al., 1996; Yamaguchi-Iwai et al., 1995; Yamaguchi-Iwai et al., 1997; Zhao and Eide, 1996). In contrast, conditions of excess metal ions in the environment result in the inhibition of expression of metal ion uptake genes as well as the induced expression of a different subset of genes (Culotta et al., 1994; Furst et al., 1988; Gralla et al., 1991). Genes activated by metal ions encode proteins that typically have protective roles (Culotta et al., 1994; Furst et al., 1988; Gralla et al., 1991).

In the past few years considerable effort has been focused on mechanisms by which cells specifically sense the intracellular copper ion concentration and transduce the signal into the physiological regulation of the cellular copper levels. Copper is an essential nutrient and yet excess accumulation of copper ions results in toxicity. Copper-induced toxicity may arise, in part, from cell damage caused by reactive oxygen intermediates. Transition metal ions, i.e. copper, can catalyze the formation of highly reactive hydroxyl radicals through Fenton chemistry (Fridovich, 1978).

Copper ions are required for at least three key enzymes in the yeast *Saccharomyces cerevisiae*. The ability of cells to grow on non-fermentable carbon sources is dependent on having an active cytochrome oxidase complex which requires Cu ions as cofactors. Oxidative growth requires defense molecules against reactive oxygen intermediates. Superoxide dismutase is a Cu-metalloenzyme that dismutes superoxide anions. A third key Cu-metal-

Copper Transport and Its Disorders, edited by Leone and Mercer.
Kluwer Academic / Plenum Publishers, New York, 1999.

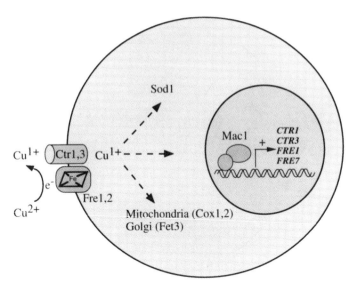

Figure 1. Response of cells cultured in medium containing <1 nM Cu(II). Under Cu deficiency conditions, genes encoding two high affinity Cu permeases Ctr1 and Ctr3 are expressed. In addition, one metalloreductase, Fre1, and a Fre1-homolog, Fre7, are expressed. Fre1 elaborates a diffusable reductant in the growth medium that assists in mobilizing Cu ions.

loenzyme is Fet3 which is a ferro-oxidase critical for uptake of Fe(II) (Askwith et al., 1994). A myriad of other oxidases and oxygenases require Cu(II) as a functional cofactor, but the presence of these enzymes in the yeast *Saccharomyces cerevisiae* is unclear.

The essentiality of Cu for normal yeast physiology is consistent with the fact that *S. cerevisiae* possesses mechanisms to ensure that adequate cellular copper levels are attained. Conditions of copper deficiency result in the derepression of a family of genes whose products are involved in cellular uptake of copper ions (Dancis et al., 1994; Hassett and Kosman, 1995; Labbe et al., 1997; Yamaguchi-Iwai et al., 1997). Derepression of Cu uptake genes occurs in *S. cerevisiae* cultured in medium containing less than 1 nM Cu(II) (Figure 1). Two of the genes encode for two high affinity plasma membrane copper ion permeases, Ctr1 and Ctr3 (Dancis et al., 1994; Labbe et al., 1997). Another gene encodes for the Fre1 metalloreductase which elaborates a diffusable reductant into the growth medium to mobilize Cu(II) ions (Lesuisse et al., 1996). A fourth gene derepressed is a homolog to Fre1, designated Fre7, whose functions remains unresolved. The expression of these genes is a cellular response to inadequate intracellular Cu levels.

The expression of these genes whose products function in the high affinity uptake system is regulated at the level of transcription through the Mac1 transcription factor (Jungmann et al., 1993). *MAC1* was originally identified as a partially dominant *MAC1* mutation, *MAC1^{up1}* (Jungmann et al., 1993). Cells harboring the *MAC1^{up1}* allele were found to be incapable of repression of Cu-uptake genes (Hassett and Kosman, 1995; Yamaguchi-Iwai et al., 1997). As a result, the cells were found to be hypersensitive to copper salts in the growth medium (Jungmann et al., 1993). The copper sensitivity of *MAC1^{up1}* cells demonstrates the critical importance of down-regulating the high-affinity uptake system in copper replete cells. The role of Mac1 is transcriptional activation of *CTR1* and *FRE1* is further highlighted by the reduced copper transport observed in *mac1* cells (Hassett and Kosman, 1995).

Copper-Regulatory Domain Involved in Gene Expression

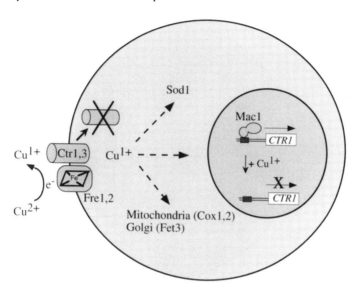

Figure 2. Response of cells cultured in medium containing >1 nM Cu(II). Under these conditions, *CTR1*, *CTR3*, *FRE1* and *FRE7* are inhibited in their expression. Furthermore, Cu-replete cells remove any pre-existing Ctr1 permease from the cell surface.

Expression of *CTR1*, *CTR3* and *FRE1* is inhibited in wild-type yeast cells cultured in medium containing greater than 1 nM Cu(II) (Figure 2). Cells cultured in such medium are in positive Cu balance and have no need for the high affinity Cu uptake system. Cu uptake can occur through low affinity permeases such as Ctr2 (Kampfenkel et al., 1995). One cellular response to a positive Cu balance is, therefore, inhibition of the transcription of genes whose products are involved in high affinity Cu uptake. A second response appears to be removal of any pre-existing Ctr1 transporters. Western analysis of epitope-tagged Ctr1 revealed that Ctr1 is targeted for degradation in Cu-replete cells (Ooi et al., 1996).

Cells exposed to an environment with high Cu(II) levels exhibit protective responses to counteract the cytotoxicity of Cu ions. Under conditions where the extracellular copper concentration exceeds 1 μM, at least three genes, *CUP1*, *CRS5* and *SOD1*, are transcriptionally activated (Culotta et al., 1994; Furst et al., 1988; Gralla et al., 1991) (Figure 3). *CUP1* and *CRS5* encode cysteinyl-rich polypeptides in the metallothionein (MT) family (Culotta et al., 1994; Jensen et al., 1996). *CUP1* is the dominant locus that confers the ability of yeast cells to propagate in medium containing copper salts (Fogel and Welch, 1982; Hamer et al., 1985; Jensen et al., 1996). Cells highly resistant to copper salts contain a *CUP1* locus with tandem arrays of genes encoding the Cup1 metallothionein (Fogel and Welch, 1982).

The Cup1 MT buffers the intracellular copper ion concentration by binding Cu(I) ions within a heptacopper-thiolate cluster (Narula et al., 1991; Peterson et al., 1996). The effectiveness of Cup1 MT in Cu ion buffering is dramatically enhanced by the cellular coupling of the biosynthesis of Cup1 to the cellular concentration of Cu(I) ions (Furst et al., 1988). The Cu-regulation of Cup1 biosynthesis occurred at the level of transcriptional activation (Butt et al., 1984; Karin et al., 1984). *S. cerevisiae* contains a second MT gene *CRS5* whose expression is also Cu-regulated (Culotta et al., 1994). *CRS5* is present as a single copy gene, unlike the tan-

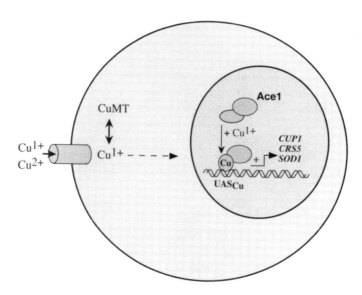

Figure 3. Response of cells cultured in medium containing > 1μM Cu(II). In Cu-replete cells Cu(I) or Cu(II) ions can be taken up through low affinity transporters. Cu-induces the expression of *CUP1* and *CRS5* which encode Cu-buffering metallothionein molecules. Cu-induced expression of these genes is mediated by the Cu-activated transcription factor, Ace1.

dem array of *CUP1* genes (Culotta et al., 1994). Targeted disruption of *CUP1* in cells with a wild-type *CRS5* locus confers hypersensitivity to Cu salts (Culotta et al., 1994). An additional targeted disruption of *CRS5* exacerbates the hypersensitivity (Culotta et al., 1994). The dominance of *CUP1* in Cu buffering arises from the amplification of the *CUP1* locus, the greater Cu-responsiveness of the *CUP1* promoter, and the greater stability of the CuCup1 complex compared to the CuCrs5 complex (Jensen et al., 1996).

Additional functions may exist for Cup1 and Crs5 as both genes are highly basally expressed. SAGE (serial analysis of gene expression) analysis revealed that *S. cerevisiae* cells contain at least 4,665 genes, with expression levels ranging from 0.3 to 200 mRNA transcripts per cell (Velculescu et al., 1997). The *CUP1* transcript level was near 75, which is a similar abundance as many ribosomal protein genes (Velculescu et al., 1997). DNA microchip analysis of yeast reveals that *CRS5* is also highly basally expressed. The high basal expression of *CUP1* and *CRS5* may imply that Cup1 and Crs5 have additional housekeeping roles.

One additional function of CuCup1 is that is possesses limited superoxide dismutase activity (Tamai et al., 1993). *CUP1* can functionally suppress the oxygen sensitivity of a *sod1*Δ strain (Tamai et al., 1993). The mechanism of superoxide dismutation by CuCup1 involves a thiyl radical species (Sievers et al., 1996). Sod1, likewise, has a secondary role of contributing to copper buffering in *S. cerevisiae* (Culotta et al., 1995). Thus, Cup1 and Sod1 each contribute to copper and oxygen radical homeostasis.

A. COPPER-ACTIVATED TRANS-ACTING FACTORS

The *trans*-acting factor that mediates Cu(I) activation in *S. cerevisiae* was identified as Ace1 (also designated Cup2) (Buchman et al., 1989; Thiele, 1988) (Figure 3). The mechanism of Cu(I) activation through Ace1 was shown to be Cu-dependent Ace1 binding to 5' promoter elements (designated UAS_{Cu}) within the *CUP1* gene (Buchman et al., 1990; Furst et al., 1988). Ace1 formed a specific complex with UAS_{Cu} DNA only in the presence of Cu(I) or Ag(I) ions (Furst et al., 1988). The DNA-binding domain of Ace1 was shown to map to the N-terminal 122 residues (Furst et al., 1988). Twelve cysteinyl residues are present in the N-terminal 122 residues of Ace1, eleven of which are critical for Cu-induced expression of *CUP1* (Hu et al., 1990). Ten of the eleven critical cysteinyl residues are present in Cys-$X_{1,2}$-Cys sequence motifs that are commonly found in metal-binding proteins such as metallothionein. These results led Furst et al. to postulate in 1988 that Cu(I) binding to Ace1 triggered a conformational change to a fold that was poised for DNA binding (Furst et al., 1988).

Footprinting analyses of Ace1 binding to UAS_{Cu} revealed major groove base contacts at the two ends of UAS_{Cu} and minor groove contacts in the middle A/T-rich region (Buchman et al., 1990; Dobi et al., 1995). The prediction was made that Ace1 lies atop the minor groove contacting the major groove on both sides (Buchman et al., 1990).

Additional insights on the mechanism of copper metalloregulation in yeast comes from studies on a related system in the yeast *Candida glabrata*.

C. glabrata .

C. glabrata contains a family of MT genes as is observed in *S. cerevisiae* (Mehra et al., 1989). The MT genes of *C. glabrata* are specifically copper-responsive in their expression (Mehra et al., 1989; Mehra et al., 1992). The *trans*-acting factor that mediates Cu-induced expression of MT genes in *C. glabrata* is Amt1 (Zhou and Thiele, 1991).

Ace1 and Amt1 are structurally homologous. The N-terminal half of Amt1 is 50% identical to the corresponding region of Ace1 with complete conservation of sequence positions of the eleven critical cysteinyl residues if one gap is added in the sequence alignment in Ace1. Homology between Ace1 and Amt1 ends after residue 100 in Ace1. The C-terminal region (residues 120–225) of Ace1 has been shown to contain the transactivation domain (Munder and Furst, 1992). Thus, Ace1 and Amt1, like many yeast transcription factors, are modular in nature with a specific DNA-binding domain and a separate and independently acting transactivation domain (Hahn, 1993).

B. PRESENCE OF A POLYCOPPER-THIOLATE CLUSTER IN Ace1 AND Amt1

The DNA-binding activity of Ace1 was shown to be dependent on the presence of copper ions (Dameron et al., 1991; Furst et al., 1988). The addition of Cu(I) to an Ace1 peptide consisting of residues 1–122 resulted to a specific protein:DNA complex (Dameron et al., 1991; Furst et al., 1988). The only metal ions that facilitated protein:DNA complex formation were Cu and Ag ions (Dameron et al., 1991; Furst et al., 1988). Expression of recombinant Ace1 in bacteria enabled the isolation of a Cu_4Zn_1Ace1 complex (Dameron et al., 1991; Farrell et al., 1996). Biophysical analyses of the Cu_4Zn_1Ace1 and Cu_4Zn_1Amt1 complexes revealed ultraviolet transitions consistent with Cu-thiolate charge-transfer bands, and luminescence in aqueous solutions indicative of trigonal Cu(I) binding in an environment shielded from solvent interactions (Dameron et al., 1991; Farrell et al., 1996; Thorvaldsen et al., 1994). Copper reconstitution studies of apo-Amt1 revealed an all-or-nothing formation of a tetracopper species (Thorvaldsen et al., 1994).

X-ray absorption spectroscopy (Cu K-edge EXAFS) confirmed the trigonal coordination of Cu(I) ions in Ace1 and Amt1 (Pickering et al., 1993; Thorvaldsen et al., 1994). In addition, EXAFS revealed the existence of a short Cu-Cu distance of 2.7 Å which is diagnostic of a Cu(I) polynuclear cluster. The CuS cluster in Ace1 and Amt1 resembled the properties of a synthetic tetracopper cluster with thiolphenolate ligands (Dance, 1986; Pickering et al., 1993).

The polycopper cluster in Ace1 and Amt1 resembles the polycopper cluster in metallothioneins in certain regards. The Cup1 metallothionein consisting of 53 residues enfolds a single heptacopper thiolate cluster (Narula et al., 1991; Peterson et al., 1996). The Cu(I) cluster in Cup1 spectroscopically resembles the cluster in Cu,ZnAce1 (Pickering et al., 1993). Thus, the Ace1/Cup1 system in yeast is curious in that the regulatory factor (Ace1), like the product of the pathway (Cup1), contains a polycopper cluster.

C. DISSECTION OF Ace1 AND Amt1 INTO FUNCTIONAL DOMAINS

Mapping studies were carried out on Amt1 and Ace1 to determine which cysteinyl thiolates serve as ligands for Zn(II) and the tetracopper cluster (Farrell et al., 1996). Site-directed mutagenesis of *AMT1* revealed that the single Zn(II) ion is coordinated within the N-terminal 40 residues (Farrell et al., 1996). Ligands for the bound Zn(II) ion include the thiolates of Cys11, Cys14, Cys23 and the imidazole of His25 (Farrell et al., 1996; Posewitz et al., 1996). Synthetic peptides encompassing residues 1–42 of Amt1 and Ace1 form stably folded complexes in the presence of 1 mol eq. Zn(II).

The tetracopper domain of Amt1 is enfolded by residues 41–110 (Graden et al., 1996). Expression of a truncated Amt1 (residues 37–110) in bacteria resulted in the isolation of a Cu(I)-containing complex with 4 mol eq. Cu(I) bound (Graden et al., 1996). Spectral analyses of the truncated Cu$_4$Amt1 complex suggested that the four Cu(I) ions were bound similarly as in the intact Cu,ZnAmt1 complex. The truncated CuAmt1 complex exhibits high affinity and specific DNA binding (Graden et al., 1996). Thus, Amt1 and Ace1 appear to consist of three separate domains: residues 1–40, the Zn(II) domain; residues 41–110 (100 in Ace1), the tetracopper domain; and residues 110-C-terminal end, the transactivation domain. A model for the organization of Ace1 and Amt1 is shown in Figure 4.

The activation of Amt1 and Ace1 to become DNA-binding proteins occurs by formation of the tetracopper cluster. Thus, the tetracopper domain is functionally a Cu-regulatory domain (CuRD). The CuRD contains 8 cysteinyl residues. Only six thiolates are needed to form a tetracopper cluster based on the known Cu_4S_6 synthetic cage clusters.

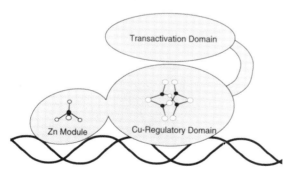

Figure 4. Model of the Cu-activated Ace1 factor. Reversible Cu(I) binding in the Cu-regulatory domain controls the ability of Ace1 to bind to DNA.

The tetracopper clusters in Ace1 and Amt1 may not contain all bridging sulfurs as is observed in the Cu_4S_6 cage clusters. One model is that the Cu-thiolate cage cluster in Amt1 and Ace1 is stabilized by four terminal and four bridging thiolates. Bridging thiolates may be the predominant stabilizing force in the integrity of the tetracopper clusters in Ace1 and Amt1.

Cu(I) activation of Amt1 or Ace1 appears to consist of conversion of this 70 residue CuRD from an apo-conformer or inactive Zn(II) conformer to a structure containing the tetracopper cluster. If the basal state of the regulatory domain is a Zn(II) conformer, then Cu-activation would occur through a metal exchange reaction shown below. A model for such metal exchange is the facile metal exchange kinetics observed in metallothioneins (Otvos et al., 1989).

$$Zn_4MT + Cu_4Zn_1Ace1 \leftrightarrow Cu_7MT + Zn_4Ace1$$

The significance of a tetracopper cluster as the structural unit within the activated transcription factors is three-fold. First, a polycopper cluster formed by eight cysteinyl residues organizes and stabilizes a larger structural unit than a single bound metal ion. A single Cu(I) site expectedly is three or four coordinate, and therefore, would anchor the polypeptide in only three or four places rather than eight anchor sites in the candidate Cu-regulatory domain of Ace1 and Amt1.

A second significant aspect of a tetracopper cluster in the Cu-regulatory domain is that a polycopper cluster provides metal ion specificity. Polymetal clusters are also known for Zn(II) and Cd(II) ions, but these clusters are structurally distinct from the polycopper clusters (Dance, 1986).

The third important feature is the observed cooperativity in cluster formation (Casas-Finet et al., 1992; Dameron et al., 1991; Thorvaldsen et al., 1994). Cooperativity in cluster formation may be significant in that it permits a direct coupling of the intracellular exchangeable Cu ion concentration to transcriptional activation of *CUP1* and to a lesser extent *CRS5* and *SOD1*. Cells can respond to small increases in copper ion concentration to activate Ace1 and therefore enhance MT biosynthesis.

D. COPPER REGULATION OF Mac1 FUNCTION

A polycopper cluster may likewise be important in the Cu-dependent inhibition of gene expression in *S. cerevisiae*. As mentioned earlier, Cu-dependent repression of *CTR1* and *FRE1* occurs at the level of transcription in an Ace1-independent manner (Dancis et al., 1994; Hassett and Kosman, 1995; Yuan et al., 1995). The *trans*-acting factor responsible for Cu-repression of *CTR1* and *FRE1* is Mac1 (Jungmann et al., 1992; Yamaguchi-Iwai et al., 1997). Mac1 contains two cysteine-rich motifs consisting of CxCxxxxCxCxxCxxH sequence repeats that resembles Cu(I) binding cysteinyl sequence motifs found in Ace1 and metallothioneins (Jungmann et al., 1992).

The Cys-rich motifs in Mac1 are adjacent to candidate transactivation domains (TAD) in Mac1. The proximity of the Mac1 Cys-rich motifs to a candidate acidic TAD led us to postulate that the activity of Mac1 TAD(s) may be copper-regulated. We confirmed that Mac1 possesses a Cu-regulated TAD by using a Gal4/Mac1 fusion protein in which the DNA-binding domain of Gal4 (residues 1–147) was fused in frame to residues 42–417 of Mac1 (Graden and Winge, 1997). The first 40 residues of Mac1 were deleted since they may likely contribute to DNA-binding. The first 40 residues of Mac1 are homolgous to the

Zn modules of Ace1 and Amt1. The Gal4 fragment does not contain a TAD, therefore, it fails to trans-activate expression of reporter genes cloned downstream of a promoter element containing Gal4 binding sites. Expression of the Gal4/Mac1 fusion protein in a cell containing p*GAL4*/lacZ resulted in expression of ß-galactosidase implying that residues 42–417 of Mac1 contain transcriptional activation activity (Graden and Winge, 1997). Expression of ß-galactosidase was Cu-regulated (Graden and Winge, 1997). Under conditions of limited copper ion uptake into cells, ß-galactosidase activity was markedly elevated; whereas cells from copper-supplemented medium exhibited low ß-galactosidase levels. Cu(I) ions bind to Mac1 in Cu(I)-thiolate coordination similar to that observed in Ace1 and MT. Reversible Cu(I) binding may be important in regulating the activity of the TAD domain(s) in Mac1. Regulation of a transactivation domain is a novel form of cellular regulation. Our working model is that Mac1 Cu-specific inactivation occurs through formation of a polycopper cluster.

The C-terminal segment of Mac1 contains two Cys-rich motifs. Only the first Cys-rich motif appears to be important in Cu-modulation of TAD activity in Mac1 (Graden and Winge, 1997). It is not clear at this time whether Cu attenuates other functions of Mac1. Our preliminary data suggest that Mac1 is a Cu-binding protein and that Mac1 binds multiple Cu(I) ions raising the possibility of a polycopper center. The intriguing question is by what mechanism Cu-represses the function of Mac1 at an extracellular Cu concentration of 1 nM, whereas Ace1 requires Cu levels in excess of 1 μM for optimal activity.

Mac1 appears to be fully functional only under Cu-deficient growth conditions. Cells exposed to inadequate copper growth conditions would result in derepression of Mac1 and subsequent enhanced expression of genes whose products are involved in copper ion uptake. Thus, activation of Mac1 by copper limitation is a copper starvation response.

In summary, copper ion homeostasis in yeast is maintained through regulated expression of genes involved in copper ion uptake, Cu(I) sequestration and defense against reactive oxygen intermediates. Positive and negative copper ion regulation is observed, and both effects are mediated by Cu(I)-sensing transcription factors. The mechanism of Cu(I) regulation is distinct for transcriptional activation verses transcriptional repression.

Cu(I) activation of gene expression in *S. cerevisiae* and *C. glabrata* occurs through Cu-regulated DNA binding. The activation process involves Cu(I) cluster formation within a regulatory domain in Ace1 and Amt1. In contrast, Cu(I) inhibition of gene expression in *S. cerevisiae* appears to occur through Cu-dependent repression of Mac1 function. One function that is clearly repressed is the TAD activity of Mac1 (Graden and Winge, 1997).

Several major questions concerning intracellular copper homeostasis persist that will stimulate future research. One question in yeast is how Cu(I) ions are routed to the nucleus for Ace1 and Mac1 binding. A second unresolved question is elucidation of the structural basis of how Cu(I) binding to Mac1 represses Mac1 function. A third question is why Mac1 is Cu-regulated in cells exposed to nM Cu(II), whereas Ace1 is only activated in cells exposed to μM Cu(II). The identification of Cu ion sensors in animal cells is a significant future goal.

REFERENCES

Askwith, C., Eide, D., Van Ho, A., Bernard, P. S., Li, L., Davis-Kaplan, S., Sipe, D. M., and Kaplan, J. (1994). The *FET3* gene of *S. cerevisiae* encodes a multicopper oxidase required for ferrous iron uptake. Cell 76, 403–410.

Buchman, C., Skroch, P., Dixon, W., Tullius, T. D., and Karin, M. (1990). A single amino acid change in CUP2 alters its mode of DNA binding. Mol. Cell. Biol. *10*, 4778–4787.

Buchman, C., Skroch, P., Welch, J., Fogel, S., and Karin, M. (1989). The CUP2 gene product, regulator of yeast metallothionein expression, is a copper activated DNA-binding protein. Mol. Cell. Biol. *9*, 4091–4095.

Butt, T. R., Sternberg, E. J., Gorman, J. A., Clark, P., Hamer, D., Rosenberg, M., and Crooke, S. T. (1984). Copper metallothionein of yeast, structure of the gene, and regulation of expression. Proc. Natl. Acad. Sci. U.S.A. *81*, 3332–3336.

Casas-Finet, J. R., Hu, S., Hamer, D., and Karpel, R. L. (1992). Characterization of the copper- and silver-thiolate clusters in N- terminal fragments of the yeast ACE1 transcription factor capable of binding to its specific DNA recognition sequence. Biochemistry *31*, 6617–26.

Culotta, V. C., Howard, W. R., and Liu, X. F. (1994). *CRS5* encodes a metallothionein-like protein in *Saccharomyces cerevisiae*. J. Biol. Chem. *269*, 1–8.

Culotta, V. C., Joh, H. D., Lin, S. J., Slekar, K. H., and Strain, J. (1995). A physiological role for *Saccharomyces cerevisiae* copper/zinc superoxide dismutase in copper buffering. J Biol Chem *270*, 29991–29997.

Dameron, C. T., Winge, D. R., George, G. N., Sansone, M., Hu, S., and Hamer, D. (1991). A copper-thiolate polynuclear cluster in the ACE1 transcription factor. Proc. Natl. Acad. Sci. USA *88*, 6127–6131.

Dance, I. G. (1986). The structural chemistry of metal thiolate complexes. Polyhedron *5*, 1037–1104.

Dancis, A., Haile, D., Yuan, D. S., and Klausner, R. D. (1994). The *Saccharomyces cerevisiae* copper transport protein (Ctr1p). J. Biol. Chem. *269*, 25660–25667.

Dobi, A., Dameron, C. T., Hu, S., Hamer, D., and Winge, D. R. (1995). Distinct regions of Cu(I):ACE1 contact two spatially resolved DNA major groove sites. J. Biol. Chem. *270*, 10171–10178.

Farrell, R. A., Thorvaldsen, J. L., and Winge, D. R. (1996). Identification of the Zn(II) site in the copper-responsive yeast transcription factor, AMT1: A conserved Zn module. Biochemistry *35*, 1571–1580.

Fogel, S., and Welch, J. W. (1982). Tandem gene amplification mediates copper resistance in yeast. Proc. Natl. Acad. Sci. USA *79*, 5342–5346.

Fridovich, I. (1978). The biology of oxygen radicals. Science *201*, 875–880.

Furst, P., Hu, S., Hackett, R., and Hamer, D. (1988). Copper activates metallothionein gene transcription by altering the conformation of a specific DNA binding protein. Cell *55*, 705–717.

Graden, J. A., Posewitz, M. C., Simon, J. R., George, G. N., Pickering, I. J., and Winge, D. R. (1996). Presence of a copper(I)-thiolate regulatory domain in the copper-activated transcription factor Amt1. Biochemistry *35*, 14583–14589.

Graden, J. A., and Winge, D. R. (1997). Copper-mediated Repression of the Activation Domain in the Yeast Mac1p Transcription Factor. Proceedings of the National Academy of Sciences, USA *94*, 5550–5555.

Gralla, E. B., Thiele, D. J., Silar, P., and Valentine, J. S. (1991). ACE1, a copper-dependent transcription factor, activates expression of the yeast copper, zinc superoxide dismutase gene. Proc Natl Acad Sci U S A *88*, 8558–62.

Hahn, S. (1993). Structure (?) and function of acidic transcription activators. Cell *72*, 481–483.

Hamer, D. H., Thiele, D. J., and Lemontt, J. E. (1985). Function and autoregulation of yeast copperthionein. Science *228*, 685–690.

Hassett, R., and Kosman, D. J. (1995). Evidence for Cu(II) reduction as a component of copper uptake by *Saccharomyces cerevisiae*. J. Biol. Chem. *270*, 128–134.

Hu, S., Furst, P., and Hamer, D. (1990). The DNA and Cu binding functions of ACE1 are interdigitated within a single domain. New Biologist *2*, 544–555.

Jensen, L. T., Howard, W. R., Strain, J. J., Winge, D. R., and Culotta, V. C. (1996). Enhanced effectiveness of copper ion buffering by *CUP1* metallothionein compared with *CRS5* metallothionein in *Saccharomyces cerevisiae*. J. Biol. Chem. *271*, 18514–18519.

Jungmann, J., Reins, H. A., and Jentsch, S. (1992). The S. Cerevisiae MFT1 Gene Encodes an ACE1-Regulated Putative Transcription Factor Involved in the Regulation of the Ubiquitin System. In 16th Int. Conf. On Yeast Genetics and Molecular Biology: John Wiley & Sons Ltd.).

Jungmann, J., Reins, H. A., Lee, J., Romeo, A., Hassett, R., Kosman, D., and Jentsch, S. (1993). MAC1, a nuclear regulatory protein related to Cu dependent transcription factors is involved in Cu/Fe utiltization and stress resistance in yeast. EMBO J. *12*, 5061–5056.

Kampfenkel, K., Kushnir, S., Babiychuk, E., Inze, D., and Van Montagu, M. (1995). Molecular characterization of a putative Arabidopsis thaliana copper transporter and its yeast homologue. J Biol Chem *270*, 28479–28486.

Karin, M., Najarian, R., Haslinger, A., Valenzuela, P., Welch, J., Fogel, S., and lec (1984). Primary structure and transcription of an amplified genetic locus: the *CUP1* locus of yeast. Proc. Natl. Acad. Sci. USA *81*, 337–341.

Labbe, S., Zhu, Z., and Thiele, D. J. (1997). Copper-specific Transcriptional Repression of Yeast Genes Encoding Critical Components in the Copper Transport Pathway. J. Biol. Chem. 272, 15951–15958.

Lesuisse, E., Casteras-Simon, M., and Labbe, P. (1996). Evidence for the Saccharomyces cerevisiae Ferrireductase System Being a Multicomponent Electron Transfer Chain. Journal of Biological Chemistry 271, 13578–13583.

Mehra, R. K., Garey, J. R., Butt, T. R., Gray, W. R., and Winge, D. R. (1989). Candida glabrata Metallothioneins: Cloning and Sequence of the Genes and Characterization of Proteins. J. Biol. Chem. 264, 19747–19753.

Mehra, R. K., Thorvaldsen, J. L., Macreadie, I. G., and Winge, D. R. (1992). Disruption analysis of metallothionein-encoding genes of Candida glabrata. Gene 114, 75–80.

Munder, T., and Furst, P. (1992). The Saccharomyces cerevisiae CDC25 gene product binds specifically to catalytically inactive ras proteins in vivo. Mol Cell Biol 12, 2091–9.

Narula, S. S., Mehra, R. K., Winge, D. R., and Armitage, I. M. (1991). Establishment of the metal-to-cysteine connectivities in silver substituted yeast metallothionein. J. Amer. Chem. Soc. 113, 9354*9358.

Ooi, C. E., Rabinovich, E., Dancis, A., Bonifacino, J. S., and Klausner, R. D. (1996). Copper-dependent degradation of the Saccharomyces cerevisiae plasma membrane copper transporter Ctr1p in the apparent absence of endocytosis. The EMBO Journal 15, 3515–3523.

Otvos, J. D., Petering, D. H., and Shaw, C. F. (1989). Structure-Reactivity Relationships of Metallothionein, a Unique Metal-Binding Protein. Comments in Inorganic Chemistry 9, 1–35.

Peterson, C. W., Narula, S. S., and Armitage, I. M. (1996). 3D solution structure of copper and silver-substituted yeast metallothioneins. Febs Lett 379, 85–93.

Pickering, I. J., George, G. N., Dameron, C. T., Kurz, B., Winge, D. R., and Dance, I. G. (1993). X-ray Absoption Spectroscopy of Cuprous-Thiolate Clusters in Proteins and Model Systems. Journal American Chemical Society 115, 9498–9505.

Posewitz, M. C., Simon, J. R., Farrell, R. A., and Winge, D. R. (1996). Role of the conserved histidines in the Zn module of the copper-activated transcription factors in yeast. Journal of Bioinorganic Chemistry 1, 560–566.

Sievers, C., Deters, D., Hartmann, H., and Weser, U. (1996). Stable Thiyl Radicals in Dried Yeast Cu(I)6-Thionein. Journal of Inorganic Biochemistry 62, 199–205.

Stearman, R., Yuan, D. S., Yamaguchi-Iwai, Y., Klausner, R. D., and Dancis, A. (1996). A permease-oxidase complex involved in high-affinity iron uptake in yeast. Science 271, 1552–1557.

Tamai, K. T., Gralla, E. B., Ellerby, L. M., Valentine, J. S., and Thiele, D. J. (1993). Yeast and mammalian metallothioneins functionally substitute for yeast copper-zinc superoxide dismutase. Proc Natl Acad Sci U S A 90, 8013–7.

Thiele, D. J. (1988). ACE1 regulates expression of the Saccharomyces cerevisiae metallothionein gene. Mol. Cell. Biol. 8, 2745–2752.

Thorvaldsen, J. L., Sewell, A. K., Tanner, A. M., Peltier, J. M., Pickering, I. J., George, G. N., and Winge, D. R. (1994). Mixed Cu(I), Zn(II) coordination in the DNA binding domain of AMT1 transcription factor from Candida glabrata. Biochemistry 33, 9566–9577.

Velculescu, V. E., Zhang, L., Zhou, W., Vogelstein, J., Basrai, M. A., Bassett, D. E., Hieter, P., Volgelstein, B., and Kinzler, K. W. (1997). Characterization of the Yeast Transcriptome. Cell 88, 243–251.

Yamaguchi-Iwai, Y., Dancis, A., and Klausner, R. D. (1995). AFT1: a mediator of iron regulated transcriptional control in Saccharomyces cerevisiae. Embo J 14, 1231–1239.

Yamaguchi-Iwai, Y., Serpe, M., Haile, D., Yang, W., Kosman, D. J., Klausner, R. D., and Dancis, A. (1997). Homeostatic Regulation of Copper Uptake in Yeast via Direct Binding of MAC1 Protein to Upstream Regulatory Sequences of FRE1 and CTR1. J. Biol. Chem. 272, 17711–17718.

Yuan, D. S., Stearman, R., Dancis, A., Dunn, T., Beeler, T., and Klausner, R. D. (1995). The Menkes/Wilson disease gene homologue in yeast provides copper to a ceruloplasmin-like oxidase required for iron uptake. Proc Natl Acad Sci U S A 92, 2632–2636.

Zhao, H., and Eide, D. (1996). The yeast ZRT1 gene encodes the zinc transporter protein of a high-affinity uptake system induced by zinc limitation. Proc. Natl. Acad. Sci. USA 93, 2354–2458.

Zhou, P., and Thiele, D. J. (1991). Isolation of a metal-activated transcription factor gene from Candida glabrata by complementation in Saccharomyces cerevisiae. Proc. Natl. Acad. Sci. USA 88, 6112–6116.

22

INTRACELLULAR PATHWAYS OF COPPER TRAFFICKING IN YEAST AND HUMANS

Valeria Cizewski Culotta,[1] Su-Ju Lin,[1] Paul Schmidt,[1] Leo W. J. Klomp,[2] Ruby Leah B. Casareno,[2] and Jonathan Gitlin[2]

[1]Department of Environmental Health Sciences
Johns Hopkins University School of Public Health
615 N. Wolfe St.
Baltimore, Maryland 21205
[2]Edward Mallinckrodt Department of Pediatrics
Washington University School of Medicine
St. Louis, Missouri 63110-1014

ABSTRACT

In the bakers yeast *S. cerevisiae*, there at least four intracellular targets requiring copper ions- 1) Ccc2p and Fet3p in the secretory pathway (homologues to Menkes/Wilson proteins and ceruloplasmin); 2) cytochrome oxidase in the mitochondria; 3) copper transcription factors in the nucleus; and 4) Cu/Zn superoxide dismutase (SOD1) in the cytosol. We have discovered a small soluble copper carrier that specifically delivers copper ions to the secretory pathway. This 8.2 kDa factor known as Atx1p, exhibits striking homology to the MERp mercury carrier of bacteria and contains a single MTCXXC metal binding site also found in the Menkes/Wilson family of copper transporting ATPases. Our studies show that Atx1p is cytosolic and facilitates the delivery of copper ions from the cell surface copper transporter to Ccc2p and Fet3p in the secretory pathway; furthermore, it is not involved in the delivery of copper ions to the mitochondria, the nucleus or cytosolic SOD1, implicating specific signals directing Atx1p to the secretory pathway. Homologues to Atx1p have been found in invertebrates, plants and humans, and the human gene is abundantly expressed in all tissues. In addition to Atx1p, we have recently uncovered an additional metal trafficking protein that appears to specifically deliver copper ions to SOD1. Mutants in the corresponding gene (*lys7*) are defective for SOD1 activity, and are unable to incorporate copper into SOD1, while there is no obvious impairment in copper delivery to cytochrome oxidase or Fet3p. The encoded 27 kDa protein contains a single MHCXXC consensus copper binding sequence and close homologues have been identified in a wide array of eukaryotic species including humans.

BACKGROUND

There are least four biological targets of copper trafficking that have been identified in the bakers yeast *Saccharomyces cerevisiae* that receive copper from the cell surface copper transporters Ctr1 or Ctr3 (Dancis et al., 1994; Knight et al., 1996): 1) Fet3p a multi copper oxidase involved in iron uptake (Askwith et al., 1994; De-Silva et al., 1995; Yuan et al., 1995; Kaplan and O'Halloran, 1996; Stearman et al., 1996); 2) cytochrome oxidase of the respiratory chain; 3) copper containing transcription factors (e.g., Ace1p and Mac1p) in the nucleus (Furst et al., 1988; Thiele, 1988; Buchman et al., 1989; Jungmann et al., 1993); and 4) copper and zinc requiring superoxide dismutase (SOD1). These four copper requiring targets are harbored in four distinct cellular locations in the cell. Fet3p must obtain its copper as it passes through a late Golgi compartment (Yuan et al., 1997), cytochrome oxidase is assembled in the mitochondria, the copper transcription factors require nuclear copper, and SOD is predominantly a cytosolic enzyme. Consequently, it is important that sufficient copper is delivered to diverse locations throughout the cell. In certain instances, intracellular membranes pose a barrier to copper delivery. This problem is solved in part through the evolution of intracellular copper transporters. For example, a P-type copper transporting ATPase in the secretory pathway, known as yeast Ccc2p, is important for the delivery of copper into the late Golgi compartment for activation of Fet3p (Fu et al., 1995; Yuan *et al.*, 1995; Kaplan and O'Halloran, 1996; Yuan *et al.*, 1997). The functional homologues to Ccc2p include the human copper transporting ATPases affected in Wilson and Menkes diseases (Bull et al., 1993; Chelly et al., 1993; Mercer et al., 1993; Nulpe et al., 1993; Yamaguchi et al., 1993). The Sco1p and Sco2p mitochondrial membrane proteins may be important for proper delivery of copper to cytochrome oxidase across the mitochondrial membranes (Glerum et al., 1996).

In addition to the plasma membrane and intracellular membrane transporters for copper, there appears to be a third tier of intracellular copper transport. This involves the action of a family of low molecular weight, soluble, copper binding proteins. We refer to this class of proteins as "copper chaperones". These proteins appear to act as personal escorts for the metal and guide the copper ions across the cytosol to their proper cellular destination. The first of these to be described, Cox17p, was discovered by Glerum and Tzagoloff. This cysteine rich protein is important for the delivery of copper to cytochrome oxidase in the mitochondria, but does not deliver copper to other cellular locations such as Fet3p in the secretory pathway (Glerum et al., 1996). In this report, we describe the identification and characterization of two additional chaperones for copper: Atx1p that transports copper to Ccc2p and Fet3p; and Lys7p that delivers copper to SOD1.

RESULTS AND DISCUSSION

The *ATX1* Gene

Saccharomyces cerevisiae ATX1 was originally identified as a gene which when present in multiple copies, would overcome oxidative damage in yeast lacking SOD1. *sod1Δ* mutants of yeast exhibit many defects related to oxygen, including auxotrophies for the amino acids lysine and methionine when grown in air (Bilinski et al., 1985; Chang and Kosman, 1990; Liu et al., 1992). The molecular nature of these defects has not been completely resolved. The methionine auxotrophy appears to reflect a lowering of cellular NADPH levels (Slekar et al., 1996), yet the aerobic growth requirement for lysine remains

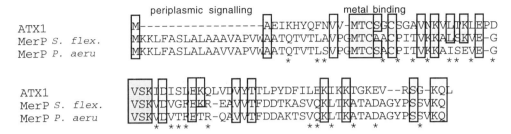

Figure 1. The sequence of the Atx1p polypeptide comparatively aligned against the MerP polypeptides of *Shigella flexneri* (*S. flex*) and *Pseudomonas aeruginosa* (*P. aeru*). The periplasmic signal sequence of MerP is indicated and the consensus metal binding region is shaded. Boxes indicate regions of amino acid identity and asterisks mark similar residues.

enigmatic. Never the less, these aerobic auxotrophies have served as excellent biomarkers for oxidative stress in yeast lacking SOD1. *ATX1* was originally isolated in a screen for genes which bypassed the lysine requirement of *sod1Δ* yeast; subsequent studies demonstrated that *ATX1* in fact overcomes all the markers of oxidative stress in these yeast including a sensitivity to the O_2^- generating agent, paraquat (Lin and Culotta, 1995). We therefore postulated that *ATX1* was a new anti-oxidant factor.

Sequence analysis of *ATX1* revealed that the gene encodes a 8.2 kDa protein with a striking resemblance to MerP (Summers, 1986; Sahlman and Skarfstad, 1993), a periplasmic mercury carrier in bacteria (Fig. 1). With the exception of the MerP periplasmic signaling sequence, the homology between Atx1p and MerP extends throughout the corresponding polypeptides. Like MerP, Atx1p contains in its amino terminal segment, a single copy of the well conserved MTCXXC metal binding motif known to bind mercury (in the case of MerP) and copper (in the case of the copper transporting ATPases). We failed to observe any role of yeast Atx1p in mercury detoxification. However, Atx1p function was found to be completely dependent upon copper ions. The ability of multi-copy *ATX1* to guard against oxidative damage was reversed in yeast cells starved for copper by treatment with batho cuprione sulfate (BCS). The anti-oxidant function of *ATX1* was also abolished in cells missing the cell surface transporter, Ctr1p (Lin and Culotta, 1995).

Our subsequent studies indicated that physiological levels of Atx1p function in the delivery of copper to Fet3p in the secretory pathway (Lin et al., 1997). First, cells containing an *atx1Δ* mutation were found to be defective in high affinity iron uptake (Lin *et al.*, 1997). Secondly, these mutants exhibit defective incorporation of ^{64}Cu into Fet3p in a metabolic labeling study (Klomp et al., 1997). Our localization studies demonstrated that Atx1p is cytosolic and together with genetic epistasis studies, we have established a pathway of copper transport that flows from Ctr1p to Atx1p to Ccc2p and Fet3p (Lin *et al.*, 1997).

To address whether the delivery of copper to Ccc2p and Fet3p was responsible for the apparent anti-oxidant activity of Atx1p, we tested whether multi-copy *ATX1* could suppress oxidative damage in a *sod1Δ* mutant also containing null mutations in either *CCC2* or *FET3*. We observed that *ATX1* could still suppress *sod1Δ*-related oxygen toxicity in the absence of a functional Ccc2p or Fet3p (Lin *et al.*, 1997). Hence, *ATX1* appears to function in two independent pathways of copper trafficking: 1) the delivery of copper to the secretory pathway, and 2) the delivery of copper to a novel target that uses copper in an anti-oxidant capacity.

Although Atx1p has dual functions, it is highly specific for its target(s) and does not deliver copper to proteins in the mitochondria, nucleus or to SOD1 in the cytosol. What is the nature of this remarkable specificity? Our studies show that a direct protein-protein interaction between Atx1p and its target Ccc2p may mediate this specificity. Protein interaction was demonstrated through the use of the yeast two-hybrid system. The entire *ATX1* coding region was fused in frame to the DNA binding domain of the GAL4 trans-activator, and 4 independent segments of Ccc2p were fused in frame to the trans-activation domain of GAL4. These domains represented the four postulated cytoplasmic loops of Ccc2p as predicted from the topology map of this transporter (Fu *et al.*, 1995). The most amino terminal of these domains contains the two MTCXXC copper binding motifs. When the Atx1p containing fusions were co-transformed with each of the four Ccc2-containing fusions, a positive signal for interaction was only obtained with the combination of Atx1p and the most amino terminal region of Ccc2p containing the copper binding consensus sequences (Pufahl et al., 1997). This interaction appeared weak and transient compared to signals typically obtained with stably interacting proteins, which is precisely what one might predict with proteins functioning in a copper transfer reaction. Furthermore, this reaction is copper dependent, as treatment with the copper chelator BCS abolished the interaction signal between Atx1p and Ccc2p (Pufahl *et al.*, 1997). Therefore, Atx1p interacts transiently and in a copper dependent fashion with the amino terminal region of Ccc2p containing the MTCXXC copper binding sites. The ability of Atx1 to specifically recognize and interact with Ccc2p presumably mediates the remarkable specificity of this copper chaperone.

Atx1p is not unique to yeast, and homologues have been identified in plants, *C. elegans*, and humans (Lin and Culotta, 1995; Klomp *et al.*, 1997). When expressed in yeast, the human homologue (HAH1 or ATOX1) is able to complement the yeast *atx1* mutant and restore delivery of copper to Ccc2p and Fet3p (Klomp *et al.*, 1997). Like Atx1p, expression of HAH1 in a yeast *sod1*Δ mutant circumvents oxidative damage; therefore, HAH1 exhibits anti-oxidant function (Klomp *et al.*, 1997). It is conceivable that the mammalian targets of HAH1 include the human Wilson and Menkes copper transporting ATPases and a distinct target responsible for the anti-oxidant function of this copper chaperone.

The Copper Chaperone for SOD1

Since neither Cox17p nor Atx1p were responsible for copper delivery to Cu/Zn SOD, we utilized a yeast genetic approach to identify the chaperone responsible for inserting copper into this target molecule. We predicted that a yeast mutant for this chaperone would harbor an inactive SOD1 molecule and would therefore exhibit the same phenotypes as a *sod1*Δ mutant. We noted that the *lys7*Δ mutant of *S. cerevisiae* was not only auxotrophic for lysine, but for methionine as well (Culotta et al., 1997). Furthermore, as is the case with *sod1*Δ mutants, these phenotypes were completely reversed under anaerobic conditions (Culotta *et al.*, 1997). To test whether the *lys7*Δ mutation affected SOD1 activity, we monitored the level of SOD1 polypeptide and SOD1 O_2^- scavenging activity in a *lys7*Δ mutant. We found that despite normal levels of SOD1 protein, *lys7*Δ mutants were devoid of Cu/Zn SOD1 activity. Furthermore, metabolic labeling of cells with ^{64}Cu demonstrated that the SOD1 polypeptide in *lys7*Δ mutants was devoid of copper (Culotta *et al.*, 1997). Therefore, Lys7p is necessary for copper insertion into SOD1 in yeast. Lys7p however is not needed for copper incorporation in cytochrome oxidase, for trans-activation of Ace1p in the nucleus or for copper delivery to Ccc2p and Fet3p in the secretory pathway (Culotta *et al.*, 1997).

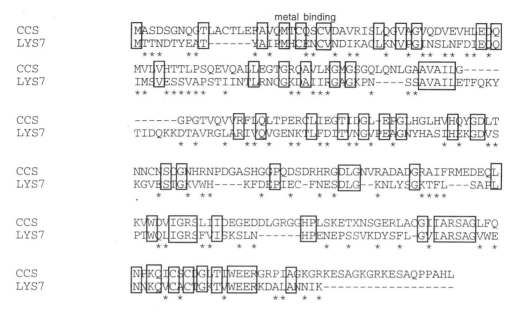

Figure 2. The sequence of the human (CCS) and yeast (LYS7) copper chaperones for superoxide dismutase. The presumed copper binding site is indicated in the shadowed box. Boxes indicate regions of amino acid identity and asterisks mark similar residues.

The human homologue of *LYS7* was identified by inspection of EST data bases and has been denoted as CCS (copper chaperone for SOD1). Upon sequence alignment, nearly 28% identity is shared among the yeast and human gene (Fig. 2). Both genes encode proteins of approximately 30 kDa in mass. The amino terminal region of each protein contains a single copy of the same copper binding motif (MT/HCXXC) that is found in Atx1p and in the copper transporting ATPases Ccc2p, and the Wilson and Menkes proteins. Like Atx1p, we found that Lys7 is soluble and appears to be uniformly distributed throughout the cytosol (Culotta *et al.*, 1997). Other than the metal binding domain, Atx1p and Lys7p share little identity, which presumably accounts for the remarkable specificity of these two chaperones for their cognate targets. Notably, Lys7p exhibits a small degree of homology to SOD1 that has been described as a dimerization domain of the SOD (Carlsson et al., 1996). Furthermore, the mid-segment of CCS exhibits vast homology to human SOD1, and contains three out of the four histidines that serve as copper binding ligands (the fourth is replaced by a glutamate). This suggested that CCS may represent a bi-functional molecule that is both a copper chaperone and a SOD.

To test whether CCS is a SOD, the cDNA was placed under the control of a strong constitutive *S.cerevisiae* promoter and was expressed in a *sod1Δ* mutant. Expression of CCS failed to complement the *sod1Δ* mutation and furthermore, no SOD activity was associated with this expressed CCS molecule (Culotta *et al.*, 1997). However, expression of CCS in this manner fully complemented a *lys7Δ* mutation and restored aerobic growth on medium lacking lysine and methionine. Furthermore, CCS was active for incorporating ^{64}Cu into SOD1 in a strain lacking Lys7p (Culotta *et al.*, 1997). Therefore, CCS is fully functional as a copper chaperone for SOD1, but exhibits no detectable SOD activity. Presumably the strong homology to SOD promotes transient heterodimer formation between CCS and SOD1, facilitating the copper transfer reaction.

The identification of the copper chaperone for SOD1 has important implications with regard to amyotrophic lateral sclerosis (ALS). This fatal, adult onset disorder causes rapid and progressive degeneration of the motor neurons, and certain familial cases have been attributed to dominant mutations in Cu/Zn SOD1 (Deng et al., 1993; Rosen et al., 1993). These mutations (nearly 50 have been described to date) occur throughout the polypeptide and are believed to confer a new toxic property to the SOD1 polypeptide that is responsible for motor neuron death (Gurney et al., 1994; McCord and Gurney, 1994; Brown, 1995; Ripps et al., 1995). Although the mechanism of this toxicity is not completely clear, evidence to date implicates the bound copper ion (Beckman et al., 1993; Wiedau-Pazos et al., 1996; Yim et al., 1996; Bruijin et al., 1997; Yim et al., 1997). Hence, it is predicted that SOD1-linked ALS would be ameliorated by agents that block copper entry into the polypeptide. Our identification of CCS will now provide the tools necessary to test this hypothesis. Knock out mice for CCS are currently being constructed and by mating these mice to transgenic mouse models for SOD1-linked ALS (Gurney *et al.*, 1994; McCord and Gurney, 1994; Ripps *et al.*, 1995; Bruijin *et al.*, 1997), one can directly test whether the metal is responsible for motor neuron degeneration. As such, these studies may lead to the discovery of novel therapeutic targets for SOD1-linked ALS.

OVERVIEW

The incorporation of copper into polypeptides in vivo is a complex process and requires the action of multiple chaperones that bind and escort the metal to its proper destination. Studies thus far indicate that these chaperones are highly specific for their targets and there is no apparent cross reactivity. Atx1p is specific for Ccc2p and this specificity appears to be maintained through defined protein-protein interactions between Atx1p and the amino terminal region of Ccc2p. Lys7p is specific for SOD1. Although chaperone-SOD interactions have not been demonstrated, it is conceivable, based on the homology between the human CCS and SOD1, that a transient heterodimer may form to dictate specificity and facilitate copper transfer.

Why the need for chaperones? In cell free systems, copper containing proteins such as SOD1 and Fet3p are readily activated by copper in the absence of additional factors (Beem et al., 1974; Yuan *et al.*, 1995). However, in an intact cell the presence of multiple metal binding proteins can limit copper availability, creating the need for specific chaperones. Furthermore, as copper is also a potentially toxic element, the chaperones may circumvent deleterious reactions of the metal inside the cell. Although our studies have provided strong evidence for copper-specific chaperones, this phenomenon may not be unique to copper. It is highly conceivable that intracellular trafficking of other biologically active trace metals such as iron or zinc is mediated through the action of another family of metal chaperones.

ACKNOWLEDGMENTS

This work was funded by the JHU NIEHS Center and by NIH grants to DK 44464 (to J. D. G.) and GM 50016 (to V.C.C.). J. D. G. is a recipient of a Burroughs Wellcome Fund Scholar Award and P. S. was supported by NIEHS training grant ES 07141.

REFERENCES

Askwith, C., Eide, D., V-Ho, A., Bernard, P. S., Li, L., Davis-Kaplan, S., Sipe, D. M. and Kaplan, J. (1994) The *FET3* gene of *S. cerevisiae* encodes a multicopper oxidase required for ferrous iron uptake. Cell 76, 403–410.

Beckman, J. S., Carson, M. C., Smith, C. D. and Koppenol, W. H. (1993) ALS, SOD and peroxynitrite. Nature 364, 584.

Beem, K. M., Rich, W. E. and Rajagopalan, K. V. (1974) Total reconstitution of copper-zinc superoxide dismutase. J. Biol. Chem. 249, 7298–7305.

Bilinski, T., Krawiec, Z., Liczmanski, L. and Litwinska., J. (1985) Is hydroxyl radical generated by the fenton reaction in vivo? Biochem. Biophys. Res. Comm. 130, 533–539.

Brown, R. H. (1995) Amyotrophic lateral sclerosis: recent insights from genetics and transgenic mice. Cell 80, 687–692.

Bruijin, L. I., Becher, M. W., Lee, M. K., Anderson, K. L., Jenkins, N. A., Copeland, N. G., Sisodia, S. S., Rothstein, J. D., Borchelt, D. R., Price, D. L. and Cleveland, D. W. (1997) ALS-linked SOD1 mutant G85R mediates damage to astrocytes and promotes rapidly progressive disease with SOD1 containing inclusions. Neuron 18, 327–338.

Buchman, C. P., Skroch, P., Welch, J., Fogel, S. and Karin., M. (1989) The *CUP2* gene product, regulator of yeast metallothionein expression, is a copper-activated DNA binding protein. Mol. Cell. Biol. 9, 4091–4095.

Bull, P. C., Thomas, G. R., Rommens, J. M., Forbes, J. R. and Cox, D. W. (1993) The Wilson disease gene is a putative copper transporting P-type ATPase similar to the Menkes gene. Nature Genet. 5, 327–337.

Carlsson, L. M., Marklund, S. L. and Edlund, T. (1996) The rat extracellular superoxide dismutase dimer is converted to a tetramer by the exchange of a single amino acid. Proc. Natl. Acad. Sci. USA 93, 5219–5222.

Chang, E. and Kosman, D. (1990) O_2-dependent methionine auxotrophy in Cu,Zn Superoxide dismutase deficient mutants of *Saccharomyces cerevisiae*. J. Bacteriol. 172, 1840–1845.

Chelly, J., Tumer, Z., Tonnesen, T., Petterson, A., Ishikawa-Brush, Y., Tommerup, N., Horn, N. and Monaco, A. P. (1993) Isolation of a candidate gene for Menkes disease that encodes a potential heavy metal binding protein. Nature Genetics 3, 14–19.

Culotta, V. C., Klomp, L., Strain, J., Casareno, R., Krems, B. and Gitlin, J. D. (1997) The copper chaperone for superoxide dismutase. J. Biol. Chem. 272, 23469–23472.

Dancis, A., Yuan, S., Haile, D., Askwith, C., Eide, D., Moehle, C., Kaplan, J. and Klausner, R. (1994) Molecular characterization of a copper transport protein in *S. cerevisiae*; an unexpected role for copper in iron transport. Cell 76, 393–402.

De-Silva, D. M., Askwith, C. C., Eide, D. and Kaplan, J. (1995) The *FET3* gene product required for high affinity iron transport in yeast is a cell surface ferroxidase. J. Biol. Chem. 270, 1098–1101.

Deng, H. X., Hentati, A., et al. (1993) Amyotrophic lateral sclerosis and structural defects in Cu,Zn superoxide dimutase. Science 261, 1047–1051.

Fu, D., Beeler, T. J. and Dunn, T. M. (1995) Sequence, mapping, and disruption of *CCC2*, a gene that cross-complements the Ca^{2+}-sensitive phenotype of *csg1* mutants and encodes a P-type ATPase belonging to the Cu^{2+}-ATPase subfamily. Yeast 11, 283–293.

Furst, P., Hu, S., Hackett, R. and Hamer, D. (1988) Copper activates metallothionein gene transcription by altering conformation of a specific DNA binding protein. Cell 55, 705–717.

Glerum, D. M., Shtanko, A. and Tzagoloff, A. (1996) Characterization of *COX17*, a yeast gene involved in copper metabolism and assembly of cytochrome oxidase. J. Biol. Chem. 271, 14504–14509.

Glerum, D. M., Shtanko, A. and Tzagoloff, A. (1996) *SCO1* and *SCO2* act as high copy suppressors of a mitochondrial copper recruitment defect in *Saccharomyces cerevisiae*. J. Biol. Chem. 34, 20531–20535.

Gurney, M. E., Pu, H., Chiu, A. U., Canto, M. C. D., Polchow, C. Y., Alexander, D. D., Caliendo, J., Hentati, A., Kwon, Y., Deng, H. S., Ehen, W., Zhai, P., Sufit, R. L. and Siddique, T. (1994) Motor neuron degeneration in mice that expresss a human Cu,Zn superoxide dismutase mutation. Science 264, 1772–1775.

Jungmann, J., Reins, H., Lee, J., Romeo, A., Hassett, R., Kosman, D. and Jentsch, S. (1993) MAC1, a nuclear regulatory protein related to Cu-dependent transcription factors is involved in Cu/Fe utilization and stress resistance in yeast. EMBO J. 13, 5051–5056.

Kaplan, J. and O'Halloran, T. V. (1996) Iron metabolism in eukaryotes: Mars and Venus at it again. Science 271, 1510–1512.

Klomp, L. W. J., Lin, S. J., Yuan, D., Klausner, R. D., Culotta, V. C. and Gitlin, J. D. (1997) Identification and functional expression of HAH1, a novel human gene involved in copper homeostasis. J. Biol. Chem. 272, 9221–9226.

Knight, S., Labbe, S., Kwon, L. F., Kosman, D. J. and Thiele, D. J. (1996) A widespread transposable element masks expression of a yeast copper transport gene. Genes and Develop. 10, 1917–1929.

Lin, S. and Culotta, V. C. (1995) The *ATX1* gene of *Saccharomyces cerevisiae* encodes a small metal homeostasis factor that protects cells against reactive oxygen toxicity. Proc. Natl. Acad. Sci. USA 92, 3784–3788.

Lin, S. J., Pufahl, R., Dancis, A., O'Halloran, T. V. and Culotta, V. C. (1997) A role for the *Saccharomyces cerevisiae ATX1* gene in copper trafficking and iron transport. J. Biol. Chem. 272, 9215–9220.

Liu, X. F., Elashvili, I., Gralla, E. B., Valentine, J. S., Lapinskas, P. and Culotta, V. C. (1992) Yeast lacking superoxide dismutase: isolation of genetic suppressors. J. Biol. Chem. 267, 18298–18302.

McCord, J. M. and Gurney, M. E. (1994) Mutant mice, Cu, Zn superoxide dismutase, and motor neuron degeneration. Science 266, 1586–1587.

Mercer, J. F. B., Liningston, J., Hall, B., Paynter, J. A., Begy, C., Chandrasekharappa, S., Lockhart, P., Grimes, A., Bhave, M., Siemieniak, D. and Glover, T. W. (1993) Isolation of a partial candidate gene for Menkes disease by positional cloning. Nature Genetics 3, 20–25.

Nulpe, C., Levinson, B., Whitney, S., Packman, S. and Gitschier, J. (1993) Isolation of a candidate gene for Menkes disease and evidence that it encodes a copper-transporting ATPase. Nature Genetics 3, 7–13.

Pufahl, R., Singer, C., Peariso, K. L., Lin, S. J., Schmidt, P., Fahrni, C., Culotta, V. C., Penner-Hahn, J. E. and O'Halloran, T. V. (1997) Metal ion chaperone function of the soluble Cu(I) receptor Atx1. Science 278, 853–856.

Ripps, M. E., Huntley, G. W., Hof, P. R., Morrison, J. H. and Gordon, J. W. (1995) Transgenic mice expressing an altered murine superoxide dismutase gene provide an animal model of amyotrophic lateral sclerosis. Proc. Natl. Acad. Sci. USA 92, 689–693.

Rosen, D. R., Siddique, T., et al. (1993) Mutations in Cu/Zn superoxide dismutase gene are associated with familial amyotrophic lateral sclerosis. Nature 362, 59–62.

Sahlman, L. and Skarfstad, E. G. (1993) Mercuric ion binding abilities of MerP variants containing only one cysteine. Biochem. Biophys. Res. Commun. 196, 583–588.

Slekar, K. H., Kosman, D. and Culotta, V. C. (1996) The yeast copper/zinc superoxide dismutase and the pentose phosphate pathway play overlapping roles in oxidative stress protection. J. Biol. Chem. 271, 28831–28836.

Stearman, R., Yuan, D., Yamaguchi-Iwan, Y., Klausner, R. D. and Dancis, A. (1996) A permease-oxidase complex involved in high-affinity iron uptake in yeast. Science 271, 1552–1557.

Summers, A. O. (1986) Organization, expression, and evolution of genes for mercury resistance. Ann. Rev. Microbiol. 40, 607–634.

Thiele, D. (1988) Ace1 regulates expression of the *Saccharomyces cerevisiae* metallothionein gene. Mol. Cell. Biol. 8, 2745–2752.

Wiedau-Pazos, M., Goto, J., Rabizadeh, S., Gralla, E. B., Roe, J. A., Lee, M. K., Valentine, J. S. and Bredesen, D. E. (1996) Altered reactivity of superoxide dismutase in familial amyotrophic lateral sclerosis. Science 271, 515–518.

Yamaguchi, Y., Heiny, M. E. and Gitlin, J. D. (1993) Isolation and characterization of a human liver cDNA as a candidate gene for Wilson disease. Biochem. Biophys. Res. Commun 197, 271–277.

Yim, H., Kang, J., Chock, P. B., Stadtman, E. R. and Yim, M. B. (1997) A familial amyotrophic lateral sclerosis-associated A4V Cu,Zn-superoxide dismutase mutant has a lower K_m for hydrogen peroxide. J. Biol. Chem. 272, 8861–8863.

Yim, M. B., Kang, J. H., Yim, H. S., Kwak, H. S., Chock, P. B. and Stadtman, E. R. (1996) A gain-of-function of an amyotrophic lateral sclerosis-associated Cu,Zn superoxide dismutase mutant: An enhancement of free radical formation due to a decrease in Km for hydrogen peroxide. Proc. Natl. Acad. Sci. USA 93, 5709–5714.

Yuan, D. S., Dancis, A. and Klausner, R. D. (1997) Restriction of copper export in *Saccharomyces cerevisiae* to a late golgi or post-golgi compartment in the secretory pathway. J. Biol. Chem. 272, 25787–25793.

Yuan, D. S., Stearman, R., Dancis, A., Dunn, T., Beeler, T. and Klausner, R. D. (1995) The Menkes/Wilson disease gene homologue in yeast provides copper to a ceruloplasmin-like oxidase required for iron uptake. Proc. Natl. Acad. Sci., USA 92, 2632–2636.

23

COPPER HOMEOSTASIS IN *ENTEROCOCCUS HIRAE*

Haibo Wunderli-Ye and Marc Solioz*

Department of Clinical Pharmacology, University of Berne
3010 Berne, Switzerland

1. INTRODUCTION

Copper acts as a redox-active cofactor in over 30 enzymes by means of its two oxidation states, Cu^+ and Cu^{2+}. But copper can also be very toxic to cells by its ability to form radicals. Thus, copper must be carefully controlled by all cells (Vulpe and Packman, 1995; Linder and Hazegh Azam, 1996). Two key elements of most, if not all, copper homeostatic mechanisms have only recently been discovered and are milestones in field of trace element research: the copper ATPases and the copper chaperones. Copper ATPases were first discovered in the Gram-positive bacteria *Enterococcus hirae* (Odermatt et al., 1992). Similar ATPases were later identified in humans as underlying the copper metabolic defects of Menkes and Wilson disease, respectively. More recently, copper ATPases were cloned from yeast and other organisms and over two dozen putative copper ATPases have been described today (Lutsenko and Kaplan, 1995; Solioz and Vulpe, 1996). In fact, copper ATPase genes have been found in every genome that has been completely sequenced, suggesting that these enzymes are ubiquitous.

The other novel components of copper homeostasis are the copper chaperone. These 8 kDa copper binding proteins, termed CopZ in *E. hirae*, ATX1 in yeast and HAH1 in humans, appear to serve in the intracellular shuttling of copper between components of the copper homeostatic system. The sequences of copper chaperones are highly conserved between species, again suggesting a universal mechanism. Multiple copies of the conserved sequence domains of copper chaperones are also present in the N-termini of larger heavy metal binding proteins, such as copper ATPases, cadmium ATPases, and mercuric reductases. The role of these putative heavy metal binding sites in these proteins is not yet clear.

---

*Mailing address: Marc Solioz, Dept. of Clinical Pharmacology, University of Berne, Murtenstrasse 35, 3010 Berne, Switzerland, Tel. +41 31 632 3268, Fax. +41 31 632 4997, E-mail: marc.solioz@ikp.unibe.ch

Copper Transport and Its Disorders, edited by Leone and Mercer.
Kluwer Academic / Plenum Publishers, New York, 1999.

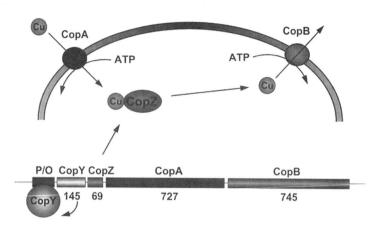

Figure 1. Schematic drawing of the *cop*-operon and model of copper homeostasis in *E. hirae*. CopY is a copper-responsive repressor which binds to the operator/promoter (P/O) region and regulates the expression of itself and that of the other *cop* genes. CopZ is a small copper binding protein that probably functions in the intracellular shuttling of copper between copper transporters and copper requiring proteins. CopA and CopB are copper transporting ATPases that serve in the uptake of copper under copper limiting conditions, and its excretion when copper is in excess, respectively.

In the Gram-positive bacterium *Enterococcus hirae*, the *cop*-operon appears to be essential and sufficient for copper homeostasis. The operon encodes four genes: a copper responsive repressor, CopY, the copper chaperone, CopZ, a copper uptake ATPase, CopA, and a copper exporting ATPase, CopB. Fig. 1 schematically illustrates the *cop*-operon, its gene products and a model of how they effect the control of copper in *E. hirae*. In the following sections, the experimental evidence for the functions of the four *cop*-gene products is discussed, starting with the function of the copper pumps CopA and CopB.

2. COPPER HOMEOSTASIS IN ENTEROCOCCUS

2.1. The CopA Copper Accumulating ATPase

CopA (as well as CopB) belongs to the class of P-type ATPases, classically represented by the Na^+K^+-ATPases and the Ca^{2+}-ATPases of higher cells. The term "P-type" has been adopted because of the phosphorylated intermediate that these enzymes form in the course of the reaction cycle (Pedersen and Carafoli, 1987a; Pedersen and Carafoli, 1987b). In all cases looked at, the phosphorylated residue was found to be an aspartic acid in the conserved sequence DKTGT (given in the one-letter amino acid code, used throughout this article). This sequence, also called the aspartyl kinase domain, is present in all of the more than 100 P-type ATPases sequenced today (Moller et al., 1996). Other motifs common to P-type ATPases are the so called phosphatase domain of consensus sequence TGES, and the ATP binding domains of consensus GCGINDAP (Fagan and Saier, Jr. 1994). While all the putative copper and other heavy metal ATPases appear to be members of the P-type ATPase family, they exhibit some important differences to non-heavy metal ATPases, and evolutionary analysis places them in a separate branch. Heavy metal ATPases exhibit the following striking features not found in non-heavy metal ATPases: (i)

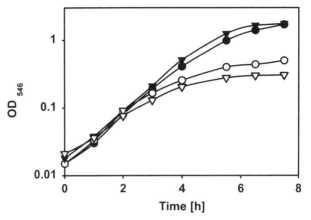

Figure 2. Growth of wild-type and *cop* deletion mutants in copper depleted media. Rich growth media (original copper content: 10 μM) was supplemented with 500 μM 8-hydroxyquinoline and growth followed spectrophotometrically at 546 nm. ●, wild-type; ▼, *cop*B deleted strain; ○, *cop*A deleted strain; ◁, *cop*A and *cop*B deleted strain.

one to six repeat elements with sequence similarity to copper chaperones comprise the N-terminal region, (ii) they have a conserved intramembranous CPC or CPH motif, (iii) they have a conserved HP motif 34 to 43 amino acids C-terminal to the CPC motif, and (iv) they have a different membrane topology. They thus form a distinct subclass of the P-type ATPases (Solioz, 1998; Solioz and Vulpe, 1996; Lutsenko and Kaplan, 1995). Whether the heavy metal ATPases also differ mechanistically from non-heavy metal ATPases will have to be seen.

Current evidence suggests that CopA is responsible for copper uptake under copper limiting conditions. Cells with a disrupted *cop*A gene grow poorly in media reduced in copper with 8-hydroxyquinoline or *o*-phenanthroline (Fig. 2). This growth inhibition could be overcome by adding copper back to the media. The same growth behavior was also displayed by strains deleted in both, CopA and CopB function, while no reduction of growth in copper limited media was observed with cells lacking functional CopB only. Interestingly, Ag$^+$ ions could apparently also enter the cells *via* CopA: cells deleted in CopA function could grow in 5 μM AgNO$_3$, a concentration that completely inhibited the growth of wild-type cells (Odermatt et al., 1993). Since silver can only exist in the oxidation state Ag(I) in solution, it indicates that Cu(I) rather than Cu(II) is transported. Indeed, Cu(I) and Ag(I) have similar ionic radii (1.2 and 1.33 Å, respectively), making transport by the same ATPase feasible.

2.2. The CopB Copper Excretion ATPase

The CopB ATPase of *E. hirae* was found to be required for copper resistant growth of the cells. Wild-type cells can grow in the presence of up to 6 mM CuSO$_4$. Null mutants in *cop*B became sensitive to copper, while null-mutation of *cop*A had no significant effect on copper tolerance. This suggested that the CopB ATPase is a copper export ATPase, extruding excess copper from the cytoplasm of *E. hirae* and thus conferring copper resistance.

Copper transport by CopB was also directly demonstrated. In native inside-out vesicles, CopB was shown to catalyze ATP-driven accumulation of ^{64}Cu$^+$ or ^{110m}Ag$^+$. Accumulation of copper by these vesicles, which would correspond to extrusion of copper by whole cells, was only observed under reducing conditions, indicating that the transported species was copper(I). Use of null-mutants in either *cop*A, *cop*B, or *cop*A and *cop*B made it possible to attribute the observed transport to the activity of the CopB ATPase, as exem-

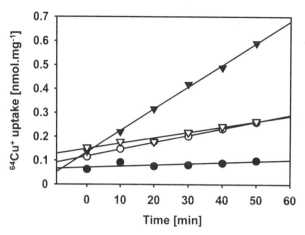

Figure 3. Copper-64 accumulation by native membrane vesicles. The complete reactions contained 0.2 mg/ml of *E. hirae* membrane vesicles in one ml of 100 mM Tris-Cl, 50 mM NaCl, 5 mM MgSO$_4$, 5 mM DTT. At time 0, transport was initiated by the addition of 0.5 mM ATP and followed by filtering and washing 0.1 ml aliquots on 0.45 μm nitrocellulose filters, followed by counting of the retained radioactivity. The copper concentration was 50 nM. Complete reaction (▼), or reaction without ATP (●), or without DTT (○) were measure with delta *copA* membranes; complete reaction measured with delta *copB* membranes (◁).

plified in Fig. 3. Copper transport exhibited an apparent K_m for Cu^+ of 1 μM and a V_{max} of 0.07 nmol/min/mg of membrane protein. $^{110m}Ag^+$ was transported with similar affinity and rate (Solioz and Odermatt, 1995). However, the K_m values must be considered with caution since Cu^+ and Ag^+ were not free in solution but complexed to Tris-buffer and dithiothreitol under the experimental conditions. The results obtained with membrane vesicles were further supported by evidence of $^{110m}Ag^+$ extrusion from whole cells pre-loaded with this isotope. Again, transport depended on the presence of functional CopB (Odermatt et al., 1994).

Vanadate showed an interesting bi-phasic pattern of inhibition of ATP-driven copper and silver transport: maximal inhibition of Cu^+ transport was observed at 40 μM VO_4^{3-} and of Ag^+ transport at 60 μM VO_4^{3-}. Higher concentrations relieved the inhibition of transport. This behavior is unexplained at present. It may relate to the complex chemistry of vanadate that involves several oxidation states (Pope and Dale, 1968).

Taken together, these results suggest that CopA functions as a Cu^+/Ag^+-ATPase for the import of Cu^+ and Ag^+, and CopB as a Cu^+/Ag^+-ATPase for the excretion of Cu^+ and Ag^+. Silver transport by CopA and CopB are probably fortuitous and have no physiological significance: Ag^+ is very toxic and has no known biological role in living cells. In wild-type cells, Ag^+ imported by CopA cannot be extruded efficiently enough by CopB to prevent cell death and therefore constitutes a suicide mechanism.

2.3. The CopY Repressor

Expression of the *E. hirae* *cop*-operon exhibits a unique bi-phasic regulation: induction is lowest in standard growth media (average copper content = 10 μM). In higher media copper, the *cop*-operon is induced, reaching maximal expression (50-fold induction) at 2 mM extracellular copper. Ag^+ or Cd^{2+} at μM concentrations also fully induced expression (Odermatt et al., 1993). The induction by these heavy metal ions is probably fortuitous, since the *cop* genes products do not appear to confer resistance to these toxic metals. Interestingly, strong induction of the *cop*-operon is also observed in copper depleted media (Fig. 4). Since both, the copper accumulating CopA ATPase and the copper excreting CopB ATPase are apparently under the same control, induction by excessive as well as limiting copper would be expected. But why did this co-regulation of CopA and CopB evolve? Conceivably, this is a safety mechanism: if cells would express only the import

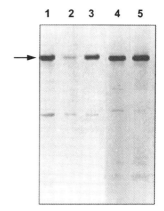

Figure 4. Figure expression of CopB under various copper concentrations. Logarithmically growing cells were challenged with copper or a copper chelator for one hour, followed by lysis and Western blot analysis of CopB expression. The *arrow* indicates the band corresponding to CopB. Standard growth media were supplemented with 100 μM o-phenanthroline (*lane 1*), nothing (*lane2*), 0.1 mM CuSO$_4$ (*lane 3*), 2 mM CuSO$_4$ (*lane 4*), and 6 mM CuSO$_4$ (*lane 5*).

ATPase under copper-limiting conditions, they would become highly vulnerable to copper poisoning by a sudden increase in available copper, as could be induced by a drop in pH.

By chromosome crawling, two genes, *copY* and *copZ*, were identified upstream of the *copA* and *copB* genes (Ochman et al., 1988). They encode predicted proteins of 145 and 69 amino acids, respectively, Disruption of the *copY* gene in *E. hirae* resulted in constitutive overexpression of the *cop*-operon, suggesting that *Cop*Y encodes a repressor protein (Odermatt and Solioz, 1995). The N-terminal half of CopY exhibits around 30% sequence identity to the bacterial repressors of β-lactamases, MecI, PenI and BlaI (Himeno et al., 1986; Suzuki et al., 1993; Hackbarth and Chambers, 1993). In the best studied of these, PenI, the N-terminal portion appears to be the domain that binds DNA (Wittman and Wong, 1988). In the C-terminal half of CopY, there are multiple cysteine residues, arranged as CXCX$_4$CXC (Fig. 5). Such a consensus motif is also found in two

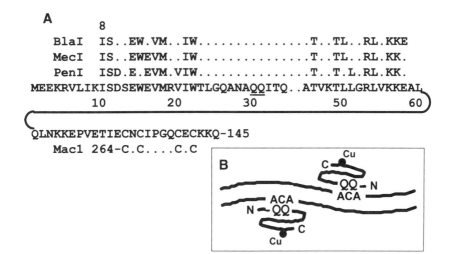

Figure 5. Schematic representation of the CopY repressor. A, the N-terminal sequence similarity with the β-lactamase repressors BlaI, MecI, and PenI of Gram-positive organisms is indicated, as are the C-terminal cysteine patterns that also occur in Mac1p. B, model of the interaction of two CopY monomers with the inverted repeat sequence of the *cop*-operon.. From the X-ray structure of the lambda 434 suppresser, it is known that a pair of glutamine residues forms a tight interaction with an ACA base triplet (Anderson et al., 1987). See text for other details.

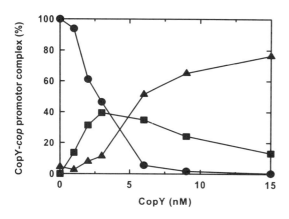

Figure 6. Interaction of the CopY repressor with promoter DNA. A 316 bp ^{32}P-labeled *cop*-promoter DNA fragment (0.15 nM) was incubated with the indicated concentrations of purified CopY repressor, followed by electrophoresis on a 6% non-denaturing polyacrylamide gel as described (Strausak and Solioz, 1997). Free promoter DNA (●), retarded form I (■), and retarded form II (▲) were quantified by phosphoimaging.

cysteine-rich clusters in the C-terminal half of Mac1p. This yeast transcription factor regulates the expression of several genes in response to copper. (Jungmann et al., 1993). It was shown that mutation of the cysteine residues of cluster 1 shown in Fig. 5 abolished copper repression by Mac1p (Graden and Winge, 1997). This would suggest that the corresponding region of CopY interacts with copper, while the N-terminal half of the protein is the DNA binding domain.

Isolated CopY was found to be a stable dimer in solution, as ascertained by crosslinking and gel exclusion chromatography (Strausak and Solioz, 1997). By DNaseI footprinting, CopY could be shown to bind to two regions of a promoter DNA fragment *in vitro*. The binding sites corresponded to the two halves of an inverted repeat that flank the site of initiation of transcription. The interaction of CopY with the promoter was also analyzed in band shift experiments and was found to occur in two steps: at low DNA:repressor ratios, the free DNA fragments were converted to an intermediate slow form I, while at higher repressor concentrations, form I disappeared and an intermediate form II became apparent (Fig. 6). Intermediate forms I and II presumably correspond to the binding of one and two CopY dimers to the promoter DNA. The interaction of CopY with promoter DNA was completely abolished by 50 μM copper. These *in vitro* findings would explain the induction of the *cop*-operon by high copper concentrations, but leave open the mechanism that leads to induction of the *cop*-operon by low copper. This aspect is currently under investigation.

Based on sequence features of CopY, it appears conceivable that the structure of its N-terminal DNA binding domain resembles that of the phage 434 repressor. Both of these repressor proteins possess a glutamine pair at position 30/31 and 29/30, respectively. X-ray structure analysis of the 434 repressor has revealed that this glutamine pair has an exceptionally tight fit with an ACA triplet in the DNA. ACA triplets are present in both CopY binding sites of the *cop*-promoter. It was found that mutation of these ACA triplet would nearly abolished binding of CopY to the respective binding site (Strausak and Solioz, 1997). A double mutant with both ACA triplets mutated to TCA and thus with only very weak binding of CopY to the promoter was analyzed *in vivo*, using a reporter gene. The *cop*-operon of such a double mutant was induced at very low copper concentrations that did not lead to incuction of the wild-type promoter (Fig. 7). This confirms the *in vitro* observations that release of CopY from the promoter turns on expression of the *cop*-operon.

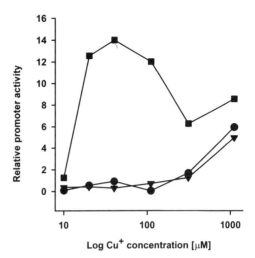

Figure 7. Copper induction of wild-type and mutant *cop*-promoters. Expression of a reporter protein under the control of the *cop*-promoter on a plasmid was analyzed in *E. hirae* wild-type cells. ●, wild-type promoter; ▼, A-30T mutant promoter; ■, A-30T/A-61T double mutant promoter.

2.4. The CopZ Copper Chaperone

CopZ is a 69 amino acid protein that belongs to the new family of copper chaperones that includes MerP, ATX1 and HAH1. CopZ-like motifs are also found in one to six copies in proteins involved in heavy metal metabolism, such as mercuric reductases, or copper and cadmium ATPases. Fig. 8 shows a sequence alignment of CopZ with other copper chaperones and CopZ-like modules of larger proteins. Common to all the sequences is a CXXC motif, but other amino acids are also conserved or conservatively replaced. For MerP, a bacterial periplasmic mercury binding protein, a solution structure is available. The protein has a βαβ-βαβ fold with the two alpha helices overlaying a four-strand antiparallel beta sheet. Mercury(II) is complexed bicoordinately with the Cys side chain ligands of the CXXC motif (Steele and Opella, 1997). Based on this structure, the three-dimensional structure of CopZ was modeled by 'Swiss-Model' (Peitsch, 1996). The resultant structure closely resembles that of MerP, featuring a βαβ-βαβ fold (Fig. 9).

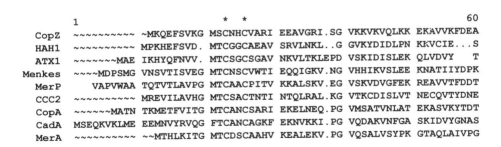

Figure 8. Sequence alignment of CopZ with related metal binding proteins or motifs. EMBL/GenBank accession numbers are given in brackets. *CopZ*, copper chaperone of *E. hirae* (Z46807); *HAH1*, human copper chaperone (U70660); *ATX1*, yeast copper chaperone (L35270); *Menkes*, copper binding motif of human Menkes ATPase (L06133); *MerP*, periplasmic mercury binding protein (P04129); *CCC2*, copper binding motif of yeast CCC2 copper ATPase (L36317); *CopA*, copper binding motif of *E. hirae* CopA copper ATPase (L13292); *CadA*, cadmium binding motif of *Staphylococcus aureus* cadmium ATPase (J04551); *MerA*, mercury binding motif of mercuric reductase (A00406). The *asterisks* denote the universally conserved cysteine residues.

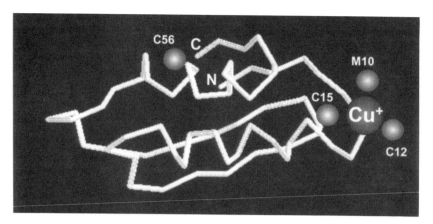

Figure 9. Model of the three-dimensional structure of CopZ. Two α-helices lie on top of a four-stranded antiparallel β-sheet, formed by the N- and C-termini, and a loop between the two α-helices. The calculated positions of the sulfur atoms of M9, C11, C13 and C55 and of a bound copper ion are indicated. The three N-terminal and the C-terminal amino acids of CopZ could not be modeled and were omitted from the structure.

CopZ, ATX1 and HAH1 are members of the new family of intracellular copper carriers, termed 'copper chaperones' (Lin and Culotta, 1995; Klomp et al., 1997). ATX1 and HAH1 were shown to bind copper(I) a the site of entry into the cytoplasm and deliver it to the vesicular CCC2 copper ATPase of yeast. With the yeast two-hybrid system, it could be demonstrated that direct interaction of the copper chaperone with the CopZ-like elements in the N-terminus of the CCC2 copper ATPase occurs (Pufahl et al., 1997). Another chaperone, called LYS7 in yeast and CCS in humans, was found to deliver copper specifically to superoxide dismutase 1 (Pufahl et al., 1997). Finally, COX17 was identified as a yeast protein that specifically delivers copper to cytochrome *c* oxidase in the mitochondria (Glerum et al., 1996). It thus appears that in eukaryotes, a whole family of proteins is required to assure secure delivery of copper to the sites of utilization. CopZ of *E. hirae* most likely takes a similar role. Whether CopZ is the only copper chaperone acting in this bacterium remains to be seen.

3. CONCLUSION

Copper in biological systems presents a formidable problem due to the high reactivity of this transition metal. The discovery of a new subclass of P-type ATPases that can pump copper across biological membranes in either direction has only recently set the stage to study copper homeostasis. Over two dozen putative copper ATPases have been described today. Due to the inherent difficulties of working with copper, the function of most of these ATPases could not yet been demonstrated and much of the evidence for their function remains indirect. ATPases involved in the transport of other heavy metals have recently also joined the heavy metal subclass of P-type ATPases. On an other level, the discovery of different specific copper chaperones that deliver copper intracellularly to the sites of utilization has revealed an unexpected complexity of the copper homeostatic mechanism. It is likely that similar mechanisms also operate for the transport and intracellular distribution of other trace elements and study of these systems may offer many more surprises.

ACKNOWLEDGMENT

The work described here was supported in part by grant 3200-046804 from the Swiss National Foundation.

REFERENCES

Anderson, J.E., Ptashne, M., and Harrison, S.C. (1987). Structure of the repressor-operator complex of bacteriophage 434. Nature *326*, 846–852.
Fagan, M.J. and Saier, M.H., Jr. (1994). P-type ATPases of eukaryotes and bacteria: sequence analyses and construction of phylogenetic trees. J. Mol. Evol. *38*, 57–99.
Glerum, D.M., Shtanko, A., and Tzagoloff, A. (1996). Characterization of *COX17*, a yeast gene involved in copper metabolism and assembly of cytochrome oxidase. J. Biol. Chem. *271*, 14504–14509.
Graden, J.A. and Winge, D.R. (1997). Copper-mediated repression of the activation domain in the yeast Mac1p transcription factor. Proc. Natl. Acad. Sci. USA *94*, 5550–5555.
Hackbarth, C.J. and Chambers, H.F. (1993). *bla*I and *bla*R1 regulate β-lactamase and PBP 2a production in methicillin-resistant *Staphylococcus aureus*. Antimicrob. Agents Chemother. *37*, 1144–1149.
Himeno, T., Imanaka, T., and Aiba, S. (1986). Nucleotide sequence of the penicillinase repressor gene penI of *Bacillus licheniformis* and regulation of *penP* and *penI* by the repressor (published erratum appears in J Bacteriol 1987 Jul; 169(7):3392). J. Bacteriol. *168*, 1128–1132.
Jungmann, J., Reins, H.A., Lee, J.W., Romeo, A., Hassett, R., Kosman, D., and Jentsch, S. (1993). MAC1, a nuclear regulatory protein related to Cu-dependent transcription factors is involved in Cu/Fe utilization and stress resistance in yeast. EMBO J. *12*, 5051–5056.
Klomp, L.W., Lin, S.J., Yuan, D.S., Klausner, R.D., Culotta, V.C., and Gitlin, J.D. (1997). Identification and functional expression of HAH1, a novel human gene involved in copper homeostasis. J. Biol. Chem. *272*, 9221–9226.
Lin, S.J. and Culotta, V.C. (1995). The ATX1 gene of Saccharomyces cerevisiae encodes a small metal homeostasis factor that protects cells against reactive oxygen toxicity. Proc. Natl. Acad. Sci. USA *92*, 3784–3788.
Linder, M.C. and Hazegh Azam, M. (1996). Copper biochemistry and molecular biology. Am. J. Clin. Nutr. *63*, 797S-811S.
Lutsenko, S. and Kaplan, J.H. (1995). Organization of P-type ATPases: Significance of structural diversity. Biochemistry *34*, 15607–15613.
Moller, J.V., Juul, B., and Le Maire, M. (1996). Structural organization, ion transport, and energy transduction of P-type ATPases. Biochim. Biophys. Acta Rev. Biomembr. *1286*, 1–51.
Ochman, H., Gerber, A.S., and Hartl, D.L. (1988). Genetic applications of an inverse polymerase chain reaction. Genetics *120*, 621–623.
Odermatt, A., Suter, H., Krapf, R., and Solioz, M. (1993). Primary structure of two P-type ATPases involved in copper homeostasis in *Enterococcus hirae*. J. Biol. Chem. *268*, 12775–12779.
Odermatt, A., Krapf, R., and Solioz, M. (1994). Induction of the putative copper ATPases, CopA and CopB, of *Enterococcus hirae* by Ag^+ and Cu^{2+}, and Ag^+ extrusion by CopB. Biochem. Biophys. Res. Commun. *202*, 44–48.
Odermatt, A. and Solioz, M. (1995). Two *trans*-acting metalloregulatory proteins controlling expression of the copper-ATPases of *Enterococcus hirae*. J. Biol. Chem. *270*, 4349–4354.
Pedersen, P.L. and Carafoli, E. (1987a). Ion motive ATPases. I. Ubiquity, properties, and significance to cell function. Trends Biochem. Sci. *12*, 146–150.
Pedersen, P.L. and Carafoli, E. (1987b). Ion motive ATPases. II. Energy coupling and work output. Trends Biochem. Sci. *12*, 186–189.
Peitsch, M.C. (1996). ProMod and Swiss-Model: Internet-based tools for automated comparative protein modelling. Biochem. Soc. Trans. *24*, 274–279.
Pope, M.T. and Dale, B.W. (1968). Isopoly-vanadates, -niobates, and tantalates. Rev. Chem. Soc. *22*, 527–545.
Pufahl, R.A., Singer, C.P., Peariso, K.L., Lin, S., Schmidt, P.J., Fahrni, C.J., Culotta, V.C., and Penner-Hahn, J.E. (1997). Metal ion chaperone function of the soluble Cu(I) receptor Atx1 [see comments]. Science *278*, 853–856.
Solioz, M. (1998). Copper homeostasis by CPX-type ATPases, the subclass of heavy metal P-type ATPases. In Ion pumps. J.P. Andersen, ed. (London: JAI Press, Inc.),

Solioz, M. and Odermatt, A. (1995). Copper and silver transport by CopB-ATPase in membrane vesicles of *Enterococcus hirae*. J. Biol. Chem. *270*, 9217–9221.

Solioz, M. and Vulpe, C. (1996). CPx-type ATPases: a class of P-type ATPases that pump heavy metals. Trends Biochem. Sci. *21*, 237–241.

Steele, R.A. and Opella, S.J. (1997). Structures of the reduced and mercury-bound forms of MerP, the periplasmic protein from the bacterial mercury detoxification system. Biochemistry *36*, 6885–6895.

Strausak, D. and Solioz, M. (1997). CopY is a copper-inducible repressor of the *Enterococcus hirae* copper ATPases. J. Biol. Chem. *272*, 8932–8936.

Suzuki, E., Kuwahara Arai, K., Richardson, J.F., and Hiramatsu, K. (1993). Distribution of mec regulator genes in methicillin-resistant Staphylococcus clinical strains. Antimicrob. Agents Chemother. *37*, 1219–1226.

Vulpe, C.D. and Packman, S. (1995). Cellular copper transport. Annu. Rev. Nutr. *15*, 293–322.

Wittman, V. and Wong, H.C. (1988). Regulation of the penicillinase genes of *Bacillus licheniformis*: interaction of the *pen* repressor with its operators. J. Bacteriol. *170*, 3206–3212.

INDEX

*ace*1, 224
   metal binding motif, 155
ACE1/CUP2, 231, 240
   DNA binding actvity, 241
   metal binding motif, 240
   polycopper clusters, 241–243
   X-ray absorption spectroscopy, 241
   zinc binding, 241
Albumin, 1, 4, 30
Alpha antitrypsin deficiency, 141
Alpha-2-macroglobulin, 5
Alzheimer's disease, 183
Ammonia
   concentration, 149
Ammonium tetrathiomolybdate, 140, 143
Amt1, 241
*amt*1, 241
   metal binding motif, 155
   X-ray absorption spectroscopy, 241
Amyloid Aβ protein, 183
Amyloid precursor protein (APP), 183
   aging, 190
   APLP1 gene, 184
   APLP2 gene, 184
   cell growth, 183
   copper binding motif, 184
   Cu(I) complex, 190
   electron transfer reactions, 186
   extracellular matix, 184
   genes, 183
   isoforms, 184
   metal binding sites, 184
   oxidative modifications, 186, 189
   reduction properties, 184, 186
   sulphydryl groups, 187
   trafficking, 188, 189
   zinc binding motif, 184
Amyotrophic lateral sclerosis, 193
AP1, 234
AREI, acceptable range of exposure/intake, 19, 20, 23, 25

Ascorbate oxidase, 175
Ascorbic acid, 220
ATPases, P-type, 33, 153, 255, 256
   localisation, 34
   copper, 99
ATP7A, 54, 84, 110
ATP7A, 40, 84
atp7a, 84
ATP7B, 84, 110
ATP7B, 33, 84
ATX1, 248–250
ATX1, 135, 255, 261
   copper binding motif, 250
Atx1, 154, 161, 166, 247, 250

Bedlington terrier, 131, 141
   copper, brain, 149
   disease forms, 141
   liver histology, 146
Biliary atresia, 140
Blotchy mouse mutant, 101
Brindled mouse mutant, viable, 100, 101
Brefeldin A, 75

Ca^{2+}-ATPase, 156
CaCo-2, 216, 217, 247, 249
   ascorbic acid, 220
   permeability, 219, 220, 221
   pH, 219, 220, 221
Cadmium, 218
CCC2, 78
Ccc2, 60, 262
ccc2, 60
Ceruloplasmin, 1, 6, 7, 175
   antioxidant defense, 7
   copper binding, 175, 176
   copper delivery, 180
   copper uptake, 8
   domains, 179
   ferroxidase activity, 178
   functions, 7, 8

Ceruloplasmin (cont.)
  Indian Childhood Cirrhosis, 135
  inflammation, 8
  iron oxidation, 178
  milk, 9
  redox potential, 178
  redox site, 175
  structure, 175
  Wilson's disease, 140
Chaperone, 33, 135, 250, 261, 262
CopA, 40, 155, 256, 257
copA, 257
CopB, 34, 256, 257
cop B, 257
CopY repressor, 258, 259, 260
copZ, 259
CopZ, 154, 161, 255, 259, 261
Copper
  absorption, 1
  biliary excretion in dogs, 141
  biokinetics, 20
  biomarkers, 20, 21
  chaperone, 33, 135, 250, 261, 262
  chronic exposure, 19
  concentrations, brain, 148
  concentrations, Bedlington terrier, 149
  concentrations, cerebrospinal fluid, 148
  concentrations, infantile cirrhosis, 130
  concentratons, toxic milk mutant, 148
  concentrations, Wilson's disease, 139
  deficiency, 21, 32
  efflux, 155
  enzymes, 166, 237, 255
  excess, 21
  exposure, 18
  glutathione complex, 33
  health risks, 17
  hepatocyte, 30
  injury, nutritional induced, 143, 146
  intake, 18
  histidine, **32**
  liver, 29, 140
  metabolism, 29, 32
  nutrition, 215
  overload, dog, 141
  overload, rodents, 142
  overload, ruminants, 142
  overload, sheep, 149
  peptide binding motifs, 55
  P type ATPases, 40, 55, 85
  resistance, 72
  therapy, 101, 105
  toxicosis in animals, 141, 142, 143
  toxicity, 18, 215
  transport, 1, 2, 31
  uptake, 228
  Wilson's disease, 140
COX17, 135, 262
CRS5, 239, 240

CTR1, 31
Ctr1p protein, 189, 238, 239, 243, 249
Ctr2, 238
Ctr3, 238
Cu(I)
  binding, 162
  luminescence, 158
CUP1, 224, 239, 240
Cystic fibrosis transmembrane protein, 68
Cytochrome c oxidase, 83, 97, 188, 247

Dappled mouse mutant, 99
Diamsar, 32
Diethyldithiocarbamate, 207
Dimethylsulphoxide, 220
Doberman pinscher, 141
Dopamine-β-hydroxylase, 98, 188
Down's syndrome, 206

Enterococcus hirae, 155, 255

Familial amyothropic lateral sclerosis (FALS), 189, 190
  calcium concentrations, 208, 209, 210
  CuZnSOD, 206
  CuZnSOD aggregation, 196
  Japanese form, 207
  lysil tRNA synthetase, 196
  mitochondrial damage, 208, 209
  neurofilament abnormalities, 196
  peroxidative activity, 197, 207
  translocon-associated protein delta, 196
Familial amyothropic lateral sclerosis (FALS), CuZn-SOD mutants
  human mutants, 195, 196
  metal binding properties, 196
  neuroblasoma cell lines, 207
  nitrotyrosine, 199
  peroxidase activity, 197, 207
  protein mutants, 206
  structural properties, 198
  subunits, 199
  transfected forms, 207, 208
  transgenic mice, 196, 197, 199
  tyrosine nitration catalyst, 197
  X-ray structure, 199
  yeast mutants, 195, 197
Fet3p, 60, 78, 238, 247, 249
Fre2, 238, 243
Fre7, 238

Gal 4, 226
Glutathione, 32
Golgi apparatus, 40

HAH1, 36, 135, 255, 261
Heat shock factors, 233
Hepatic encephalopathy, 149
Histidine, 30

Index

Indian Childhood Cirrhosis (ICC), 127, 128, 140
  animal models, 131
  ceruloplasmin, 135
  clinical criteria, 128
  environmental cause, 131
  epidemiology, 129
  genetics, 129, 131
  histological criteria, 128
  pathology, 129
Infantile cirrhosis
  high copper in water, 130
  live copper concentrations, 130
Intestinal cells
  pH, 219
  transport, 219
Iron, excretion, 20

Jax brindled mouse mutant, 102

Keyser–Fleischer ring, 139

Laccase, 175
LEC rat, 84, 142, 143
  copper in the brain, 148
  mutation, 35
Lys7, 251
Lysil oxidase, 53, 78, 83, 98

MAC 1, 238
Mac1, 243
  cysteine domains, 244
Macular mouse mutant, 100
Mannitol, 220
MBF-1, 224
Menkes disease, 39, 53, 83, 84, 109, 189
  carrier determination, 91, 92
  copper accumulation, 98
  clinical features, 97
  cytochrome c oxidase, 97
  lysil oxidase, 98
  metallothionein, 98
  mouse models, 98, 100
  mild form, 97, 98
  patient phenotype, 91
  phenotypic diversity, 105
  prenatal diagnosis, 91
  tyrosinase, 97
Menkes gene, human
  alternative splicing, 46
  cDNA, 39, 42
  chromosomal breakpoints, 86
  expression, 110
  mRNA, 40, 42, 43
  mutations, 85, 86, 88, 91
  organization, 85
  polymorphism, 92
  protein sequence, 60
  transcripts, 91

Menkes protein
  antibodies 56, 70, 79,
  carboxyl tail 61, 69,
  cDNA transcripts, 39, 42, 70
  conformation, 161
  copper titration, 159
  in BeWo cells 39, 48,
  in CHO cells, 55–58, 62, 70, 75
  in Cu resistant cells, 70, 75
  in HeLa cells, 58, 75
  in human fibroblasts, 75
  exocytic pathway, 80
  expression vector, 67
  low copy number expression vector, 68
  metal binding motif, 78, 84, 154, 155
  N-terminal domain, 157, 158, 159, 165
  overexpression, 56
  protein sequence, 54
  regulatory model, 156
  stable expression, 76,
  structure, 40
  subcellular localisation, 40, 56, 57, 58, 62, 73, 104
  subcellular localisation, copper induced, 57, 58
  trafficking, 42, 75
  transient expression, 70
  Western blot, 70
MEP 1, 225
Mercury, 154
MerP, 154, 161, 249, 261
Metal binding motifs, 85, 154, 155, 247
  ACE1, 155
  AMT1, 155
  ATX1, 250
  metallothionein, 155
  Wilson's disease protein, 166
Metal regulatory element (MRE), 224, 225
  binding factors, 224
Metalloreductase/NADH oxidase, 31
Metallothionein 32, 98, 121, 223, 245,
  metal binding motif, 155
  Indian Childhood Cirrhosis-like, 133
  sheep 1a promoter, 68
  transcription, 224, 225
Metals
  gene expression, 223
  transcription, 223
Microtubule organizing center, 75
Mottled blotchy mouse
  mutation, 90
Mottled mouse mutant, 78, 84, 99
  gene expression, 110
  mutation, 90
Mottled pewter mouse
  mutation, 90
Mottled viable brindled mouse
  mutation, 90
Mouse double mutant, 102
MTF-1, 225
  binding site, 229

MTF-1 (*cont.*)
  footprint analysis, 231
  knock outs, 229
  overexpression, 225
  zinc finger, 225

Occipital horn syndrome, 78, 84, 97, 101
  clinical features, 98
  mutation, 90
Occludin, 216

Pennicillamine, 118, 140, 148, 207
Primary biliary cirrhosis, 141
Protein sorting, 60

Reactive oxygen species (ROS), 190, 237

S-adenosyl homocysteine hydrolase, 33, 36
Sheep, 146
  copper neurological lesion, 149
Skye terrier, 141
SOD1, 239
SP1, 229
Superoxide dismutase, Cu, Zn, 83, 188, 189, 190, 193, 194, 205, 237, 247
  antioxidant functions, 194
  copper chaperon, 250
  FALS mutants, 195
  histidine residues, 194
  mutation, 193, 194
  overexpression, 194, 206
  paraquat, 195
  sod1Δ yeast 200, 201
  transgenic mouse, 189, 194

Tetrathiomolybdate, 119, 121
Therapy
  Wilson's disease, 121
  with copper in mottled mouse, 101
  with copper in Menkes disease, 105
Tight junction, 216
  cadmium, 218
  copper, 218, 219
  permeability, 218, 219
Toxic milk mutant mouse, 102, 142, 143
  copper in the brain, 148
  gene expression, 110
Transcuprein, 1, 4, 5, 30
Trans epithelial electrical resistance (TEER), 216, 217, 219
Trans Golgi network, 56, 59, 75
  localisation signal, 60
Transocytosis accessory protein, 45
Trientine, 116
Triethylene tetramine, 140
Trolox, 220
Tyrosinase, 54, 83, 97
Tyrrolean Childhood Cirrhosis (TCC), 127, 128
  animal models, 131

Tyrrolean Childhood Cirrhosis (*cont.*)
  environmental cause, 131
  epidemiology, 129
  genetics, 129, 131, 132
  pathology, 129

Vitamin E, 220

West Highland white terrier, 141
Wilson's disease, 139, 189
  animal models, 141, 143
  behavioral manifestations, 120
  ceruloplasmin, 140
  chromosomal location, 139
  clinical excurse, 120, 139
  copper concentrations, 139
  expression, 167
  follow-up, 124
  Keyser–Fleischer ring, 139
  hepatic disease, 121
  histology, brain, 147, 149
  histology, liver, 143, 146
  liver copper, 139
  metallothionein, 121
  neurological symptoms, 120, 149
  patients, pediatric, 123
  patients, pregnant, 122
  patients, presymptomatic, 122
  prognosis, 124
  serum copper, 140
  symptoms, 139
  therapy, 121, 140
  therapy with pennicillamine, 118
  therapy with tetrathiomolybdate, 119, 121
  therapy with trientine, 116
  therapy with zinc, 115, 116, 120
  urinary copper, 140
Wilson's disease gene, human
  expression, 139
  in Indian Childhood Cirrhosis, 132
  mutations, 88
  organization, 85
  poymorphisms, 133
  structure, 84
Wilson's disease gene, mouse, 84
Wilson's disease gene, rat
  expression, 110
  structure, 84
Wilson's disease gene, sheep, 142
Wilson's disease protein, 78, 54, 109
  metal binding motif, 78, 155, 166, 167
  neutron activation analysis, 167, 168
  N-terminal domain, 165
  protein sequence, 54
  structure, 40
  subcellular localisation, 60
  trafficking, 42
  zinc blotting, 167, 169, 171

Index

Zinc
    blotting, 167, 169, 171
    MTF-1, 225
    sulphate, 140
    therapy, 115, 116, 120
    ZiRF1, 228, 231

ZiRF1
    antisense, 233
    MRE binding activity, 228
    MREc/d binding site, 229
    reconstitution by zinc, 231
    SP1 binding site, 229